International Conference on Applied Sciences (ICAS 2015)

AA002746

IOP Conference Series: Materials Science and Engineering
Volume 106

Wuhan, China
3 – 5 June 2015

Editors:

Ludovic Dan Lemle
Yiwen Jiang

ISBN: 978-1-5108-1957-3
ISSN: 1757-8981

Printed from e-media with permission by:

Curran Associates, Inc.
57 Morehouse Lane
Red Hook, NY 12571

Some format issues inherent in the e-media version may also appear in this print version.

Printed by Curran Associates, Inc. (2016)

For permission requests, please contact the Institute of Physics
at the address below.

Institute of Physics
Dirac House, Temple Back
Bristol BS1 6BE UK

Phone: 44 1 17 929 7481
Fax: 44 1 17 920 0979

techtracking@iop.org

Additional copies of this publication are available from:

Curran Associates, Inc.
57 Morehouse Lane
Red Hook, NY 12571 USA
Phone: 845-758-0400
Fax: 845-758-2634
Email: curran@proceedings.com
Web: www.proceedings.com

Table of contents

Volume 106

International Conference on Applied Sciences 2015 (ICAS2015)
3–5 June 2015, Wuhan, China

Accepted papers received: 3 December 2015
Published online: 1 February 2016

Preface

Papers

IOP Conf. Series: Materials Science and Engineering **106** (2016) 011001 doi:10.1088/1757-899X/106/1/011001

International Conference on Applied Sciences ICAS2015

Wuhan, China, June 3-5, 2015

Organizers:

Military Economics Academy of Wuhan ***Politehnica University of Timişoara***

in cooperation with

Wuhan University
Huazhong University of Sciences and Technology
Naval University of Engineering
Zhongnan University of Economics and Law
"Henri Coanda" Air Force Academy of Brasov

and

Scientific Interdisciplinary Values for Education and Research Society

The **International Conference on Applied Sciences ICAS2015** took place in Wuhan, China on June 3–5, 2015 at the Military Economics Academy of Wuhan. The conference is regularly organized, alternatively in Romania and in P.R. China, by Politehnica University of Timişoara, Romania, and Military Economics Academy of Wuhan, P.R. China, with the joint aims to serve as a platform for exchange of information between various areas of applied sciences, and to promote the communication between the scientists of different nations, countries and continents.

The **topics of the conference** cover a comprehensive spectrum of issues from:
>Economical Sciences and Defense: Management Sciences, Business Management, Financial Management, Logistic, Human Resources, Crisis Management, Risk Management, Quality Control,

IOP Conf. Series: Materials Science and Engineering **106** (2016) 011001 doi:10.1088/1757-899X/106/1/011001

Analysis and Prediction, Gouvernment Expenditure, Computational Methods in Economics, Military Sciences, National Security, and others...
>*Fundamental Sciences and Engineering:* Interdisciplinary applications of physics, Numerical approximation and analysis, Computational Methods in Engineering, Metallic Materials, Composite Materials, Metal Alloys, Metallurgy, Heat Transfer, Mechanical Engineering, Mechatronics, Reliability, Electrical Engineering, Circuits and Systems, Signal Processing, Software Engineering, Data Bases, Modeling and Simulation, and others...

The conference gathered qualified researchers whose expertise can be used to develop new engineering knowledge that has applicability potential in Engineering, Economics, Defense, etc. The number of participants was 120 from 11 countries (China, Romania, Taiwan, Korea, Denmark, France, Italy, Spain, USA, Jamaica, and Bosnia and Herzegovina). During the three days of the conference four invited and 67 oral talks were delivered.

Based on the work presented at the conference, 38 selected papers have been included in this volume of *IOP Conference Series: Materials Science and Engineering*. These papers present new research in the various fields of Materials Engineering, Mechanical Engineering, Computers Engineering, and Electrical Engineering. It's our great pleasure to present this volume of *IOP Conference Series: Materials Science and Engineering* to the scientific community to promote further research in these areas. We sincerely hope that the papers published in this volume will contribute to the advancement of knowledge in the respective fields.

Special thanks should be directed to the **Organizing Committees** for their tremendous efforts in organizing the conference:

International Organizing Committee
Co-Presidents
Guan Jiefu, Military Economics Academy of Wuhan
Serban Viorel-Aurel, Politehnica University of Timisoara
Members
Dan Daniel, Politehnica University of Timisoara
Fang Zhengqi, Military Economics Academy of Wuhan
Gherman Laurian, "Henri Coanda" Air Force Academy of Brasov
Guo Zhonghou, Military Economics Academy of Wuhan
Jiang Yiwen, Military Economics Academy of Wuhan
Lemle Ludovic Dan, Politehnica University of Timisoara
Lin Pinghua, Huazhong University of Sciences and Technology
Panoiu Caius, Politehnica University of Timisoara
Popa Mircea, Politehnica University of Timisoara
Zhou Yanming, Military Economics Academy of Wuhan
Rosu Serban, Scientific Interdisciplinary Values for Education and Research Society

Local Organizing Committee
General Chair
Guan Jiefu, Military Economics Academy of Wuhan
Co-Chairs
Zhou Yanming, Military Economics Academy of Wuhan
Zhou Yezhong, Wuhan University
Wang Enke, Huazhong University of Science and Technology
Members
Fang Zhengqi, Military Economics Academy of Wuhan
Guo Zhonghou, Military Economics Academy of Wuhan
Liu Dong, Military Economics Academy of Wuhan

IOP Conf. Series: Materials Science and Engineering **106** (2016) 011001 doi:10.1088/1757-899X/106/1/011001

Wu Shaohua, Military Economics Academy of Wuhan

Secretariat

Cheng Manli, Military Economics Academy of Wuhan
Deng Li, Military Economics Academy of Wuhan

The members of the **Scientific Committee** who reviewed and selected the papers included in the Proceedings of ICAS2015 are:

Co-Chairs

Fang Zhengqi, Military Economics Academy of Wuhan
Lemle Ludovic Dan, Politehnica University of Timisoara

Sessions Co-Chairs

Economical Sciences and Defense session

Harau Carmen, Politehnica University of Timisoara
Wu Shaohua, Military Economics Academy of Wuhan

Fundamental Sciences and Engineering session

Jiang Yiwen, Military Economics Academy of Wuhan
Pater Lucian Flavius, Politehnica University of Timisoara

Members

Bordeasu Ilare, Politehnica University of Timisoara
Chen Hua, Wuhan University
Cioaca Catalin, "Henri Coanda" Air Force Academy of Brasov
Constantinescu George, "Henri Coanda" Air Force Academy of Brasov
Deaconu Sorin, Politehnica University of Timisoara
Fang Guojia, Wuhan University
Gao Liang, Huazhong University of Science and Technology
Guan Zailin, Huazhong University of Science and Technology
Herban Sorin, Politehnica University of Timisoara
Huang Tianming, Military Economics Academy of Wuhan
Kiss Imre, Politehnica University of Timisoara
Lesenciuc Adrian, "Henri Coanda" Air Force Academy of Brasov
Li Baofeng, Huazhong University of Science and Technology
Li Xuewu, Military Economics Academy of Wuhan
Luculescu Doru, "Henri Coanda" Air Force Academy of Brasov
Luo Min, Military Economics Academy of Wuhan
Ma Huijun, Military Economics Academy of Wuhan
Miron Liliana Ecaterina, "Henri Coanda" air Force Academy of Brasov
Muscalagiu Ionel, Politehnica University of Timisoara
Niu Lingjun, Military Economics Academy of Wuhan
Osaci Mihaela, Politehnica University of Timisoara
Pan Chunxu, Wuhan University
Panoiu Manuela, Politehnica University of Timisoara
Popoviciu Octavian Mircea, Politehnica University of Timisoara
Ratiu Sorin, Politehnica University of Timisoara
Socalici Ana Virginia, Politehnica University of Timisoara
Wan Min, Huazhong University of Science and Technology
Wang Gengsheng, Wuhan University
Wang Xianjia, Wuhan University
Wu Shaohua, Military Economics Academy of Wuhan
Wu Jun, Huazhong University of Science and Technology
Xu Yong, Military Economics Academy of Wuhan
Ye Yonggang, Wuhan University

IOP Conf. Series: Materials Science and Engineering **106** (2016) 011001 doi:10.1088/1757-899X/106/1/011001

Yu Jiajiang, Military Economics Academy of Wuhan
Zhang Chengjian, Huazhong University of Science and Technology
Zhang Tao, Military Economics Academy of Wuhan
Zou Yunsong, Military Economics Academy of Wuhan

We thank the authors for their contributions and we would also like to express our gratitude everyone who contributed to the success of this conference.

Ludovic Dan Lemle and *Yiwen Jiang*
Editors of the Proceedings of ICAS2015

IOP Conf. Series: Materials Science and Engineering **106** (2016) 011002 doi:10.1088/1757-899X/106/1/011002

Peer review statement

All papers published in this volume of *IOP Conference Series: Materials Science and Engineering* have been peer reviewed through processes administered by the proceedings Editors. Reviews were conducted by expert referees to the professional and scientific standards expected of a proceedings journal published by IOP Publishing.

IOP Conf. Series: Materials Science and Engineering **106** (2016) 012001 doi:10.1088/1757-899X/106/1/012001

Researches upon the cavitation erosion behaviour of austenite steels

I Bordeasu[1], M O Popoviciu[2], I Mitelea[1], L C Salcianu[1], D Bordeasu[1], S T Duma[1] and A Iosif[3]

[1]Politehnica University of Timisoara, Faculty of Mechanical Engineering, Mihai Viteazul Street, Nr. 1, 300222 Timisoara, Romania
[2]Academy of Romanian Scientists, Mihai Viteazul Street, Nr. 1, 300222 Timisoara, Romania
[3]Politehnica University of Timisoara, Faculty of Civil Engineering, Traian Lalescu Street, Nr. 2, 300223 Timisoara, Romania

E-mail: ilarica59@gmail.com

Abstract. Paper analyzes the cavitation erosion behavior of two stainless steels with 100% austenitic structure but differing by the chemical composition and the values of mechanical properties. The research is based on the MDE(t) and MDER(t) characteristic curves. We studied supplementary the aspect of the eroded areas by other to different means: observations with performing optical microscopes and roughness measurements. The tests were done in the T2 vibratory facility in the Cavitation Laboratory of the Timisoara Polytechnic University. The principal purpose of the study is the identification of the elements influencing significantly the cavitation erosion resistance. It was established the effect of the principal chemical components (determining the proportion of the structural components in conformity the Schäffler diagram) upon the cavitation erosion resistance. The results of the researches present the influence of the proportion of unstable austenite upon cavitation erosion resistance. The stainless steel with the great proportion of unstable austenite has the best behavior. The obtained conclusion are important for the metallurgists which realizes the stainless steels used for manufacturing the runners of hydraulic machineries (turbines and pumps) with increased resistance to cavitation attack.

1. Introduction

As a rule, the heavy equipments of the hydraulic machineries running in cavitation conditions are manufactured from martensitic stainless steels and for repair works there are used electrodes with austenitic structure because those steels can be easy welded. This situation is a result of economic reasons, the austenitic steels containing large amounts of nickel and chromium, which are very expensive elements [1], [3], [7]. The present work analyzes the cavitation erosion behavior of austenitic steels through different methods: characteristic curves MDE, MDER, microscopic aspects and roughness measurements of the eroded areas. As austenitic materials, there were chosen the steels **X15CrNiSi25-20 şi X5CrNiMo17-12-2** frequently used in manufacturing the electrodes used for welding the damaged areas. These two steels have a complete austenitic structure but different mechanical properties because of differences in the chemical composition. For the laboratory researches was used a laboratory facility with piezoelectric crystals realized by respecting the ASTM-G32 Standard.

2. Researched materials

The materials researched are stainless steels with austenite structure. In conformity with the European Standards the symbols used for theses steels **X15CrNiSi25-20** and **X5CrNiMo17-12-2.** The researched specimens were manufactured from round rolled iron bars. The chemical composition and mechanical properties certificated by the manufacturer are:

- for the steel **X15CrNiSi25-20:** <0.15%C, 24-26%Cr, 19-21%Ni, 1.5-2.5%Si, ≤2%Mn, ≤0.045%P, ≤0.015%S, the rest being iron; Rm = (500-700) MPa, $Rp_{o,2}$ ≈ 230 MPa, Brinell hardness (HB) ≈ 223 daN/mm^2;
- for the steel **X5CrNiMo17-12-2**: <0.08%C, 16.5-18.5%Cr, 10-13%Ni, <1.0 %Si, <2.0 %Mn, ≤0.045%P, <0.03%S, 2-2.5 % Mo, 2-3 %N; Rm = (530-680) MPa, $Rp_{o,2}$ ≈ 240 MPa, Brinell hardness (HB) ≈ 191 daN/mm^2;

The recorded values for the chemical composition and mechanical properties, established in the Materials Laboratory are:

- for the steel **X15CrNiSi25-20:** 0.095%C, 25.20%Cr, 20.47%Ni, 2.45%Si, 1.67%Mn, 0.044%P, 0.013%S, 0.001% Mo, 0.002%Al, 0.001%Co, 0.001%Cu, 0.001% Ta, the rest being iron; Rm = 580MPa, $Rp_{o,2}$ = 236MPa, Brinell hardness (HB) = 228 daN/mm^2;
- for the steel **X5CrNiMo17-12-2**: 0.036%C, 16.515%Cr, 10.105%Ni, 0.689%Si, 0.591%Mn, 0.009%P, 0.015%S, 0.206% Mo, 2.59%N, 0.027%Al, 0.069%Co, 0.165%Cu, % Ta, 0.012%Ti, 0.019%Nb, 0.025%V, 0.084%W, 0.001%B, 0.0260 %As the rest being iron; Rm = 566MPa, $Rp_{o,2}$ = 241MPa, Brinell hardness (HB) = 197daN/mm^2.

The steel **X15CrNiSi25-20** has better mechanical parameters than **X5CrNiMo17-12-2**. So, it is expected a small superiority of **X15CrNiSi25-20** at cavitation erosion.

To establish the microstructure constitution, the equivalents of Chromium (CrE) and Nickel were determined by using the relations (1):

$$Cr_e=\%Cr+1{,}5x\%Si+\%Mo+0{,}5x\%\cdot(Ta+Nb)+2x\%Ti+\%W+\%V+\%Al$$
$$Ni_e=\%Ni + 30x\%C + 0{,}5x\%Mn + 0{,}5x\%Co \quad (1)$$

In Table 1 is presented the structural constitution, established from the Schäffler diagram (Figure1), taking into account the computed values for CrE and NiE [3].

To compare the cavitation erosion resistance of the tested steels we chose the Russian martensitic stainless steel **OH12NDL** used in Romania for manufacturing numerous hydraulic turbines. In the condition of the Iron Gates I Power Plant the blades manufactured from **OH12NDL** resisted more than 10000 running hours between two successive repair works, during the period 1970-2004. So, in Table 1 and Schäffler diagram is introduced also this stainless steel.

Table 1. Structural constitution of the analyzed steels

Stainless steel	CrE	NiE	Microstructure	Notation
X15CrNiSi25-20	28.879	24.156	100% austenite	A
X5CrNiMo17-12-2	17.716	11.485	100% austenite	B
OH12NDL	13.2	4.45	74% Martensite + 26% Ferrite	C

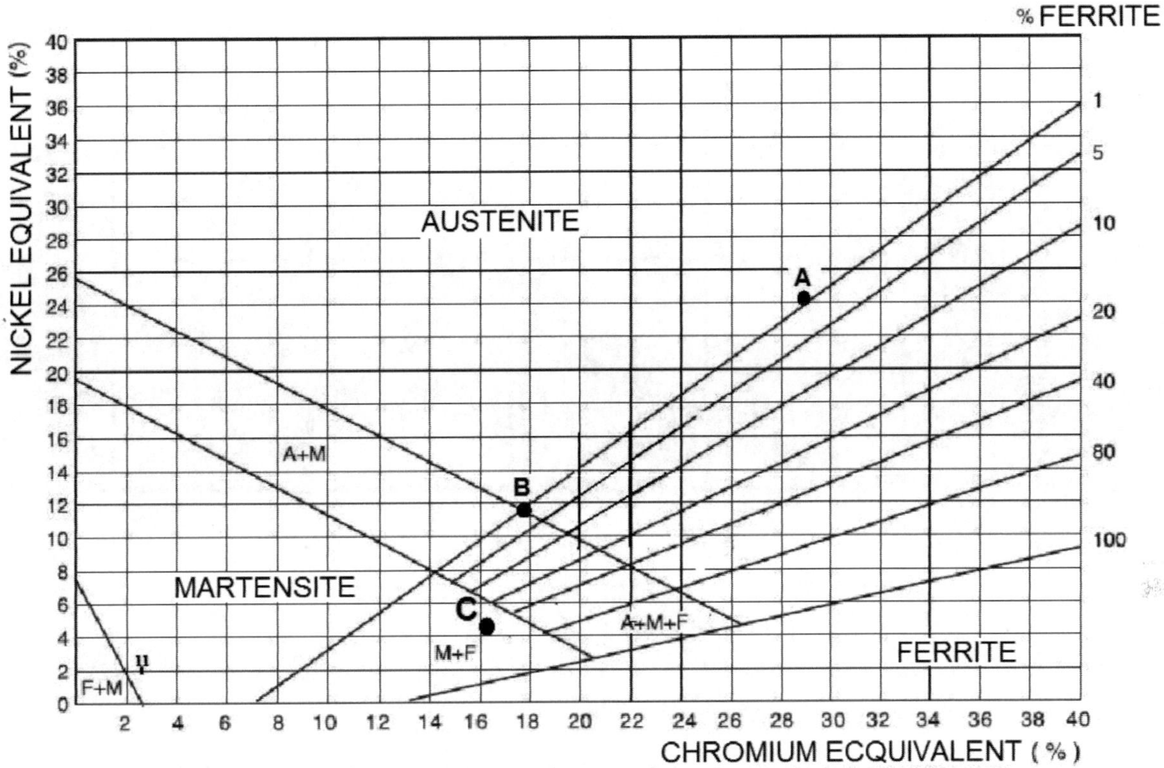

Figure 1. Position of researched stainless steels in the Schäffler diagram
A - X15CrNiSi25-20; **B** - X5CrNiMo17-12-2; **C**- OH12NDL

3. Test facility and test procedure

For tests was used the vibratory device T2 existing in the Cavitation Laboratory of the Timisoara "Politehnica" University. The facility has piezoelectric crystals [9]. The principal parameters of this facility are: vibration double amplitude 50 μm; vibration frequency 20,000 ± 2% Hz; power of ultrasonic generator 500 W; the power supply voltage 220 V/50 Hz. The device is provided with electronic facilities to maintain constant all the test parameters. In conformity of the ASTM G32-2010 Standard [10], the specimens with a diameter of 15.8 mm were tested in double distillated water at the same exposure time namely 165 minutes divided in 12 testing intervals of 5 minutes (the first one), 10 minutes (the second one) and the rest of for 15 minutes. The total exposure time was extended till the MDER curves became horizontal. Three specimens from each material were tested. The characteristic curves with the average value of those three tested values.

The testing procedure, the specimen preparations, the erosion evaluation after every exposure period and the structural analyze is specific to the laboratory bur integrally respect all the ASTM G32 recommendations [4], [5], [6], [8], [9]. The eroded areas were examined by free eye observations but also by using relevant images with optical and electronic microscopes.

4. Test results. Analyzes and discussions

In Figures 2 and 3 there are given the final results MDE(t) and MDER(t) for the cavitation erosion evolution. The points presented in the diagrams represent the average values for the three values experimentally measured and the curves were obtained by establishing a corresponding exponential equation, using the Bordeasu method [2], [4], [5], [6], [7].

IOP Conf. Series: Materials Science and Engineering **106** (2016) 012001 doi:10.1088/1757-899X/106/1/012001

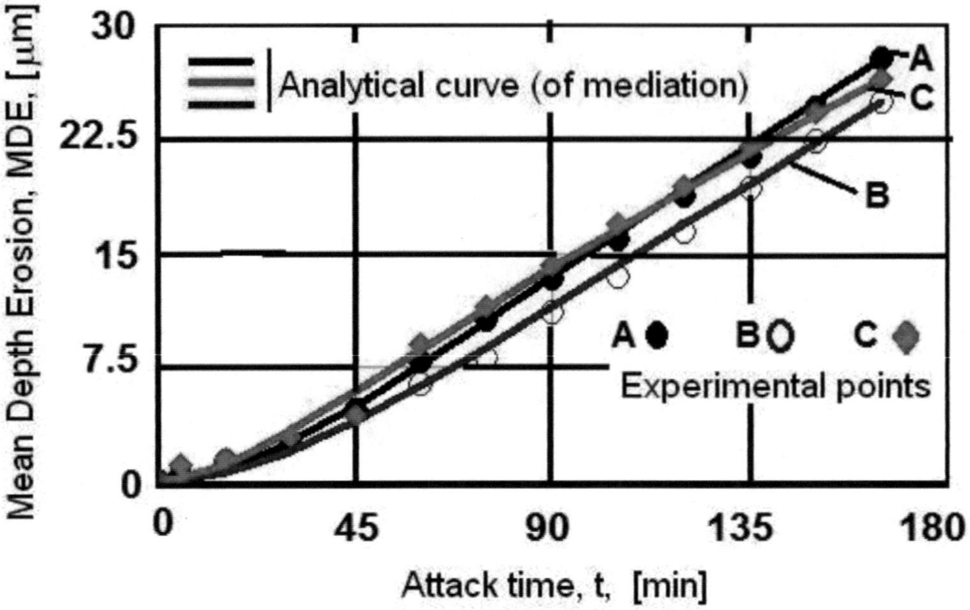

Figure 2. Time evolution of mean depth erosion
Stainless Steel - **A** (**X15CrNiSi25-20**); Stainless Steel - **B** (**X5CrNiMo17-12-2**);
Comparison Stainless Steel - **C** (**OH12NDL**)

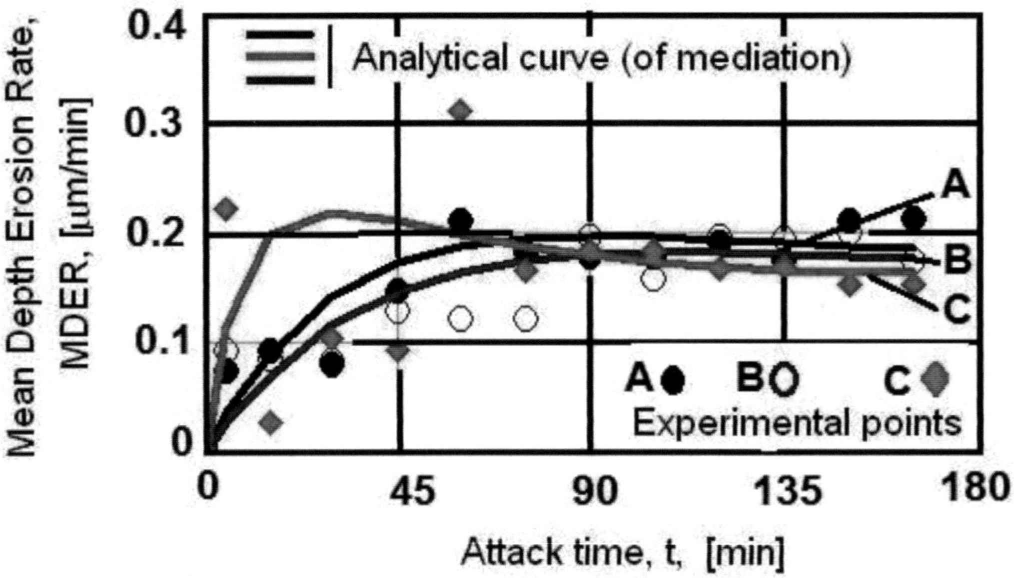

Figure 3. Time evolution of cavitation mean depth erosion rate
Stainless Steel - **A** (**X15CrNiSi25-20**); Stainless Steel - **B** (**X5CrNiMo17-12-2**);
Comparison Stainless Steel - **C** (**OH12NDL**)

The experimental results presented in Figure 2 and Figure 3 show that, in general the three stainless steels have good and very close resistance to cavitation erosion. From Figure 2 we can see that the best performance is given by the steel **B** (**X5CrNiMo17-12-2**); but from Figure 3 it results that the best performance is given by the comparison steel **C** (**OH12NDL**). This difference can be explained by the great mass losses of the steel **C** during the first 60 minutes of tests, after that the behavior is greatly improved. Our opinion is that the final conclusion must be obtained from the shape of the characteristic curve MDER(t) and not from the final value of the lost mass. Also the steel **B** presents a better resistance than **A**. This can be explained through austenitic condition, very close to the transition line to martensite (see Schäffler diagram). In this zone, the austenite is very unstable and under the impact with the collapsing bubbles is locally transformed in martensite and consequently the local resistance to cavitation is increased.

The eroded surface was analyzed also through microscopy. Supplementary, in the present research there were undertaken analyzes upon the final roughness of the eroded areas. Taking into account all the results the final conclusion was that both austenitic steels can be used for the repair works by welding of the areas eroded through cavitation. The small differences of the behavior are due to the differences of: chemical composition, mechanical properties and metallographic structure. The comparison steel with preponderant martensitic structure has the best resistance.

5. Roughness measurements

The roughness of the eroded area present a great interest because the pointed ends of the crests are easy destroyed by the implosion of the cavitation bubbles. With this purpose in mind, the roughness was measured before the beginning and after the end of the tests (165 minutes of exposure). The roughness of the exposed area before the tests is presented in Figures 4 and 5. The surface finishing operations were identical for both materials (**A** and **B**). The small differences between the roughness parameters Ra, Rt and Rz were generated by the material microstructures but can be considered without importance. As a consequence the resistance in the first exposure phase (different MDER) for the studied austenitic steels is determined in principal by the structure and the mechanical parameters and not by roughness. During the cavitation process the smoothness of the surface is modified in a complex way. There appears simultaneously deep indentation but also small degradations. Only these degradations may be considered as roughness. But the measurements can not made distinctions through these two phenomena. On the other hand the MDE(t) values increases constant in time while so roughness measurements can be done only at the end of the cavitation exposure (in our case at 165 minutes).

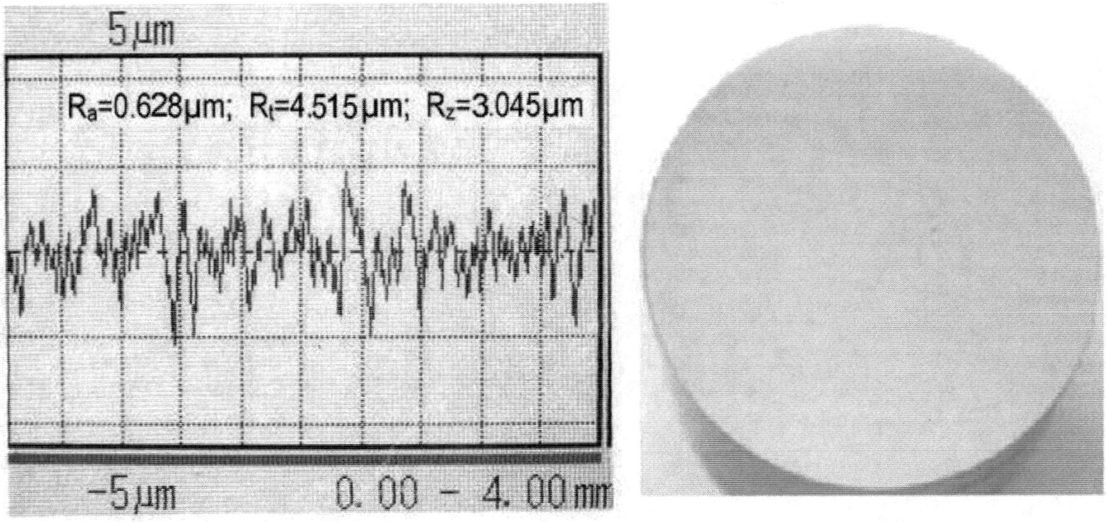

a) b)

Figure 4. Roughness measurements before cavitation exposure, Steel **A (X5CrNiMo17-12-2)**
a) roughness diagram; b) aspect of specimen surface before cavitation exposure

We measured those roughnesses along a single line, with a length of approximate 4 mm, presented in Figure 5. As it can be seen in Figures 9 and 10 the maximum depth was determined on a line with a length of about 0.2…0.5 mm. In conclusion the roughness measurements results simultaneously from the depth of cavitation indentations and the surface lake of smoothness. In order to have estimation between roughnesses and the erosion depths we made such comparisons with h_{max} the maximum depth and MDE_{165} mean depth erosion for 165 minutes of exposure. Figure 6 present the final roughness measurements and the method chosen for measuring the maximum erosion depth.

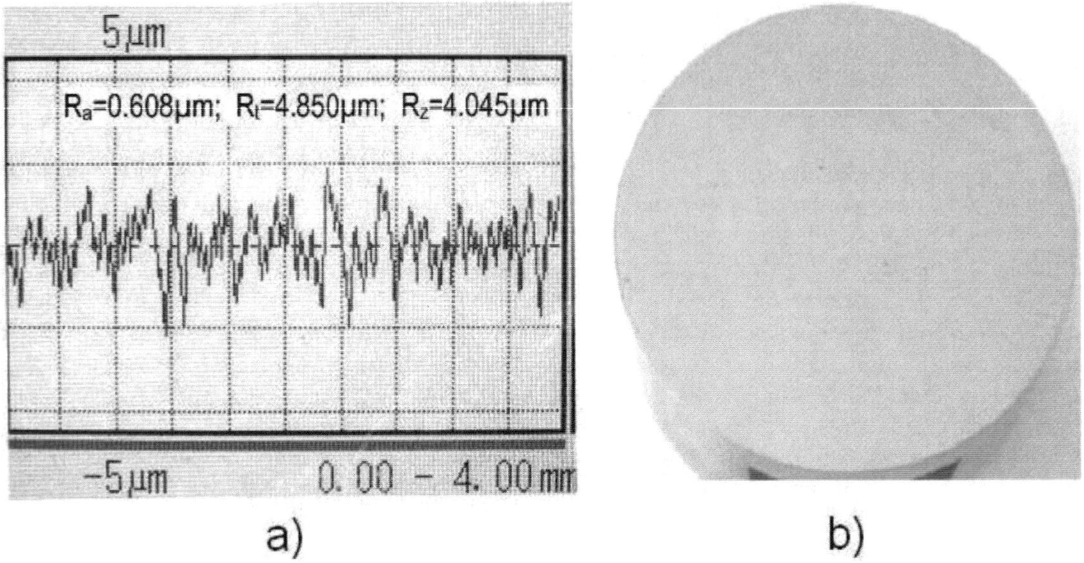

Figure 5. Roughness measurements before cavitation exposure, Steel **B (X15CrNiSi25-20)**
b) roughness diagram; b) aspect of specimen surface before cavitation exposure

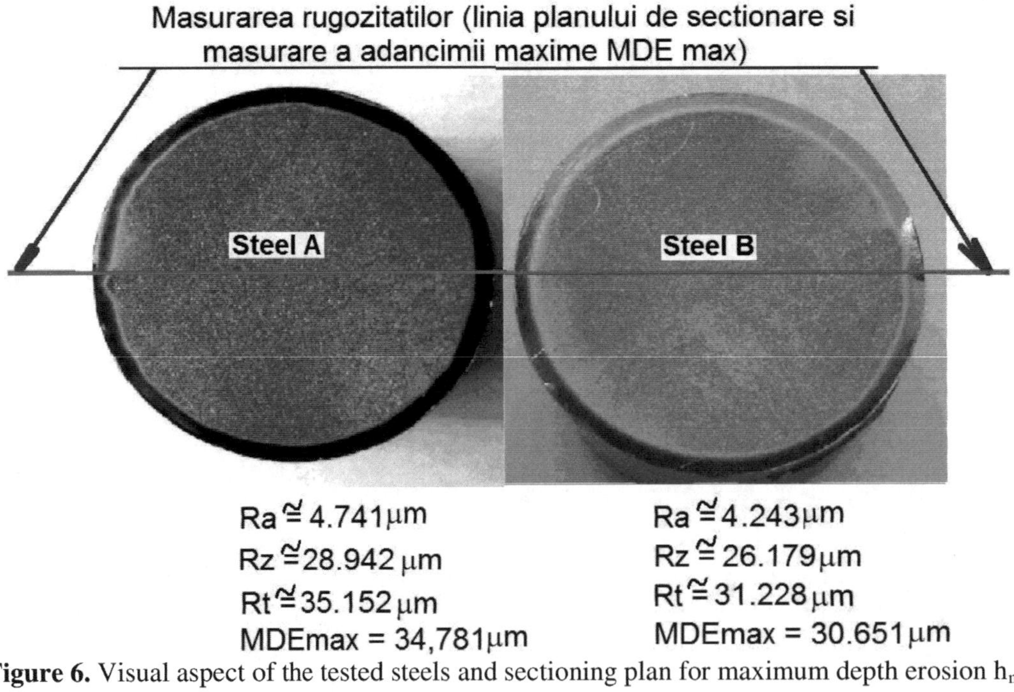

Figure 6. Visual aspect of the tested steels and sectioning plan for maximum depth erosion h_{max}

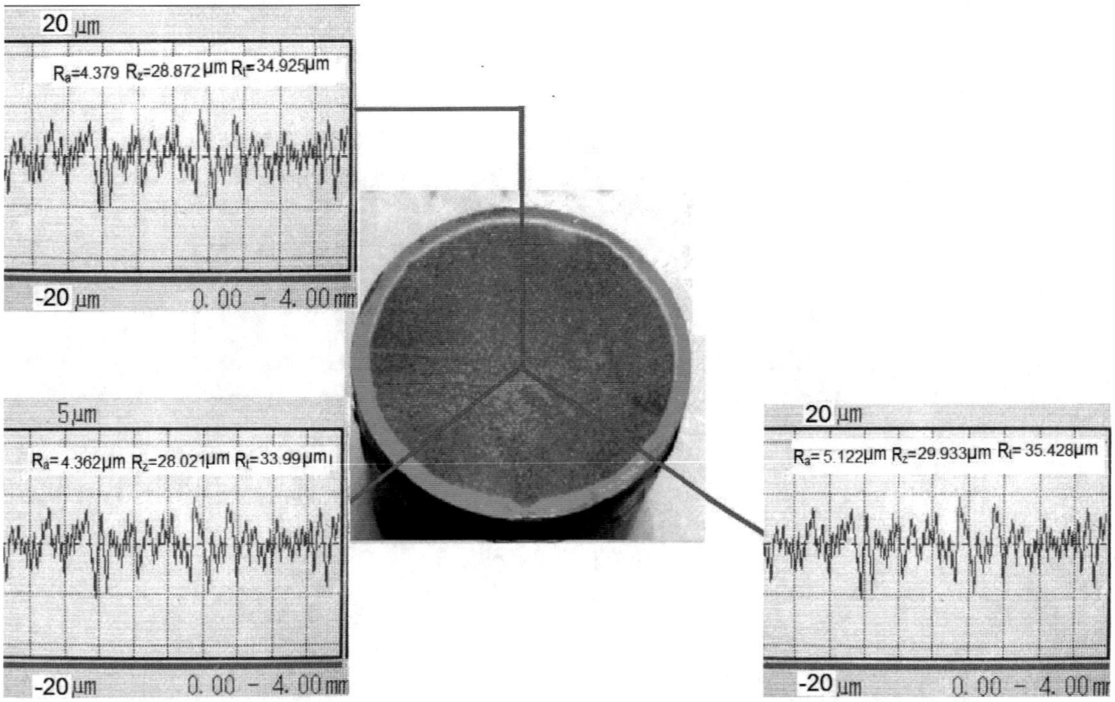

Figure 7. Ra, Rz, and Rt roughness measurements upon three directions (steel **A**)

Even form the visual aspect of the two tested steels, presented in Figure 6, it result a best behavior of the steel **B**. The histograms in Figure 7 and 8 compare the roughness parameters Rz, Rt and Ra with the values MDE_{165}, h_{max} and R_{cav}. The parameter R_{cav} represents the cavitation erosion resistance and is the inverse of the $MDER_{max}$.

It resulted that the values Rz are comparable with MDE_{165} and Rt with $h_{max.}$ It is very probable that this can be made not only for 165 minutes but for all exposure times. Small differences can occur by shifting the positions of the 4 mm measured segment along the diameter.

Regardless of those results and the facility of roughness measurements we do not consider that the use of weight measurements can be avoided and the cavitation erosion roughness can be used as a reliable method for establishing the cavitation erosion behavior.

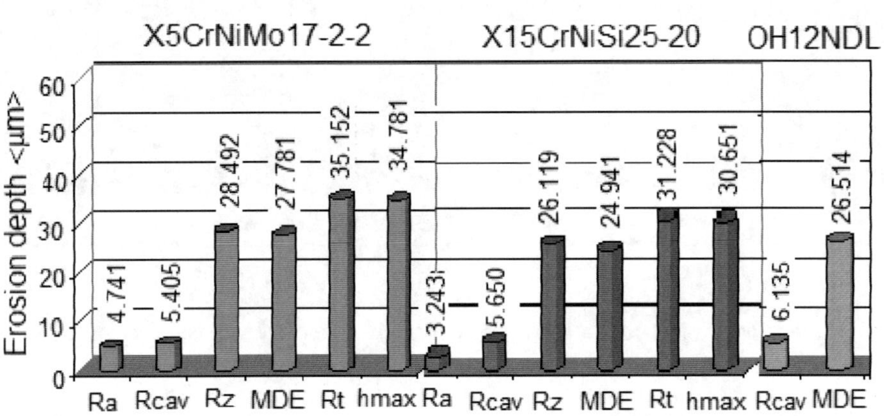

Figure 8. Comparisons between the cavitation erosion characteristic parameters after 165 minutes

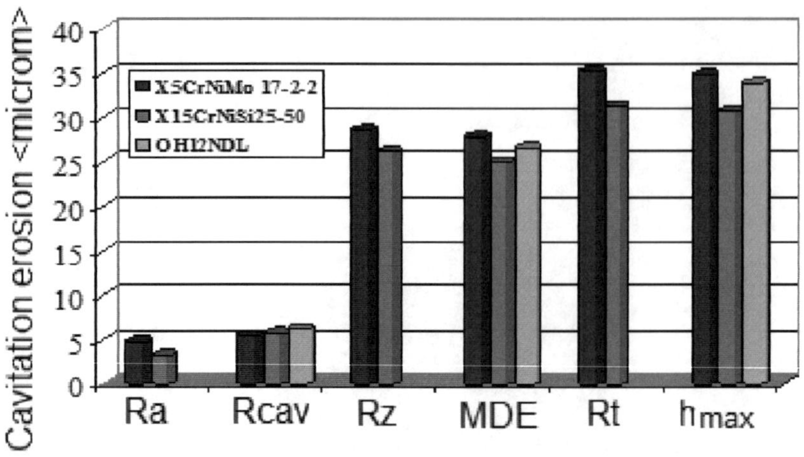

Figure 8. Comparisons regarding cavitation erosion resistance (165 min)

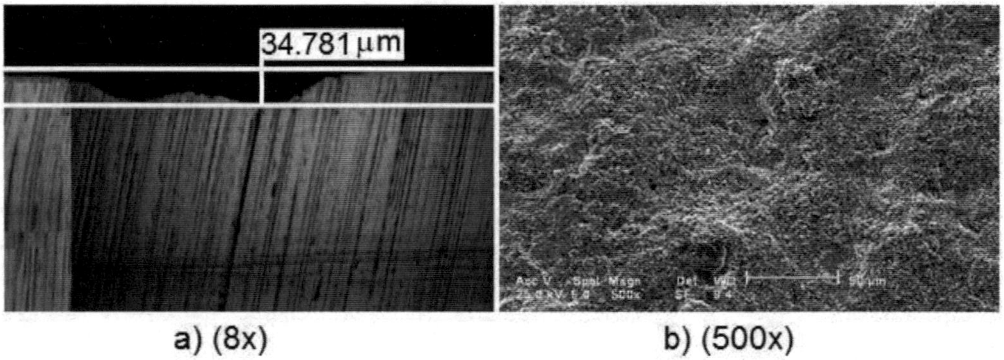

Figure 9. Degradation of the cavitation eroded surface– Specimen **A** (steel **X15CrNiSi25-20)**
a) Maximum depth erosion h_{max}; b) Image of the eroded area

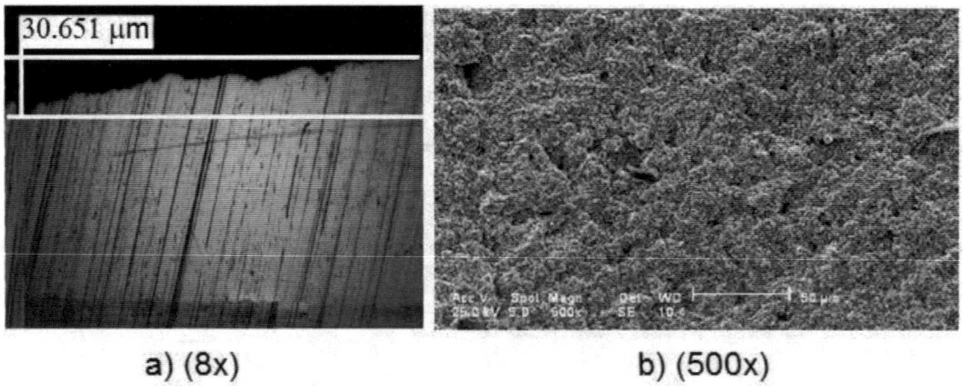

Figure 10. Degradation of the cavitation eroded surface– Specimen **B** (steel **X5CrNiMo17-12-2)**
a) Maximum depth erosion h_{max}; b) Image of the eroded area

The steel **B** (**X5CrNiMo17-12-2**) has a little better cavitation erosion resistance than the steel **A** (**X5CrNiMo17-12-2**) from all methods of comparison characteristic curves (Figures 2 and 3), roughness measurements (Figures 7 and 8) as well as maximum depth of erosion (Figures 9 and 10).

Both austenitic steels are a little weaker than the martensitic steel **C** (**OH12NDL**). Regardless of this superiority the austenitic steels, especially **B** (**X5CrNiMo17-12-2**), are recommended for repair works through welding.

6. Conclusions

a. The characteristic curves MDE(t) and MDER(t) for all three materials are very similar, with a small superiority for the martensitic steels. Regardless of this superiority the austenitic steels, especially **B** (**X5CrNiMo17-12-2**), are recommended for repair works through welding.
b. Even if the mechanical parameters (especially Rm and HB) of the steel **X15CrNiSi25-20** are greater than those for **X5CrNiMo17-12-2**, the cavitation erosion of the last steel is better. This result can be explained by the better structure of the steel **X5CrNiMo17-12-2** containing unstable austenite which under the influence of the bubble implosions is transformed suddenly in martensite.
c. For the researched steels the roughness parameters Rz and Rt, have very close values to MDE_{165} and $h_{max.}$.
d. We consider that in the future those researches must be continued also for other materials. Therw must be choosen also different position of the segment for which roughness measurements are undertaken (appoximate 4 mm).
e. We consider that the use of weight measurements (proposal of ASTM G32 Standard), especially MDE(t) can not be avoided, even if the roughness measurements are very easy to be obtained. The cavitation erosion roughness can be used as a supplementary method for confirmations.

References

[1] Anton I 1985 *Cavitation*, Tom II, Romanian Academy Publishing House, Bucharest, Romania (in Romanian)
[2] Bordeaşu I, Popoviciu M O, Balasoiu V and Patrascoiu C 2004 An Analytical Model for the Cavitation Erosion Characteristic Curves, *Scientific Bulletin "Politehnica" University of Timişoara, Transaction of Mechanics* **49**(63) 253-258
[3] Bordeaşu I 2006 *Cavitation Erosion of Materials,* Timisoara Polytechnic Publishing House, Timisoara, Romania (in Romanian)
[4] Bordeasu I, Popoviciu M O, Mitelea I, Ghiban B, Ghiban N, Sava M, Duma S T and Badarau R 2014 Correlations between mechanical properties and cavitation erosion resistance for stainless steels with 12% Chromium and variable contents of Nickel, *IOP Conf. Ser.: Mater. Sci. Eng.* **57** 012006
[5] Bordeasu I, Popoviciu M O, Mitelea I, Ghiban B and Ghiban N 2014 Cavitation erosion resistance of two steels with the same percentage of Chromium and Nickel but different Carbon content, *IOP Conf. Ser.: Mater. Sci. Eng.* **57** 012007
[6] I Bordeasu, M O Popoviciu, L M Micu, O V Oanca, D Bordeasu, A Pugna and C Bordeasu 2015 Laser beam treatment effect on AMPCO M4 bronze cavitation erosion resistance, *IOP Conf. Ser.: Mater. Sci. Eng.* **85** 012005
[7] Karabenciov A 2013 *Cercetări asupra eroziunii produse prin cavitaţie vibratorie la oţelurile inoxidabile cu conţinut constant în nichel şi variabil de crom*, Politehnica University of Timisoara, Romania, Doctoral Thesis
[8] Oanca O, Bordeasu I, Mitelea I and Crăciunescu C 2013 *Phenomenology of Degradation by Cavitation for Heat Treated CuNiAlFe Bronzes*, 22-th International Conference on Metallurgy and Materials, Brno, Czech Republic, May 15-17, pp 1561-1566
[9] Oanca O, 2014 *Tehnici de optimizare a rezistenţei la eroziunea prin cavitaţie a unor aliaje CuAlNiFeMn destinate execuţiei elicelor navale*, Politehnica University of Timisoara, Romania, Doctoral Thesis
[10] ***2010 *Standard method of vibratory cavitation erosion test*, ASTM, Standard G32-10

Pellet plant energy simulator

D Bordeasu[1], T Vasquez Pulido[2] and C Nielsen[1]

[1]VIA University College, VIA Engineering Department, str. Chr. M. Østergaards Vej, nr. 4, 8700 Horsens, Denmark
[2]KinTech Engineering, str. Calle José Anselmo Clave, nr. 37-45, 50004 Zaragoza, Spain

E-mail: dorin_craiova@yahoo.com

Abstract. The Pellet Plant energy simulator is a software based on advanced algorithms which has the main purpose to see the response of a pellet plant regarding certain location conditions. It combines energy provided by a combined heat and power, and/or by a combustion chamber with the energy consumption of the pellet factory and information regarding weather conditions in order to predict the biomass consumption of the pellet factory together with the combined heat and power, and/or with the biomass consumption of the combustion chamber. The user of the software will not only be able to plan smart the biomass acquisition and estimate its cost, but also to plan smart the preventive maintenance (charcoal cleaning in case of a gasification plant) and use the pellet plant at the maximum output regarding weather conditions and biomass moisture. The software can also be used in order to execute a more precise feasibility study for a pellet plant in a certain location.
The paper outlines the algorithm that supports the Pellet Plant Energy Simulator idea and presents preliminary tests results that supports the discussion and implementation of the system.

1. Introduction

Wood pellets are a modern form of densified biomass fuel with consistent quality (low moisture content, high energy density and homogeneous size and shape). Consistent fuel quality makes pellets a suitable fuel type for producing electricity and/or heat for all areas of application, from stoves and central house heating systems to large scale plants, with practically complete automation in all these capacity ranges [1].

Before the wood pellets leave to their final destination they have to be produced in a pellet factory.

In order to ensure the maximum profitability for a pellet factory, the wood pellets should have a quality according to international standards [2], and the pellet production should reach the nominal capacity during the entire working period.

Before erecting the pellet plant, instead of sizing the pellet factory according to the annual average of biomass moisture content, of ambient relative humidity and temperatures, instead of selecting the dryer using the average evaporation rate at average conditions of the location ambient, a better approach is to dimension the pellet factory according to historical weather data [3] from the respective location, or by using the data from a trustworthy weather forecast.

By using this weather data [3] the pellet factory can be sized better ensuring a longer working period at the nominal power. This can be possible by using the Pellet Plant energy simulator.

The idea of Pellet Plant Energy Simulator depicted in this paper combines information regarding location's historical weather data [3] (temperature, relative humidity), biomass availability, biomass moisture content [4] and pellet production [5] in order to:

- Calculate the amount of biomass consumption for the next simulating period on simulation step basis;
- Calculate the amount of pellets that can be produced for the next simulating period on simulation basis;
- Calculate the thermal energy required to evaporate the water from the biomass in order to reach the desired pellets humidity;
- Simulate the psychometric diagram [6], [7], [8] for calculating the required dryer inlet thermal energy, temperature and volumetric air flow rate;
- Calculate the biomass consumption of the combined heat and power plant and/or combustion chamber;
- Calculate the produced thermal energy and volumetric flue gas, with its temperature and enthalpy by the combined heat and power plant and/or combustion chamber;
- Calculate the electrical energy produced by combined heat and power plant (in case of using it);
- Calculate the percentage of the charcoal amount into the amount of biomass combusted in the combustion chamber (in case of using a down draft gasification plant [9] as combined heat and power plant);
- The combustion chamber working range (partial load/maximum load).

The Pellet Plant Energy Simulator can be used by investors as a tool in helping to execute a more precise feasibility study, or by pellet and dryer manufacturers for a better dimensioning of pellet factory machinery.

2. Overview of the system

The following diagram (Figure 1) depicts the overall idea of the Pellet Plant Energy Simulator.

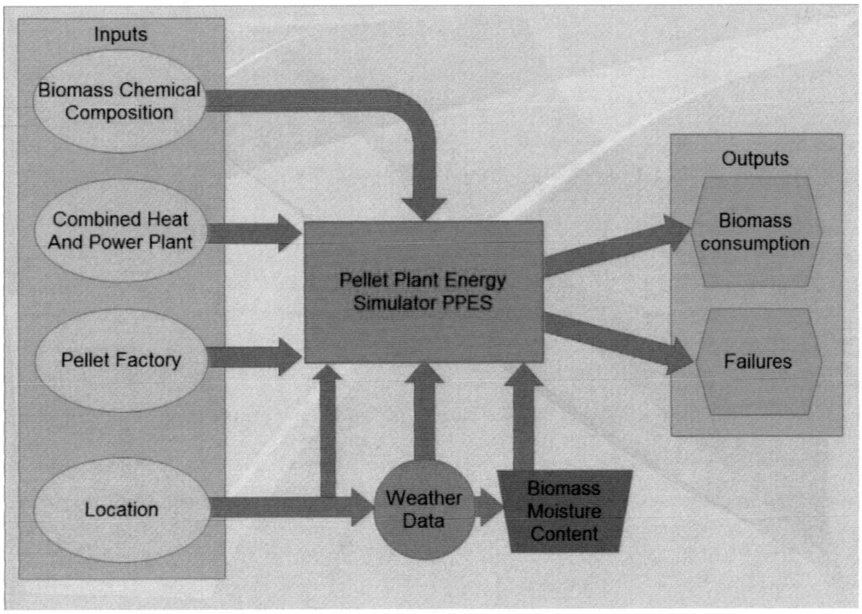

Figure 1. Pellet Plant Energy Simulator Overview

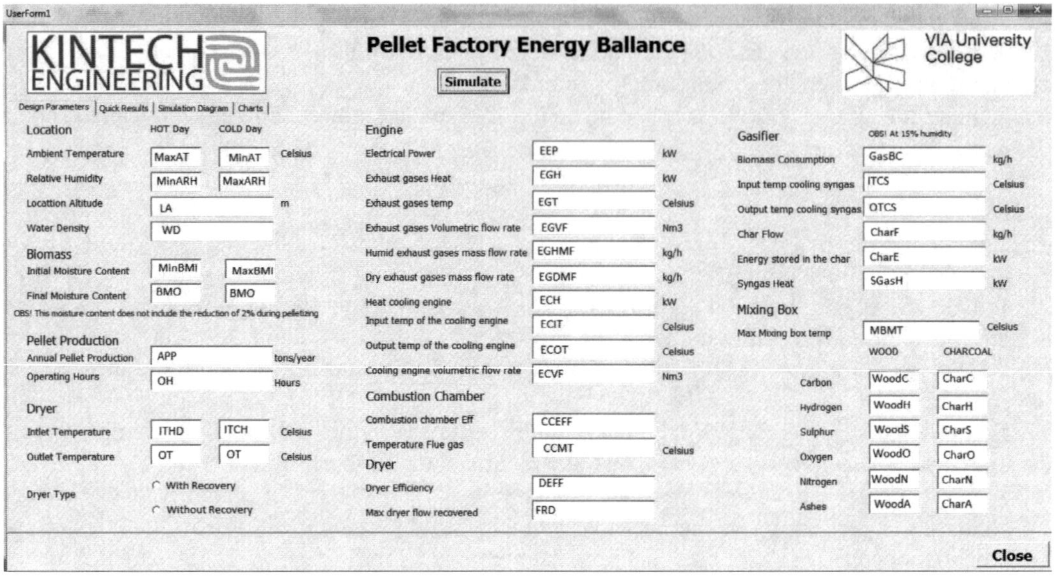

Figure 2. Pellet Plant Energy Simulator - "Design Parameters" Page

The necessary information and how it is being used can be described as follows:

1. The inputs are inserted by the user regarding:

a) Biomass chemical composition (Figure 3A) plus charcoal chemical composition (Figure 3A) in case of using a downdraft gasification plant (information obtained after Instituto de Carboquimica from Zaragoza, Spain, was analyzing the samples of biomass and charcoal) in order to calculate the biomass heating value [10];

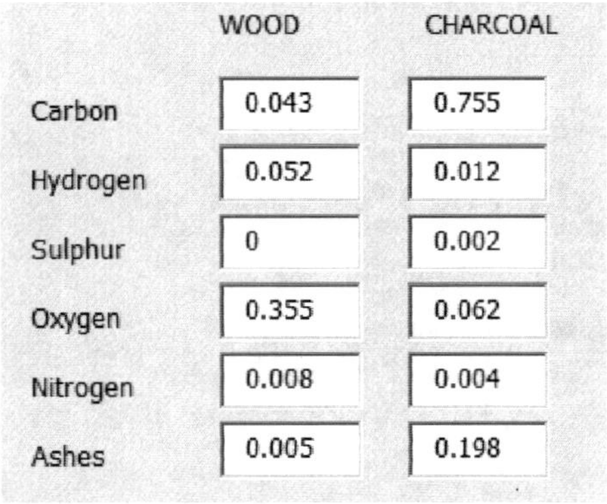

Figure 3A. Biomass and Charcoal chemical composition

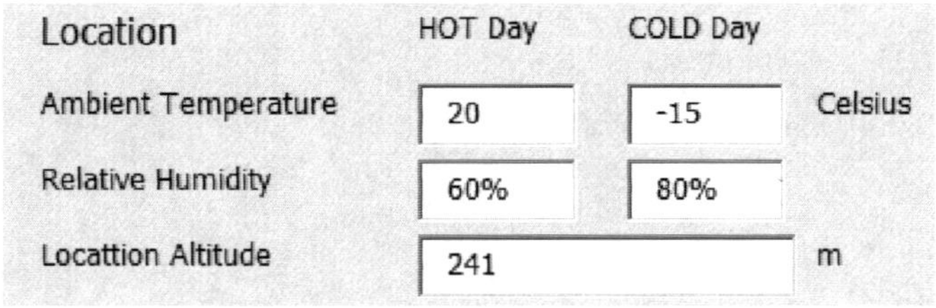

Figure 3B. Pellet Plant Location data

b) Combined heat and power plant data (Figure 4) for calculating the electrical and thermal energy produced. As a combined heat and power plant was used a downdraft gasification plant [11] combined with a gas engine [12];

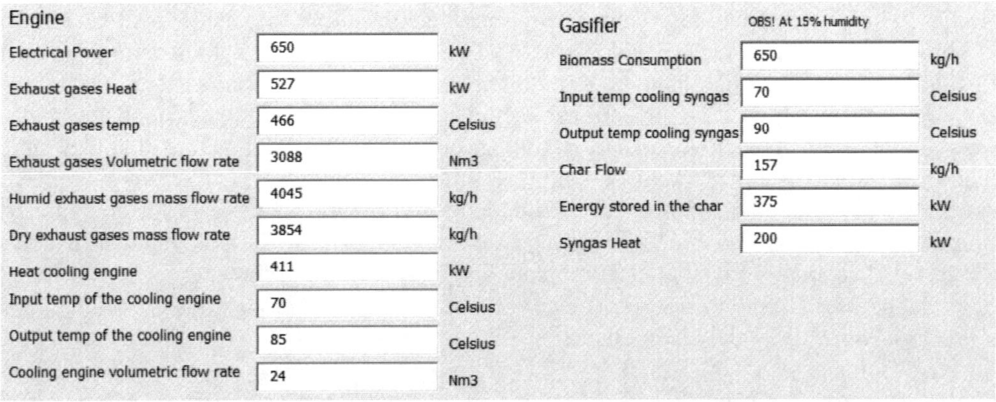

Figure 4. Combined Heat and Power Plant input data

c) Data about the pellet factory (Figure 5) (e.g. pellet production and operating hours, the drier's temperature ranges) in order to ensure that the quality of pellets is following European standards [2];

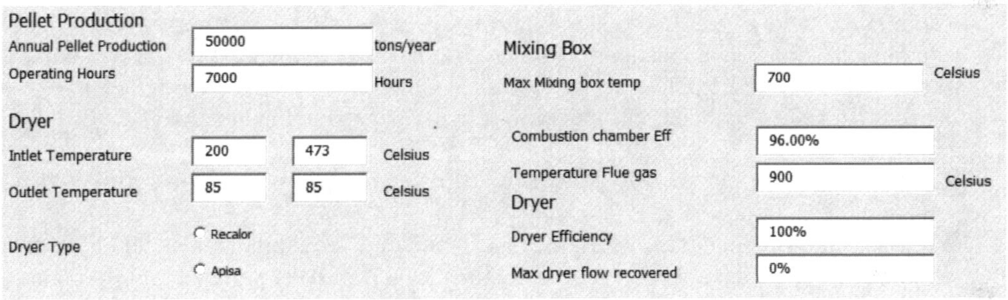

Figure 5. Pellet Factory input data

d) Location (Figure 3B).

2. The weather data (Figure 1) is either entered manually, either downloaded from a website [3] for providing the system with information regarding the weather such as temperature or humidity conditions during the simulation period in order to calculate the enthalpy of the ambient air flow and of the inlet and outlet dryer air flow [13].
3. By using the temperature and relative humidity from weather data, using the wood equilibrium moisture content formula [4] (Figure 1), the biomass moisture content [4] is estimated in order to calculate the mass flow rate of dry biomass and to calculate the heat together with its specifications required for drying the biomass.
4. The overall efficiency of the system is reported according to the annual biomass consumption of the plant (Figure 1).
5. The air flow entering in the combustion chamber for ensuring a total combustion of the biomass and charcoal, comes mainly from the syngas cooling system and from the engine cooling system. The air flow entering in the combustion chamber is adjusted with the help of outside air. If the amount of inserted outside air for adjusting the air mass flow coming from cooling the syngas and the engine is negative (described as failure in Figure 1), it means that a part of the heat recovered form syngas and engine cooling systems is given away resulting in a decrease of the overall efficiency of the system. In order to minimize the wasted heat, the inlet temperature of the dryer has to be decreased until the amount of inserted outside air becomes 0 or positive.

IOP Conf. Series: Materials Science and Engineering **106** (2016) 012002 doi:10.1088/1757-899X/106/1/012002

The Pellet Plant Energy Simulator is based on advanced algorithm that is using all the information depicted above in order to calculate and predict the pellet plant biomass consumption (pellets raw material plus biomass consumption of the combined heat and power plant and of the combustion chamber) more accurate than the conventional methods. It can also calculate which thermal energy source has to be used at each step of the simulation (only the combustion chamber at certain percentage from its maximum capacity, or both the combustion chamber at a certain percentage from its maximum capacity together with the combined heat and power plant) resulting in indicating the simulation steps when it is recommended to plan the maintenance of the combined heat and power plant (the charcoal cleaning of the gasification plant and maintenance of the engine).

In the end, the user is able to analyze the amount of biomass required in each simulation step in order to produce enough thermal energy necessary for a certain production of pellet by using the most energy efficient source.

The entire software has been developed using Visual Basic Application from Microsoft, mostly because of the good communication with Excel, a software used in feasibility studies.

The entire software has been developed using Visual Basic Application from Microsoft, mostly because of the good communication with Excel, a software used in feasibility studies.

Features of the Pellet Plant Energy Simulator:

1. Design Parameters page (Figure 2), containing 49 inputs (regarding location, biomass, pellet production, dryer, combustion chamber, biomass and charcoal chemical composition (section II.1.a.) and finally the gasification unit together with the engine) and 2 option buttons for selecting the dryer type (with or without recovering system).
2. Quick Results page (Figure 6), containing 68 results and 2 buttons:

- "Gasifier ON": When this button is activated the gasification plant is operating and when is deactivated it is not operating;
- The second button called "Show results" is displaying 68 results of a quick simulation (33 results for the hottest day, another 33 results for the coldest day), the required part load for the combustion chamber and the maximum amount of extra air flow rate at the end of the dryer, in order to keep a constant flow of the flue gas coming from the dryer so the silo can ensure lower polluting emissions in the atmosphere;
- The white boxes are displaying the results regarding the hottest day during the future simulation, and the blue boxes are displaying the results regarding the coldest day during the simulation;
- "%Chamber Min/Max" represents the required part load for the combustion chamber;
- "Cleaning ventilator Maximum air flow rate" (Figure 6) represents the maximum amount of extra air flow at the end of the dryer in order to keep a constant flow of the flue gas coming from the dryer for ensuring low amount of polluting emissions in the atmosphere.

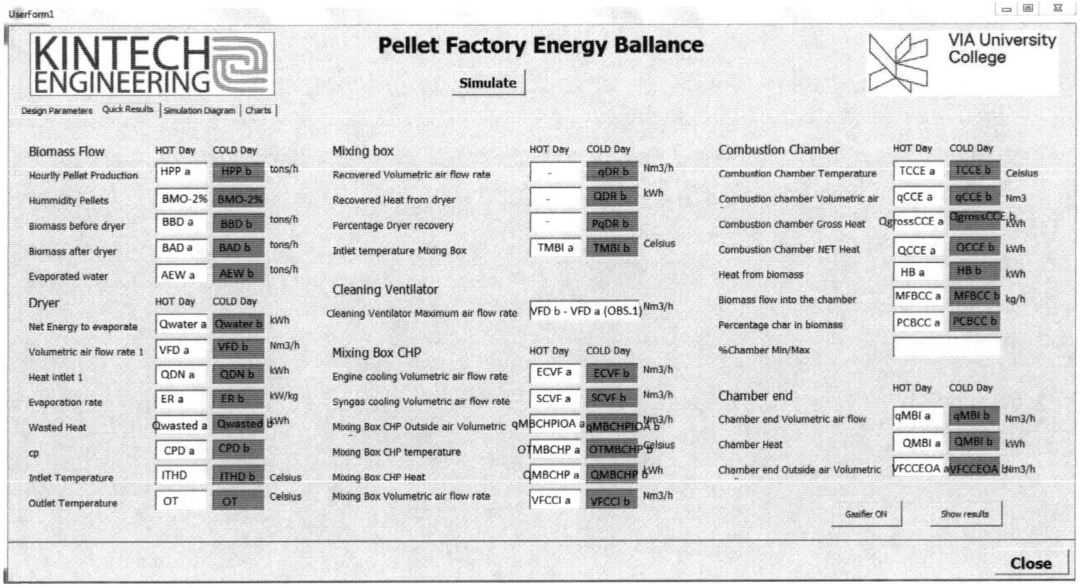

Figure 6. Pellet Plant Energy Simulator - "Quick Results" page

3. Simulation Diagram page (Figure 7):

- One Button for activating or deactivating the gasification plant operation;
- One combo box for seeing the simulated values according to the date for which the weather data was defined (See Related accessories, described above);
- One diagram showing the system and the simulated parameters for each component.

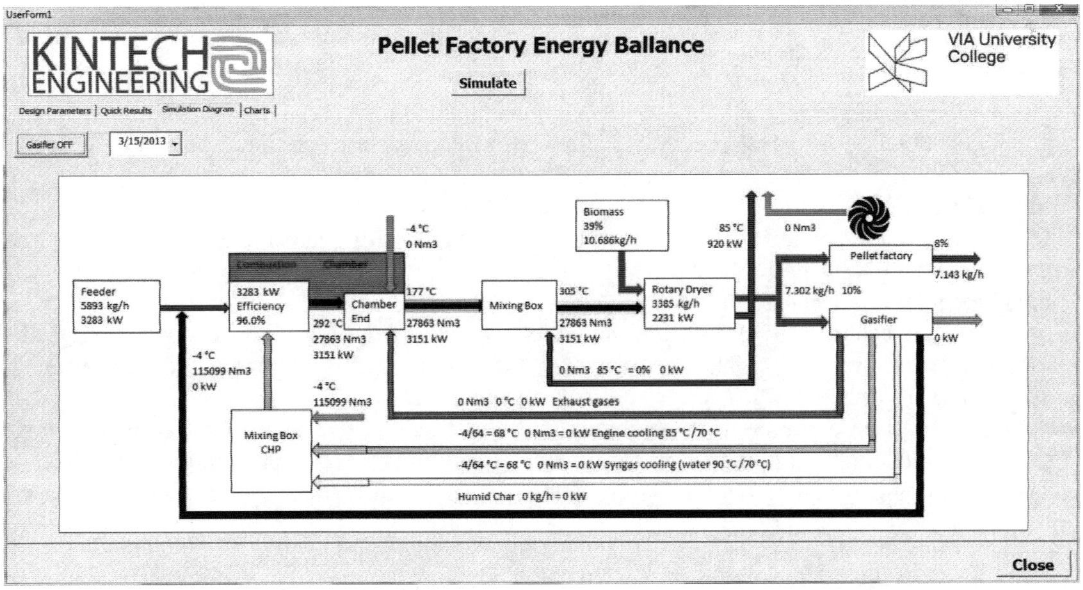

Figure 7. Pellet Plant Energy Simulator - "Simulation Diagram" page

4. Charts page (Figure 8)

- The first box called "Simulation Charts", represents the list of charts of different results according to the simulated period;

- The second box called “Range charts” contains a list of charts showing the response of the pellet factory with or without the gasification plant according to three main parameters (Ambient Temperature, Ambient Relative humidity, and Biomass Moisture Content);
- Finally the selected chart is displayed.

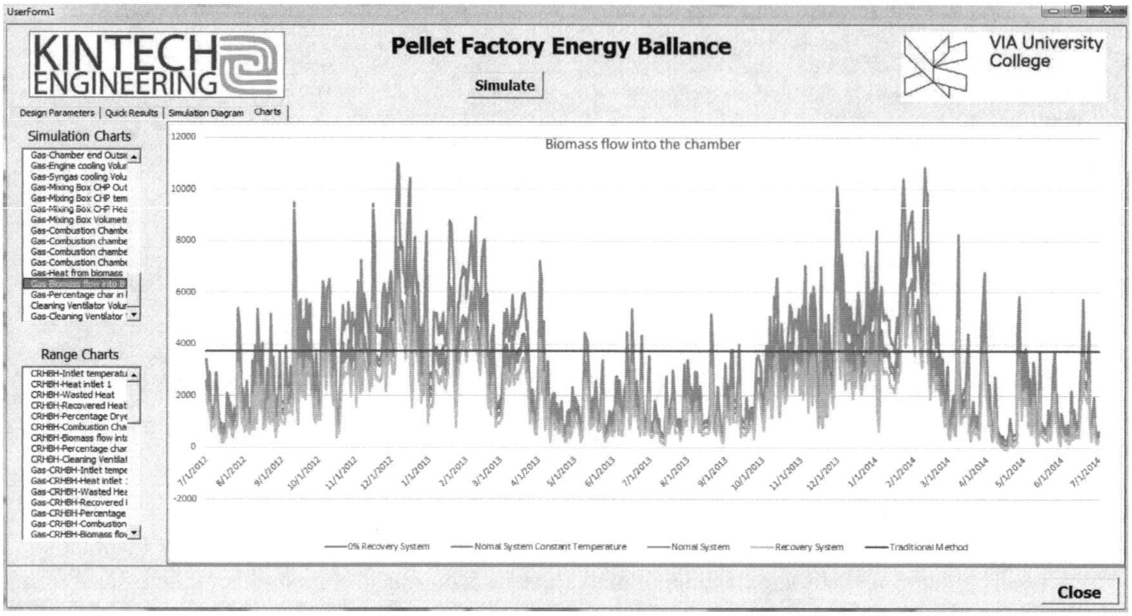

Figure 8. Pellet Plant Energy Simulator - "Charts" page

5. “Simulation” button for running the simulation according to the inputs entered on the design parameters page (point 1).

3. Software Development

The main idea behind Pellet Plant Energy Simulator is how to estimate/predict more accurate biomass consumption in a pellet plant in order to execute a more precise pellet plant feasibility study and a better management of the biomass acquisition in such plant.

After getting all information depicted above (Section II), the software can decide the best approach in calculating when it is more energy efficient to the combustion chamber with or without the combined heat and power plant unit.

First step of the calculations starts from the pellet production and the biomass consumption of the combined heat and power plant data, together with the initial and final biomass moisture content in order to calculate the mass flow rate of dry biomass through the drier.

Afterwards the absolute humidity of the biomass at the drier inlet (dry basis) is calculated from the initial biomass moisture. Similar to how the absolute humidity of biomass at the drier inlet was calculated, by using the final biomass moisture content the absolute humidity of biomass (dry basis) at the direr outlet it is calculated (eq. 1).

$$AH.biomass\ (dry\ basis) = \frac{Biomass\ moisture\ content}{100 - Biomass\ moisure\ content} \times 100 \tag{1}$$

The absolute humidities and enthalpies of drier inlet and outlet flue gases were calculated by generalizing the psychometric chart [6], [7], [8] and using the known values of ambient temperature and relative humidity together with the inlet and outlet temperatures of the drier.

IOP Conf. Series: Materials Science and Engineering **106** (2016) 012002 doi:10.1088/1757-899X/106/1/012002

By using the biomass mass flow rate, and by knowing the absolute humidity of biomass at the inlet and at the outlet of the drier, and from the absolute humidities of drier inlet and outlet flue gas, the dry air mass flow rate is calculated using a mass balance on the solvent [14] (eq. 2).

$$Mf.air \times (AH.out - AH.in) = Mf.biomass \times (AH.biomass.in - AH.biomass.out) \quad (2)$$

From the enthalpies of the drier inlet and outlet flue gas, and from the dry air mass flow rate the gross heat is calculated (3). Then the efficiency of the drier is taken into consideration (4).

$$Q.gross = Mf.air \times \Delta H \quad (3)$$

$$Q.net = Q.gross \times Eff \quad (4)$$

Next, by using the heat recovered from the gasification plant [11] (heat recovered from the syngas cooling system and the heating value contained by the charcoal) and from the gas engine [12] (heat recovered from the engine cooling system and from the exhaust gases), using first law of thermodynamics the total amount of thermal energy recovered from the combined heat and power plant is calculated (5).

$$Q.CHP = Q.charcoal + Q.syngas + Q.engine + Q.exhaust \quad (5)$$

By knowing the amount of thermal energy required by the drier and by knowing the amount of energy produced by the combined heat and power plant, the energy required to be produced by the combustion chamber it is calculated (6).

$$Q.combustion\ chamber = Q.dryer - Q.CHP \quad (6)$$

Next, according to the biomass chemical composition and the heat required to be produced by the combustion chamber, the biomass flow into the combustion chamber is calculated (7).

$$MF.biomass\ into\ combustion\ chamber = \frac{Q.combustion\ chamber}{Biomass\ Low\ Heating\ Value} \quad (7)$$

From the biomass and charcoal flow into the combustion chamber the air flow rate required for a full combustion of the biomass and charcoal is calculated (8).

$$Mf.combustion\ air = excess\ air\ ratio \times \frac{1{,}86 \times C + 5{,}55 \times H + 0{,}70 \times S - 0{,}70 \times O}{0{,}21} \quad (8)$$

Where C, H, S and O represents the biomass and charcoal chemical composition.

After comparing different configurations of the recovered heat flows, the most efficient way is found to be when most of the air flow required for a full combustion comes from the low temperature heat recovered from the combined heat and power plant. This is because the air mass flow will be reduced, resulting in higher temperatures which are offering the possibility to work also with high temperature driers.

The resultant flue gas coming from the combustion chamber is mixed with the engine exhaust gases and if it is necessary the temperature of the resultant flue gas is regulated with outside air in order to respect the working temperature ranges given by the manufacturer of the components.

4. Test Results

Four different tests were made, using the Pellet Plant Energy simulator in order to see which configuration of a pellet plant fits better in Vaggeryd, Sweden. This location was chosen due to the surrounding biomass availability and due to the relatively low ambient temperature, which increased the differences in the systems efficiencies. Afterwards the Vaggeryd weather data was downloaded and inserted into the software, and the inputs were entered (as in section 2).

Four different tests were made:

1. Pellet plant with constant dryer inlet volumetric air flow rate during the simulation period;
2. Pellet plant with constant dryer inlet temperature during the simulation period;
3. Pellet plant with a varying inlet temperature during the simulation period;
4. Pellet Plant with a dryer with recovering system up to 30% from dryer outlet volumetric air flow rate.

The test results were compared with the traditional method used in industry were the pellet plants were dimensioned according to annual average of dryer evaporation rate reported to normal ambient conditions, biomass moisture content, location relative humidity and temperature.

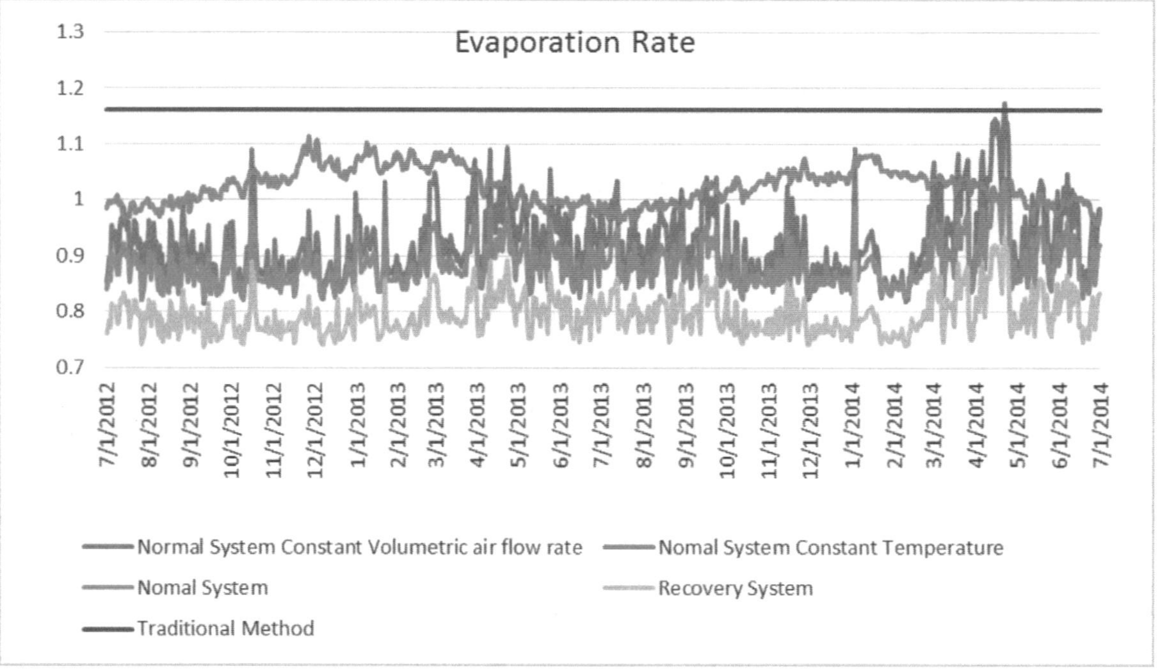

Figure 9. Comparison between simulations evaporation Rates

By comparing the results (Figure 9) of the simulations it can be concluded that the most energy efficiency configuration is when the dryer with the recovery system is used, providing the lowest evaporation rate (average: 0.80 kW/water kg) during the simulation period. The worst results were obtained by the configuration which uses the normal dryer and the same dryer inlet temperature during the entire simulation period, (average: 1.03 kW/water kg). The dryer without the recovery system, but with a variant temperature and a variant volumetric air flow at the drier inlet during the year, provides the second lowest value for the evaporation rate (average: 0.88 kW/water kg) during the simulation period. Finally the dryer without the recovery system, with a variant inlet dryer temperature and constant volumetric air flow during the year provides an evaporation rate around 0.94 kW/water kg. By comparing the results (Figure 9) of the simulations with the traditional method, it can be seen that the traditional method has a significant error.

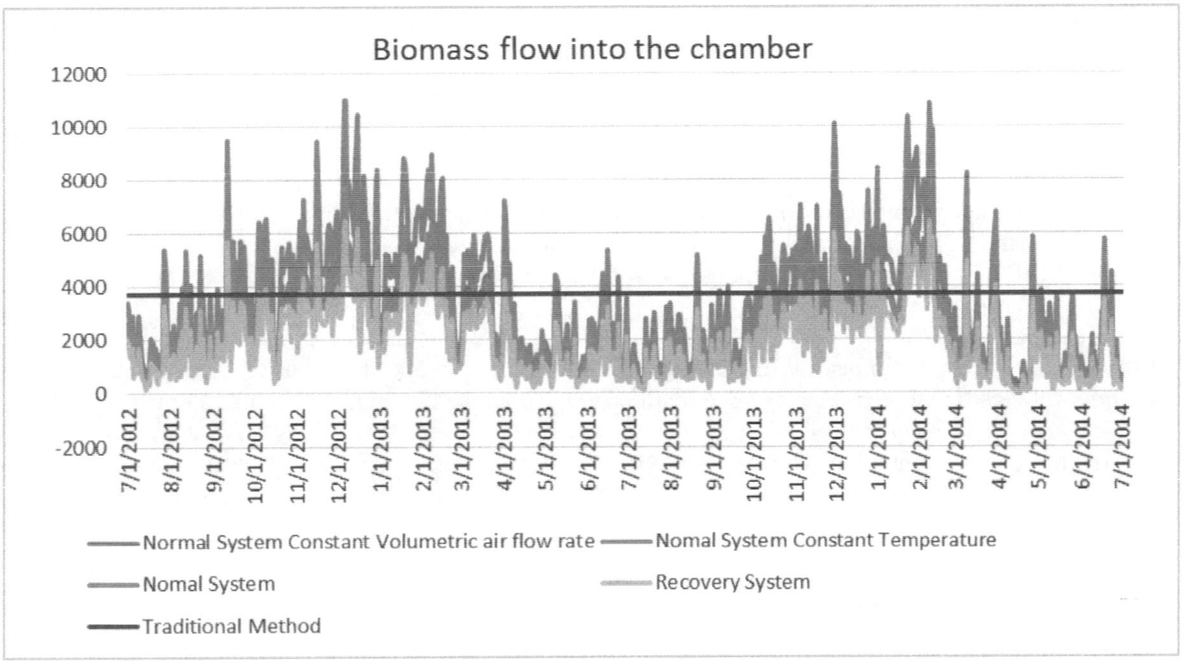

Figure 10. Comparison between simulations biomass flows into the combustion chamber

Regarding the viability of the plant, the best results (Figure 10 and Figure 11) are obtained with pellet plant with a dryer without recovery system and with the same volumetric air flow rate during the entire year, due to the reason that the partial load of the combustion chamber is down to 15% (from 5500 kg wood/h to 850 kg wood/h, representing the minimum partial load according to the manufacturer - point where all the other configurations were failing) and the maximum percentage of charcoal in biomass is 18% (from the experiments done, for a good ignition and combustion, the maximum percentage of charcoal presence in biomass was 20%).

A huge advantage in using the simulator can be seen at this point, because in the traditional method it is not taken into consideration that the combustion chamber in reality has to run at minimum partial load during the hottest days of the year, which sometimes is impossible to be reached.

Table 1. Simulations results comparison

	Normal Dryer constant inlet V.f.	*Dryer with recovering System 0%*	Normal Dryer constant inlet temperature	*Normal dryer varied inlet temperature*	Dryer with recovering system Average	*Unit*	Error
Operating hours per year	7 000	7 000	7 000	7 000	7 000	h	-
Annual pellet production	50 000	50 000	50 000	50 000	50 000	Tons	-
Annual pellet raw material	87 187	87 666	87 432	87 189	77 917	Tons	11%
Evaporation rate	0.74-0.89 (0.8)	0.82-1.10 (0.91)	0.96-1.11 (1.03)	0.82-0.99 (0.88)	1.16	kW/kg	27%
Percentage charcoal in biomass	36%	18%	18%	23%	4%	-	14% diff
Combustion chamber partial biomass load	10%	15%	12%	13%	-	-	-
Annual combustion Chamber biomass consumption	15 939	21 140	27 443	19 894	26 047	Tons	23%
Annual biomass consumption	103 126	108 806	114 875	107 083	103 964	Tons	4%

Finally it can be concluded that regarding both energy efficiency and plant viability the most suitable Pellet Plant in cogeneration with a Down Draft gasification plant regarding the Vaggeryd in Sweden should include a normal rotary dryer using the same volumetric air flow rate during the entire period in order to avoid an external ventilator at the end of the dryer for cleaning the outlet volumetric air flow rate.

Using this configuration the evaporation rate will be around 0.94 kW/evaporated kg of water, and biomass consumption in the chamber around 21 000 tons/year, which together with the pellet raw material and the CHP plant biomass consumption results in a biomass consumption of the pellet plant of around 110 000 tons/year, and a production of 50 000 pellets/year.

5. Future Work

The system offers a number of interesting areas for future work. First of all because this is a preliminary version where the preventive maintenance is planned manually, but an automated feature can be included in the next version. In the actual state, the pellet plant has to be adjusted according to the outputs of the simulation by the plant operator, but by creating an interface to support actual hardware all this can be done by a user of the interface. Further on, can extend it with a website and/or applications for mobile phones or tablets, so that the system can be monitored remotely.

6. Conclusion

This paper shows the application of leading edge technology within the area of pellets and sustainable energy, implementing an autonomous system capable of predicting the response and the amount of biomass required for operating a pellet factory in cogeneration with a combined heat and power plant based on the weather collected data and on the weather forecast. An overview of the system has been presented as well as the key characteristics of the software. In the end different test results were compared (including the traditional method used in the industry). Those results together with the comparison between them demonstrates the valid behavior of Pellet Plant Energy Simulator.

Special thanks to KinTech Engineering and Taim Weser from Zaragoza, Spain, for the good collaboration during the project, as well as to VIA University College from Horsens, Denmark, for the guidance and supervision.

References

[1] Thek G Obernberger I 2010 *The Pellet Handbook The Production and Thermal Utilisation of Pellets*, Earthscan, London, United Kingdom

[2] ***European Pelet Council 2013 *Handbook for the Certification of Wood Pellets for Heating Purposes*, Version 2.0, European Pellet Council (EPC), Brussels, Belgiun

[3] ***The National Centers for Environmental Prediction (NCEP) *Global Weather Data for SWAT*, [Online]. Available: http://globalweather.tamu.edu/home/ [Accessed 31 08 2015]

[4] Simpson W T 1998 *Equilibrium Moisture Content of Wood in Outdoor Locations in the United States and Worldwide*, United States Department of agricultural, Wisconsin, USA

[5] Francescato V, Antonini E and Bergomi L Z 2008 *Wood Fuels Handbook*, AIEL - Italian Agriforestry Energy Association, Legnaro, Italy

[6] Wilhelm L R 1976 Numerical Calculation of Psychometric Properties in SI Units, *Transactions of the ASAE* **19**(2) 318-325

[7] Ren H S 2004 Construction of a generalized psychrometric chart for different pressures, *International Journal of Mechanical Engineering Education* **32**(3) 212-222

[8] ***ASAE 1994 *Psychrometric Data* ASAE D271.2

[9] Reed T B, Das A 1988 *Handbook of Biomass Downdraft Gasifier Engine Systems,* Solar Energy Research Institute, Colorado USA

[10] Boundy B, Diegel S W, Wright L and Davis S C 2011 *Biomass Energy Data Book,* 4 Edition, Department of Energy, USA

[11] Weser T 2009 *TAIM WESER, at the forefront in biomass cogeneration solutions,* Taim Weser

SA. [Online]. Available: http://www.taimweser.com/detalle-noticia.aspx?id=29. [Accessed 31 08 2015]

[12] Jenbacher G 2015 *Technical Description Genset JGS 320 GS* GE Jenbacher. [Online]. Available: https://www.ge-distributedpower.com/products/power-generation/up-to-5mw/jenbacher-type-3. [Accessed 31 08 2015]

[13] Selivanovs J, Blumberga D, Ziemele J, Blumberga A and Barisa A 2012 Research of Woody Biomass Drying Process in Pellet Production, *Environmental and Climate Technologies* **10** 46-50

[14] Kemp I C 2012 *Fundamentals of Energy Analysis of Dryers,* In: Modern Drying Technology, Vol. 4: Energy Savings, 1st edition Tsotsas, E. and Mujumudar, A.S. eds., Wiley-VCH Verlag GmbH and Co . KGaA, Weinheim, Germany

IOP Conf. Series: Materials Science and Engineering **106** (2016) 012003 doi:10.1088/1757-899X/106/1/012003

Handling of the demilitarized zone using service providers in SAP

A Iovan[1] and R Robu[2]

[1]Politehnica University of Timişoara. Computer and Software Engineering Department, Timişoara, Romania
[2]Politehnica University of Timişoara, Automation and Applied Informatics Department, Timişoara, Romania

E-mail: adrian iovan@ymail.com

Abstract. External collaboration needs to allow data access from the Internet. In a trusted Internet collaboration scenario where the external user works on the same data like the internal user direct access to the data in the Intranet is required. The paper presents a solution to get access to certain data in the Enterprise Resource Planning system, having the User Interface on a system in the Demilitarized Zone and the database on a system which is located in the trusted area. Using the Service Provider Interface framework, connections between separate systems can be created in different areas of the network. The paper demonstrates how to connect the two systems, one in the Demilitarized Zone and one in the trusted area, using SAP ERP 6.0 with Enhancement Package 7. In order to use the Service Provider Interface SAP Business Suite Foundation component must be installed in both systems. The advantage of using the Service Provider Interface framework is that the external user works on the same data like the internal user (and not on copies). This assures data consistency and less overhead for backup and security systems.

1. Introduction

An ERP system is an integrated, configurable, and tailorable information system which plans and manages all the resources and their use in the enterprise, and streamlines and incorporates the business processes within and across the functional or technical boundaries in the organization. With ERP, an enterprise can automate its fundamental business applications, reduce the complexity and the cost of the collaboration, force the organization itself to take part in the Business Process Reengineering (BPR) to optimize its operations, and finally result in a successful business [1].

As the complexity of information systems in automation increases, so does its potential to get exploited by unauthorized access. Digital information has become an extremely important asset of organizations [2]. Current generations of factory information and automation systems follow a strictly layered hierarchical architecture. Normally an automation system and its information system is divided into three layers (automation pyramid), which are the following:

• Enterprise layer responsible for strategic decisions

• Manufacturing Execution System (MES) coordinating the factory floor by scheduling tasks, reserving resources and handling the flux of products and information

• Field control layer

SAP ERP is enterprise resource planning software developed by the German company SAP SE. SAP ERP incorporates the key business functions of an organization. The latest version (SAP ERP 6.0) was made available in 2006. The most recent Enhancement Package (EHP7) for SAP ERP 6.0 was released in 2013 [3].

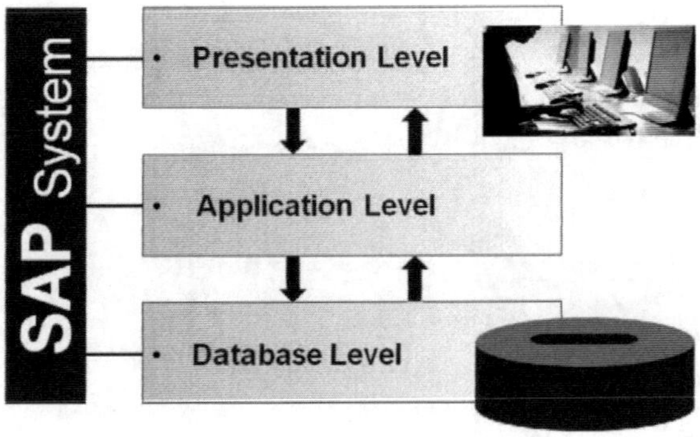

Figure 1. System Architecture [4]

The system is splitted in three layers: Presentation Level, Application Level and Database Level (see Figure 1).

In computer security, a DMZ or demilitarized zone (sometimes referred to as a perimeter network) is a physical or logical subnetwork that contains and exposes an organization's external-facing services to a larger and untrusted network, usually the Internet. The purpose of a DMZ is to add an additional layer of security to an organization's local area network (LAN); an external network node only has direct access to equipment in the DMZ, rather than any other part of the network. The name is derived from the term "demilitarized zone", an area between nation states in which military operations are not permitted [5].

The Service Provider Infrastructure (SPI) is an application and UI technology independent layer for business data exposure which is part of the SAP Business Suite Foundation (SAP_BS_FND). Its main goal is to support the application developer in building timeless software and to decouple the UI from the backend completely.

SPI handles huge amounts of data with a great performance and no additional buffering while being minimal invasive to the underlying implementation (arbitrary data repositories can easily be connected). Also dynamic backends can be accessed via the SPI, which for example do not have a static DDIC representation of their data structures or which even change their data type definitions at runtime. Every application that is using the SPI correctly can – out of the box – either run the UI and the backend on the same system or run the UI on a system in the DMZ and the backend on a second system behind the firewall. Furthermore the SPI provides a central enhancement spot which can be used by customers or partners to enhance the application backend modification free [6].

The paper presents a solution to get access to certain data in the ERP system, having the UI on a system in the DMZ and the database on a system which is located in the trusted area. The solution is based on the Service Provider Interface (SPI) framework provided by SAP AG in SAP 7.31 and Remote Function Call technology to link the systems.

In order to use this framework with the DMZ there should be at least a system with the backend and database in the trusted zone and a system with the user interface in the DMZ.

In [7] the authors evaluate cloud computing in the DMZ scenario and the conclusion is that although cloud computing has security deficiencies, the future looks less cloudy as far as more people are being attracted by this topic and pursuing research to improve on its drawbacks.

This article is structured in five chapters which include the introduction, presentation of the used technologies, the handling of the DMZ, experimental results and conclusions.

IOP Conf. Series: Materials Science and Engineering **106** (2016) 012003 doi:10.1088/1757-899X/106/1/012003

2. Service Provider Interface (SPI)

2.1. Structure of an SPI Application

An SPI application is identified by a so called Application Building Block ID (ABBID). This Application Building Block (ABB) can represent a complex application with many objects, a single object or even only a small part of an object [6].

An SPI application is identified by a so called Application Building Block ID (ABBID). This Application Building Block (ABB) can represent a complex application with many objects, a single object or even only a small part of an object [6].

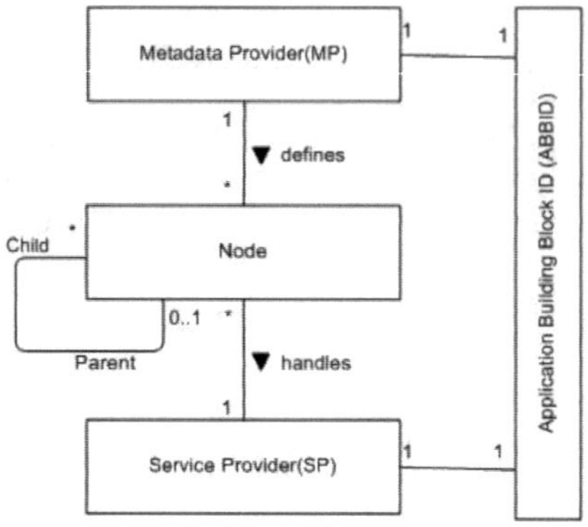

Figure 2. Application Building Block (ABB)

2.2. Consumption of an SPI Application

One central class of the SPI allows the consumption of all SPI application – the so called Connector. It implements the same access interface as the Service Providers, so it feels like calling the SP directly, when using the Connector. Internally the Connector e.g. takes care of the system switch, when running in a DMZ scenario [6].

Almost all SP methods return a direct result via export parameters, for example a table with requested node data. However there is also an indirect channel of communication between the SP and its consumer via the so called Collector. This object stores additional information like messages or field control information and thereby reduces the complexity of the SP's interface.

The complete node model of the application is described by the Metadata Provider.

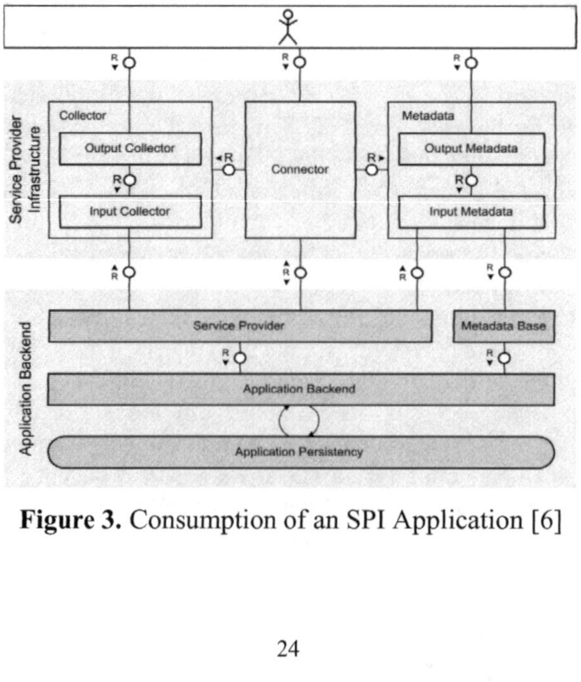

Figure 3. Consumption of an SPI Application [6]

IOP Conf. Series: Materials Science and Engineering **106** (2016) 012003 doi:10.1088/1757-899X/106/1/012003

2.3. Application Building Block

An Application Building Block (ABB) can either be a stand-alone application (e.g. material maintenance) or a smaller part of an application (e.g. long text or classification maintenance) which is integrated into the Service Provider Infrastructure (SPI). To create a dedicated Application Building Block for a smaller part of an application especially makes sense if that part can be reused by multiple applications. Otherwise it is advisable not to split an application into several Application Building Blocks. For every Application Building Block there needs to be a Service Provider and a Metadata Provider. Whenever a SPI based application is accessed, either by instantiating a Connector or by using the FPM SPI Integration (FSI) the Application Building Block ID (ABBID) is a mandatory import parameter which identifies the application that is to be instrumented.

2.4. Metadata Provider. Service Provider

A Metadata Provider (MP) describes the complete node model of a Service Provider Infrastructure (SPI) application. This includes for example the data type of nodes as well as available actions and queries that can be executed on those nodes. Beyond the definitions that are required from a technical perspective by application consumers, attributes that have a declarative/explaining character are also contained; i.e. to allow a consumer to understand what is possible with this application [6] (see Figure 3).

The main purpose of the metadata is the description/definition of:

- The application's behavior and capabilities
- The hierarchical node model and its dependencies
- The signature of generic Service Provider interface methods

The SPI offers huge flexibility in the definition of metadata:

- The metadata is provided via coding and therefore allows e.g. also the consumption (the SPI provides two interfaces for metadata consumption. One is the output interface /PLMB/IF_SPI_METADATA_OUTPUT which serves the purpose of reading and evaluating the defined metadata of an Application Building Block (ABB). This interface is returned as export parameter of the Connector instantiation method CREATE_INSTANCE and can be used by the consumer of the SP. The second interface /PLMB/IF_SPI_METADATA is handed over to the Service Provider during its instantiation. This interface includes the output interface and allows in addition manipulating the metadata at runtime, so it is supposed to be used by the SP only) of an already existing metadata model. Also the correction of coding via note is more convenient than the adjustment of customizing entries.
- The complete data type definition can be done dynamically i.e. without a DDIC representation. This allows for example to build up the node's attributes based on customizing entries or maybe even a combination of different DDIC structures.
- The metadata of a node can be modified at runtime (invalidation and re-reading), (e.g. side-effects or data type of a node).

The Service Provider (SP) functions as interface between the Service Provider Infrastructure (SPI) and the application's backend. Its main responsibility is providing access to the application data and handling requests for creating, changing or deleting the data [6] (see Figure 3).

Technically a SP is an ABAP OO class that implements the application access interface /PLMB/IF_SPI_APPL_ACCESS that consists of six node specific methods and three transactional (node-independent) methods. Additional information that is not returned via the access interface methods such as messages and field properties are handed over by the SP to the Collector.

Furthermore the SP needs to implement the initialization interface /PLMB/IF_SPI_APPL_ACCESS_INIT.

IOP Conf. Series: Materials Science and Engineering **106** (2016) 012003 doi:10.1088/1757-899X/106/1/012003

3. Handling Demilitarized Zone Using Service Provider Interface

External collaboration needs to allow data access from the internet. In a trusted internet collaboration scenario where the external users work on the same data like internal users (and not on copies) direct access to the data in the intranet is required.

This requires a special and strong security mechanism:

- No direct access from the internet to the system in the intranet is allowed
- Provide a protocol change between request and data access

A solution for this requirement is to separate UI and data in a way that there is no need to offer direct access to intranet (see Figure 4).

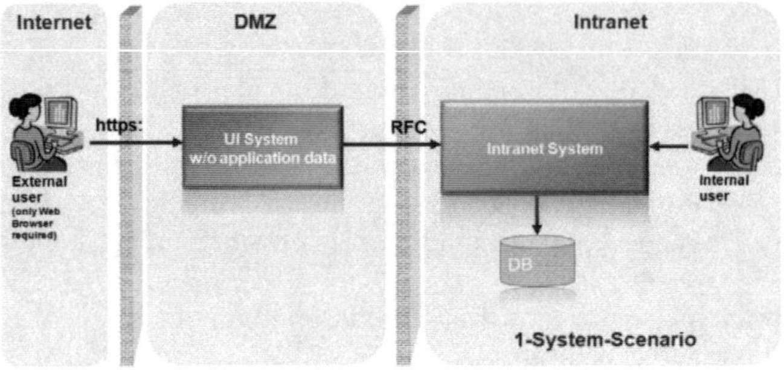

Figure 4. DMZ Scenario [6]

In the DMZ scenario the presentation layer (UI) is running on a system in the DMZ (demilitarized zone) network area which is separated from the system in the intranet where the data is stored. Therefore all data (with a few exceptions) needs to be accessed via RFC function modules. This allows a controlled access to the backend data and avoids a duplication of data needed for external collaboration.

Additional security mechanism may be needed (e.g. specialized collaboration UIs; more detailed authorization in the coding; Access control to the DMZ network area (e.g. using IP filter).

3.1. RFC Setup

User data and a few Customizing Settings (Output Formatting, RFC Connections, Language, etc) are needed in the DMZ system. Everything else has to be retrieved from the backend system by RFC.

To test the DMZ scenario internally an additional client representing the DMZ system is sufficient.

For real tests with external partners, a real DMZ system in the real SAP DMZ area is needed:

- A Security Concept which documents all accesses and connections (protocols and ports) crossing the firewalls has to be approved by the network security responsible (Security from SAP Hosting) [8].
- The DMZ system needs to be connected to the development system landscape to get all transport.
- The user ID in the DMZ system must be identical to the user ID in the intranet system.
- The authorizations for the user in the DMZ system can be reduced to a minimum to increase security.

3.1.1. Settings needed in the Backend system. Transaction SM59 would have to be used to set up a Remote Function Call (RFC) connection from the DMZ system to the backend system. A technical user like ALEREMOTE should not be used for this RFC connection. The settings needed in the Backend system are presented in Table 1.

Table 1. Settings for backend system

Attribute	Setting
RFC Destination	<Backend System Name> CLNT <client>
Connection Type	3(Connection to SAP R/3)
	Technical Settings tab page
Load Distribution	Yes
Target Host	<target host name>
	Logon/Security tab page
Trusted System	Yes
SNC	Inactive
Language	Leave this field empty
Client	<client>(Client which has the data, e.g. 002)
User	Leave this field empty
Password	Enter the following string: “is initial”
Current User	Select

Next saving RFC destination and going back to the RFC Overview screen is needed.

Next, in the RFC Overview screen it is necessary to choose from the menu bar "Extras" -> "Trusted Systems" (or Transaction SMT1) and to add the system name of the backend system as "Trusted System".

3.1.2. Registering the RFC Destination as Backend for the DMZ System. It is necessary to enter the RFC connection in the Customizing of the DMZ system (table /PLMB/RFC_DEST) IMG-Path:

Cross-Application Components -> Processes and Tools for Enterprise Applications -> Settings for BO Framework and Navigation -> BO Framework -> Define RFC Destinations.

3.1.3. Services. All data accesses using the Service Provider Infrastructure are enabled for the DMZ scenario.

Limitations of the DMZ Scenario affect the attributes with references to backend data which can't be handled in a DMZ scenario. These are:

- Object references to instances existing in the backend
- URLs (e.g. for Icons) generated on the backend

The SPI is designed for accessing Business Object (BO) like Data with CRUD services (Create, read, update and delete), but there are some use cases for data accessing which need a different approach:

- DDIC Search help
- Conversion Exits
- Customizing

3.2. DMZ Enabled Scenarios

3.2.1. DMZ Enabled DDIC Search Help. The DMZ-enabled DDIC search helps are in principle a copy of the normal DDIC search helps, which has:

- The search help exit /PLMB/SPI_SHLP_F4IF_GEN_EXIT

• An additional Search help parameter "IV_SHLP" which has as default the name of the normal DDIC search help

• Data elements which carry also DMZ-enabled DDIC Search helps

When collective search helps are needed for the DMZ they need to be build up from elementary DMZ-enabled search helps.

The search help exit /PLMB/SPI_SHLP_F4IF_GEN_EXIT should also be added to the collective DMZ-enabled search help.

Limitations of DMZ Enabled Search Help are the following:

• DDIC Search helps with programmed UIs can't be DMZ enabled.

• Packed numbers (e.g. timestamps) are not supported (limitation of internally used FM BALW_SHLP_BAPI_FILL)

• Help value is limited to char255 (help values returned in structure BAPIF4C from BALW_SHLP_BAPI_FILL)

The search help exit /PLMB/SPI_SHLP_F4IF_GEN_EXIT allows the call of application own RFC logic (available with SAP_BS_FND 7.31):

• Own function modules if the source system does not have SAP_BS_FND installed

• Own RFC determination) instead of the predefined logic

3.2.2. DMZ Enabled Conversion Exit. Conversion Exits in the DMZ system will not work if they need application data. All other Conversion Exits based on algorithm and/or customizing need no change.

For example application data is needed if an internal key is mapped to a readable key within a Conversion Exit (e.g. MATN2 for long material number)

The class /PLMB/CL_SPI_CONVEXIT is used to handle this conversion on the backend.

3.2.3. DMZ Enabled Table Provider for Customizing. Customizing which is needed in the UI (e.g. for configuration) can simply be accessed with class /PLMB/CL_FRW_TABLE_PROVIDER.

The names of tables which should be read by the table provider need to be populated in table /PLMB/FRW_TABPRO.

The limitations of table provider are that the applications tables (delivery class A) are not allowed to be used with the table provider.

3.2.4. DMZ Enabled Classes. The Table Provider class as well as the Conversion Exit class using the class /PLMB/CL_RFC_CONNECTOR for the RFC call and they are implementing the Interface /PLMB/IF_RFC_CONNECTOR.

This mechanism can also be used for application specific needs where DMZ enablement is necessary but the additional services of the SPI are not needed.

An example for this use case is the DMZ enablement of the Business Context Viewer (BCV).

In order to enable a class, the DMZ has to implement the interface /PLMB/IF_RFC_CONNECTOR which has one Method RFC_CALL_RECEIVE.

Within this method the application specific input parameters (exported with method RFC_CALL) can be imported. The requested result needs than to be exported and the export data is delivered back from Method RFC_CALL.

The data is exchanged by using the ABAP commands export and import. The format in which the data is exported and imported has to be defined by the implementing class.

It is strongly recommended that the caller method and the called method are implemented in the same class, to have the exchange protocol defined in only one class.

4. Experimental Results

For testing the connections two systems are configured: one in the DMZ Area and one in the Intranet.

The system in the DMZ Area contains only the UI and using the SPI framework and RFC connection is linked to the Intranet system which contains productive backend data (see Figure 5).

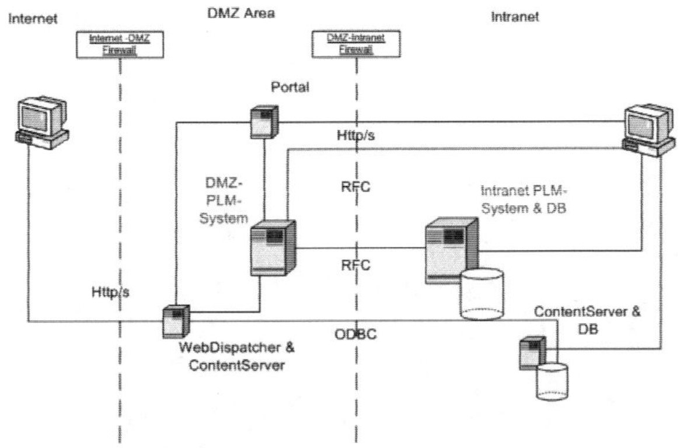

Figure 5. DMZ Systems and Connections [6]

Every application is secured using SAP Authorization objects that are assigned to users. When a user tries to access an application, the authorizations objects assigned to the application are compared with the authorizations objects assigned to the user, and if the users do not have the necessary authorization, he will not be allowed to continue.

The DMZ system contains no productive data on the database layer, so the applications are routed to get the data from the Intranet System. Specific applications only have access to specific data.

Access to unauthorized data is not permitted since the productive data is not on the same system as the application. The application gets the data through RFC connections which cannot be used by a standard user to access other data that the ones intended for the application.

5. Conclusions

Using the SPI framework to handle the DMZ makes is easy to manage external collaborations that need access to data in the company Intranet. By following the guidelines of the framework there will be no direct access from the internet to the system and a protocol will be provided to request and access data.

Another advantage is that there is no need to create separate databases for the separate systems and there is no need to administrate the replication of data from one system to another. As a result the effort and capacity to administrate and maintain the system will be low.

6. Acknowledgment

This work was partially supported by the strategic grant POSDRU/159/1.5/S/137070 (2014) of the Ministry of National Education, Romania, co-financed by the European Social Fund – Investing in People, within the Sectoral Operational Programme Human Resources Development 2007-2013.

References

[1] Wei S and Thuraisingham B 2007 Security for Enterprise Resource Planning Systems, *Information Systems Security* **16** 152–163

[2] Byres E and Lowe J 2004 *The myths and facts behind cyber security risks for industrial control systems,* Proceedings of VDE 2004 Kongress, Berlin, Germany, October, pp 213–218

[3] ***SAP Administration 2006 *Practical Guide*, SAP Press, SAP AG

[4] ***SAP Netweaver 2005 *Overview*, SAP Press, SAP AG

[5] Schmidt M, Smith M, Fallenbeck N, Picht H and Freisleben 2007 *Building a Demilitarized Zone with Data Encryption for Grid Environments*, 1st International Conference on Networks for Grid Applications GridNets07, Lyon, France, October 17-19, article no.17

[6] Schneider M and Walter W 2008 *Service Provider Infrastructure,* SAP Press, SAP AG

[7] Traian Andrei 2009 *Cloud Computing Challenges and Related Security Issues,* A project report

written under the guidance of Prof. Raj Jain, http://www1.cse.wustl.edu/~jain/cse571-09/ftp/cloud.pdf
[8] Chen B, Hanck J, P Hanck P, Hertel S, Lissarrague A and Médaille P 2015 *SAP Data Services*, SAP AG

Theoretical studies to elucidate the influence of magnetic dipolar interactions occurring in the magnetic nanoparticle systems, for biomedical applications

M Osaci[1] and M Cacciola[2]

[1]Politehnica University of Timisoara, Department of Electrical Engineering and Industrial Informatics, Vasile Parvan str., no. 1-2, 300223 Timisoara, Romania
[2]University "Mediterranea" of Reggio Calabria, DICEAM, Via Graziella Feo di Vito, I-89100 Italy

E-mail: mihaela.osaci@upt.ro

Abstract. In recent years, the study of magnetic nanoparticles has been intensively developed not only for their fundamental theoretical interest, but also for their many technological applications, especially biomedical applications, ranging from contrast agents for magnetic resonance imaging to the deterioration of cancer cells via hyperthermia treatment. The theoretical and experimental research has shown until now that the magnetic dipolar interactions between nanoparticles can have a significant influence on the magnetic behaviour of the system. But, this influence is not well understood. It is clear that the magnetic dipolar interaction intensity is correlated with the nanoparticle concentration, volume fraction and magnetic moment orientations. In this paper, we try to understand the influence of magnetic dipolar interactions on the behaviour of magnetic nanoparticle systems, for biomedical applications. For the model, we considered spherical nanoparticles with uniaxial anisotropy and lognormal distribution of the sizes. The model involves a simulation stage of the spatial distribution and orientation of the nanoparticles and their easy axes of magnetic anisotropy, and an evaluation stage of the Néel relaxation time. To assess the Néel relaxation time, we are going to discretise and adapt, to the local magnetic field, the Coffey analytical solution for the equation Fokker-Planck describing the dynamics of magnetic moments of nanoparticles in oblique external magnetic field. There are three fundamental aspects of interest in our studies on the magnetic nanoparticles: their spatial & orientational distributions, concentrations and sizes.

1. Introduction

The physics of magnetic nanoparticles systems is of special scientific interest, and has attracted much attention in recent years due to the potential for various applications, from technology to medicine. In the literature, there are many theoretical and experimental studies, often contradictory, about the influence of magnetic dipolar interactions on the magnetic behaviour of the nanoparticle systems, of which we are going to point out some of them to have an image on the matter.

Mössbauer spectroscopy measurements [1], [2], [3] on weakly interacting γ-Fe_2O_3 nanoparticles showed a decrease of the relaxation time with increasing particle interaction. Other publication [4] reported on the specific absorption rate (SAR) of a maghemite-based ferrofluid, measured at 315 K, 3 kA/m, and 109 kHz. The magnetic characterization of three different concentrations of this ferrofluid revealed the decrease of initial susceptibility values (liquid ferrofluid) and Néel relaxation times, τ_N, with increasing concentration.

In the other hand, the AC susceptibility measurements at low frequency for dynamic properties of a system constituted by small iron grains dispersed in an amorphous alumina matrix [5] show that the relaxation time increases with increasing particle concentration. The same experimental results were found in case of γ-Fe_2O_3 nanoparticles dispersed in a polymer [6] investigated by means of Mössbauer spectroscopy measurements and AC susceptibility measurements in a large frequency range ($10<\nu<104$ Hz). An increase of the effective energy barrier with increasing interaction strength has been unambiguously seen in many other cases.

The previous theoretical attempts to consider the magnetic dipolar interactions were based on the Néel-Brown model [7]. The first attempt to introduce interactions in the Néel–Brown model was made by Shtrikmann and Wohlfarth who, by using a mean field approximation, predicted a Vogel–Fulcher law for the relaxation time in the weak interaction limit [8].

A more general approach was developed by Dormann et al. [5]. Taking into account the dipole-dipole interaction, the model predicted two magnetic regimes, according to the interaction strength. Firstly, for negligible or very weak interactions, the predictions matched the Néel-Brown model of superparamagnetism. Secondly, for appreciable dipolar interactions, the individual energy barriers were modified by their strength, the case being referred to as superparamagnetic regime modified by interactions. Here, an increase of Néel relaxation time with the strength of dipolar interactions (e.g., increasing particle concentration or decreasing particle distances) was predicted.

Note that the Shtrikman and Wohlfarth model [8] led to the same conclusions as the Dorman et. al. model [5]. In a third model, proposed by Mørup and Tronc [1] for the weak interaction limit, the opposite dependence of Néel relaxation time (decrease with increasing strength of the interactions) has been predicted. In order to untangle the apparent contradiction, Mørup [1] suggested that two magnetic regimes, governed by opposite dependencies of the Néel relaxation time, occur in the systems of interacting fine particles. At high temperatures and/or for weak interactions, the Néel relaxation time decreases with increasing interactions. In contrast, at high temperatures and/or for strong interactions, a transition occurs from the superparamagnetic state to a collective state, having most of the features of typical glassy behaviour. In this case, a freezing process occurs and the Néel relaxation time increases with increasing interactions. In the last few years, this collective state and its glassy properties are some of the most studied issues regarding the assemblies of strongly interacting nanoparticles [9], [10].

The models found in the literature, which take into consideration the magnetic dipolar interactions, do not provide a rigorous theoretical treatment of the subject and are based on a series of approximations, some of them being mathematically unjustified [1]. Moreover, the problem of their contradictions [1], [5], [11], [12] is much discussed, but without a conclusion that clearly leans toward a model or another one.

Regarding the action of the external magnetic field, the Néel-Brown model (basis of the models mentioned above) considers that the nanoparticle anisotropy axes are aligned with the external magnetic field. This condition is valid only in the intense magnetic fields. In the general case, at not very intense magnetic fields, and when the nanoparticles are dispersed in solid matrix, the anisotropy axes of nanoparticles are randomly distributed and the use of Néel-Brown model is no longer justified, although it is the starting point of introducing the magnetic dipolar interactions, because the local magnetic dipolar field has usually an oblique orientation to the anisotropy axis of the nanoparticle. Coffey et al studied the external oblique magnetic field influence on the Néel relaxation time [13], succeeding to obtain a solution to the Fokker-Planck equation for nanoparticle magnetic moments dynamics, based on the Kramer's theory [13]. This solution includes the solution of Néel-Brown model as particular case.

With regard to the high degree of complexity of the problem, it seems useful to consider the "numerical experiments". For a system consisting of spherical magnetic nanoparticles with uniaxial magnetic anisotropy, we are going to start from the Coffey's solution, which we discretize to introduce the nanoparticle magnetic dipolar interactions. The 3D random distribution of the nanoparticle positions and magnetic anisotropy easy axes will be numerically simulated. The model allows obtaining the Néel relaxation time for each nanoparticle. The Néel relaxation time for the system is found through mediation. Our model is particularly suitable for predicting and understanding the Néel relaxation time trends in an interacting magnetic nanoparticle system, with a diameter of up to a few tens of nm.

2. About the magnetic dipolar interaction

A monodomain nanoparticle is in a state of uniform magnetization for any field applied. Each monodomain nanoparticle is characterized by a magnetic moment, which usually is given by [14]:

$$\vec{\mu}_i = M_s v_i \hat{\mu}_i \tag{1}$$

where M_s is the spontaneous magnetization, v_i is the particle volume and $\hat{\mu}_i$ is the unit vector of the magnetic moments.

Let $\vec{\mu}_i$ be the magnetic moment of the nanoparticle i, and $\vec{\mu}_j$ the magnetic moment of the nanoparticle j with uniaxial anisotropy. The energy of the magnetic dipolar interaction between two nanoparticles (i and j) is given by the relation [14]:

$$E_{ijd} = \frac{\mu_0}{4\pi} \cdot \frac{\mu_i \mu_j}{r_{ij}^3}\left[\hat{\mu}_i \cdot \hat{\mu}_j - 3(\hat{\mu}_i \cdot \hat{r}_{ij})(\hat{\mu}_j \cdot \hat{r}_{ij})\right] \tag{2}$$

where r_{ij} is the distance between the centres of those two nanoparticles, $\hat{r}_{ij}$ is the versor of the direction connecting the nanoparticles i and j, $\hat{\mu}_i$ and $\hat{\mu}_j$ are the unit vectors of the magnetic moments of the nanoparticles i & j, and μ_0 is the vacuum magnetic permeability.

Since the nanoparticle i has a dipole-dipole magnetostatic interaction with all the other nanoparticles, the magnetic dipolar energy of the nanoparticle i and the local dipolar magnetic field acting on the nanoparticle i can be expressed as follows:

$$E_{id} = -\mu_0 \sum_{j,j\neq i} \vec{\mu}_i \cdot \vec{H}_{ijd} \tag{3}$$

On each nanoparticle is acting a local magnetic dipolar field, given by:

$$\vec{H}_{id} = \frac{1}{4\pi} \sum_{j,j\neq i} \frac{\mu_j}{r_{ij}^3}\left[3\hat{r}_{ij} \cdot (\hat{\mu}_j \cdot \hat{r}_{ij}) - \hat{\mu}_j\right] \tag{4}$$

Due to the long range character of the magnetic dipolar interaction, the interaction effect can be significant. If on the nanoparticle system is also acting an external magnetic field, then a local magnetic field is acting on each nanoparticle, which is the vectorial sum between the external magnetic field applied ($\vec{H}_{ext}$) and the internal magnetic dipolar field $\vec{H}_{id}$ determined by the magnetic dipolar interactions among the nanoparticles.

$$\vec{H}_i = \vec{H}_{ext} + \vec{H}_{id} \tag{5}$$

It will be of great help to develop a model of the interactions between uncoated and coated nanoparticles.

3. Néel relaxation time of the nanoparticles interacting in local oblique magnetic field

The relaxation of magnetic nanoparticle systems is important for their applications. The Néel relaxation process is the solid state mechanism that occurs inside the nanoparticle, corresponding to the magnetic moment rotation between two equilibrium positions. The nanoparticle is fixed in space.

For the relaxation time assessment, we adapt an analytical Coffey solution for external magnetic fields [15], [13] - mean field method (eq. 5 and 4), in order to take into account the magnetic dipolar field on a nanoparticle. Below, we will use the notations:

$$h_i = \frac{\mu_0 M_s H_i}{2K_{ieff}} \quad \text{and} \quad \sigma_i = \frac{K_{ieff} v_i}{k_B T} \tag{6}$$

where K_{ieff} is the effective magnetic anisotropy constant of the ith nanoparticle, and H_i is assessed using the equations (4) and (5). The free diffusion time of magnetisation τ'^{i}_{ON} for low damping [16] is:

$$\tau'^{i}_{ON} = \frac{v_i M_s}{2\gamma\alpha k_B T} \tag{7}$$

where α is the damping constant, and γ is the gyromagnetic ratio.

For most systems of ferro- and ferrimagnetic nanoparticles, the magnetic damping constant α ranges around the low limit (α<<1) [17].

The Coffey analytical calculation is based on Kramer's theory [13] and in agreement with the numerical calculation and experimental results [18].This calculation shows the relaxation time dependence on the magnetic damping constant α.

In these conditions, the relaxation time relation for an oblique magnetic field is [15]:

$$\tau^{i}_{N(oblique\_field)} = \frac{4\pi\tau'^{i}_{ON}\left(S_{i1}^{-1} + S_{i2}^{-1}\right)}{\left(\sqrt{c_{i1}^{(1)}c_{i2}^{(1)}} \cdot e^{-\Delta V_{i12}} + \sqrt{c_{i1}^{(2)}c_{i2}^{(2)}} \cdot e^{-\Delta V_{i21}}\right)} \tag{8}$$

where ΔV_{i12} and ΔV_{i12} are the normalized energy barriers for magnetic moment re-orientations. In eq. (8):

$$c_{i1}^{(p)} = 2\sigma_i(\cos 2\theta_{ip} + h_i\cos(\theta_{ip} - \psi_i), c_{i2}^{(p)} = 2\sigma_i(\cos^2\theta_{ip} + h_i\cos(\theta_{ip} - \Psi_i), \text{ with p=1,2} \tag{9}$$

where θ_{ip} are the solutions of the transcendental equation $\sin 2\theta_i = 2h_i \sin(\Psi_i - \theta_i)$

If $h_i<h_{ic}(\Psi_i)<1$ (Ψ_i being the angle between $\vec{H}_i$ and the easy anisotropy axis of the ith nanoparticle). [15]

$$\cos\theta_{i1,2} = \pm 1 \mp \frac{h_i^2}{2}\sin^2\psi_i + h_i^3\sin^2\psi_i\cos\psi_i \mp \frac{h_i^4}{16}(13+11\cos 2\psi_i)\sin^2\psi_i + \frac{h_i^5}{2}(3+\cos 2\psi_i)\sin^2\psi_i\cos\psi_i \mp$$
$$\mp \frac{h_i^6}{64}(183+156\cos 2\psi_i - 19\cos 4\psi_i)\sin^2\psi_i + \ldots \tag{10}$$

$$\Delta V_{i12} = \sigma_i[1 - 2h_i(\sin\Psi_i - \cos\Psi_i) + h_i^2 + \frac{h_i^3}{2}\sin 2\Psi_i(\cos\Psi_i - \sin\Psi_i) + \frac{h_i^4}{2}\sin^2 2\Psi_i +$$
$$+ \frac{h_i^5}{32}\sin 2\Psi_i(7\cos\Psi_i - 3\cos 3\Psi_i - 7\sin\Psi_i - 3\sin 3\Psi_i) + \frac{h_i^6}{2}\sin^2 2\Psi_i + \ldots] \tag{11}$$

$$\Delta V_{i21} = \sigma_i[1 - 2h_i(\sin\Psi_i + \cos\Psi_i) + h_i^2 + \frac{h_i^3}{2}\sin 2\Psi_i(\cos\Psi_i + \sin\Psi_i) + \frac{h_i^4}{2}\sin^2 2\Psi_i +$$
$$+ \frac{h_i^5}{32}\sin 2\Psi_i(7\cos\Psi_i - 3\cos 3\Psi_i + 7\sin\Psi_i + 3\sin 3\Psi_i) + \frac{h_i^6}{2}\sin^2 2\Psi_i + \ldots] \tag{12}$$

$$S_{i1,2} = \sigma_i\sqrt{h_i\sin\Psi_i}[16 - \frac{104}{3}h_i\sin\Psi_i + h^2(1 - 21\cos 2\Psi_i) + \frac{h_i^3}{2}\sin\Psi_i(45 + 51\cos 2\Psi_i) + \ldots]\pm$$
$$\pm 2\pi\sigma_i h_i^2\sin 2\Psi_i(4 - 3h_i\sin\Psi_i - 2h_i^2\sin^2\Psi_i + \ldots) \tag{13}$$

It is defined the normalized effective energy barriers for nanoparticle magnetic moment re-orientations

$$\Delta V_{i,eff} = \frac{\Delta V_{i12} + \Delta V_{i21}}{2} \tag{14}$$

The relation (8) can be written:

$$\frac{1}{\tau^{i}_{N(oblique\_field)}} = \frac{S_{i1} \cdot S_{i2}\sqrt{c^{(1)}_{i1}c^{(1)}_{i2}}}{4\pi\tau^{i}_{0N}(S_{i1}+S_{i2})}e^{-\Delta V_{i12}} + \frac{S_{i1} \cdot S_{i2}\sqrt{c^{(2)}_{i1}c^{(2)}_{i2}}}{4\pi\tau^{i}_{0N}(S_{i1}+S_{i2})}e^{-\Delta V_{i21}} = f_{12}e^{-\Delta V_{i12}} + f_{21}e^{-\Delta V_{i21}} \quad (15)$$

where f_{12} and f_{21} are the attempt frequency factors.

The equations (8)-(13) indicate that the dipolar interactions have a complex effect, which leads to the misalignment of the moments with respect to the local easy axis.

4. Results and discussions

We simulate the random 3D arrangement of the nanoparticles in a face-centred cubic lattice. Among all the possible positions, the particles have been located in randomly selected suitable locations, according to the ratio between the volumes of the considered nanoparticles and the whole volume of the simulation box. The periodic boundary conditions have been applied on all the sides of the simulation box. A random 3D distribution of the anisotropy easy axes is also used. We are going to study a system of 1000 uncoated magnetite nanoparticles, with uniaxial magnetic anisotropy, effective magnetic anisotropy constant of $2.5 \cdot 10^4$ J/m^3 and easy axes of anisotropy, with random distribution. We consider spontaneous magnetization of $4.46 \cdot 10^5$ A/m, simulation temperature T = 298.15 and external magnetic field intensity of 15 kA/m, typical value for magnetic hyperthermia applications.

As can be seen from the equation (4), the magnetic dipolar interactions can have an influence determined by their nanoparticle concentrations and sizes.

4.1 The long range and anisotropic character of magnetic dipolar interaction

The long range and anisotropic character of the magnetic dipolar interaction is reflected in the fact that the magnetic dipolar field, respectively the magnetic dipolar energy of a nanoparticle, depends on its neighbours' contributions. To highlight this character, we will study how the strength of the average magnetic dipolar field depends on the number of nanoparticles found in the system, at the constant parameters of the simulation. The local magnetic dipolar field on the nanoparticle (4) will be simulated either using the Ewald method [19] or by direct summation. We are going to consider nanoparticles of 12 nm with the volume fraction f = 0.24, in the absence and in the presence of external magnetic field. The results are shown in the Figures 1 and 2.

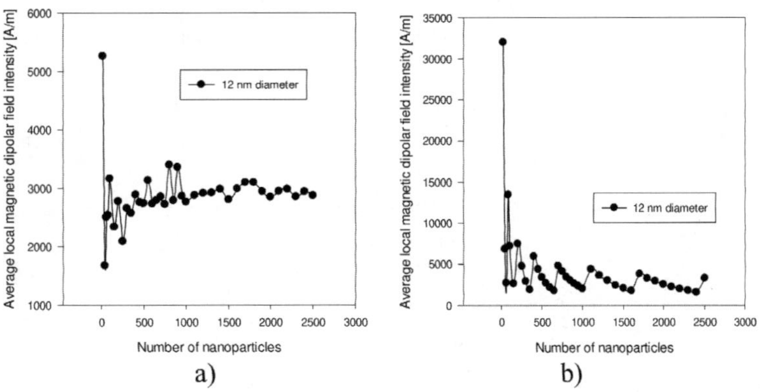

Figure 1. Average local magnetic dipolar field intensity versus number of nanoparticles
a) by direct summation, b) by applying the Ewald summation method

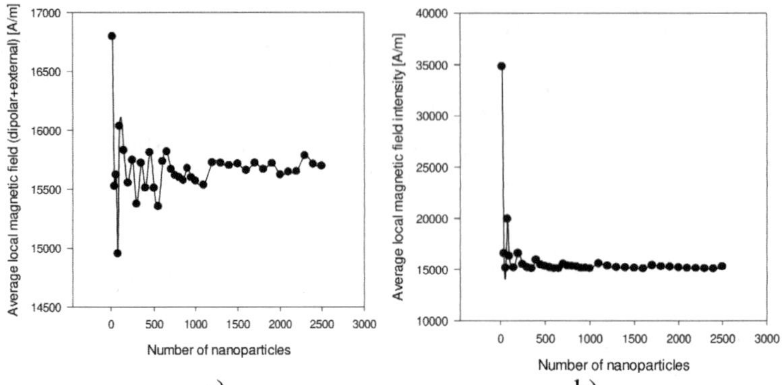

a) b)

Figure 2. Average local total magnetic field intensity (dipolar + applied) versus number of nanoparticles
a) by direct summation, b) by applying the Ewald summation method

In Figure 1, we can see that the average local magnetic dipolar field intensity varies with the number of nanoparticles in the system. If over the magnetic dipolar field is superimposed an external magnetic field (whose magnitude, in this case, is 15 kA/m, typical for magnetic hyperthermia), it can be seen that the magnetic dipolar field influences are still visible – Figure 2.

4.2 The influence of magnetic dipolar interactions on Néel relaxation time

We take into account three types of nanoparticles, i.e. small-sized (6 nm), medium-sized (12 nm) and large-sized (18 nm). In the first instance, we consider that the sizes have no dispersion. We will assess the Néel relaxation time on each nanoparticle, using the relations (8)-(13), at different values of the nanoparticle volume fraction, ranging from 0.02 to 0.3. We obtain an average value of the Néel relaxation time. The results are shown in Figure 3. It is noted that, for small and large sizes (Figures 3a and 3c), the average Néel relaxation time increases with increasing concentration, whereas, for medium-sized nanoparticles, the average Néel relaxation time decreases with increasing concentration. At small sizes of the nanoparticles, this increase is very sensitive, so that the Néel relaxation time can be considered to be approximately constant. At large sizes, the increase should be taken into account.

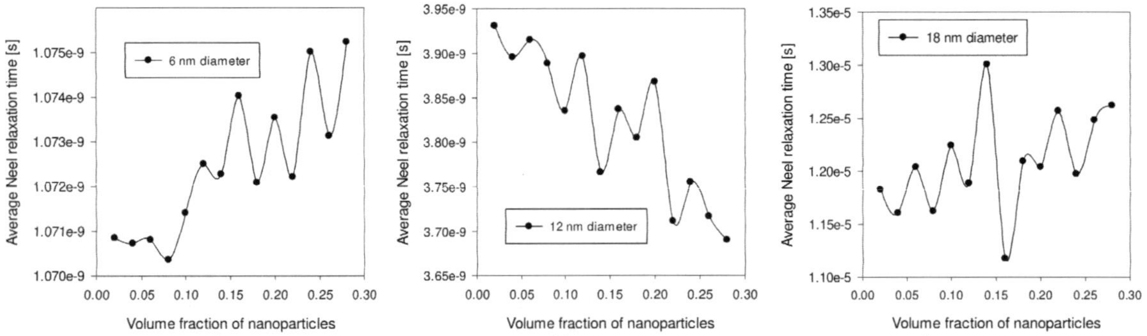

a) b) c)

Figure 3. Average Néel relaxation time versus volume fraction of nanoparticles without radii dispersion
a) small-sized; b) medium-sized; c) high-sized nanoparticles

To understand this dependence, we will simulate the local magnetic field on the nanoparticle and its average throughout the system, as well as the average normalized energy barriers and the effective average normalized energy barriers of nanoparticle reorientation. In Figure 4, we can see that the average values of the local magnetic dipolar field a) and the local magnetic dipolar field b) increase with increasing concentration of nanoparticles for all three categories of nanoparticle sizes.

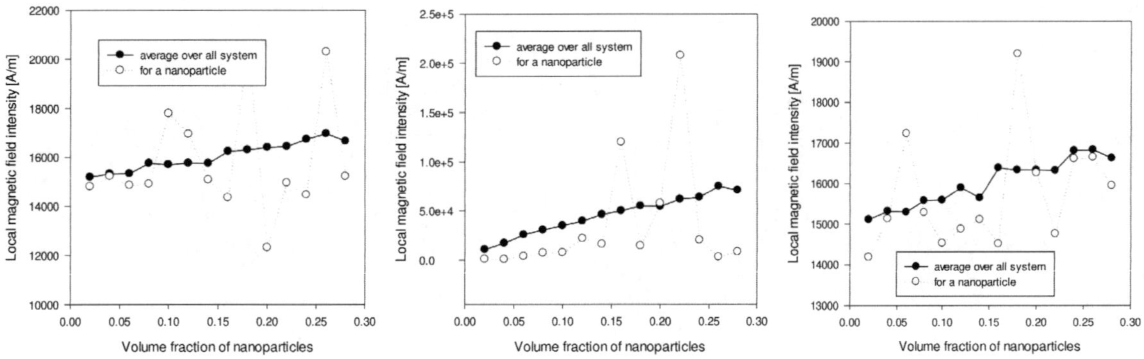

a) b) c)

Figure 4. Average local magnetic dipolar intensity versus the volume fraction of nanoparticles a) small-sized; b) medium-sized; c) high-sized nanoparticles

Regarding the dependence of the average normalized energy barriers and the effective average normalized energy barriers on the volume fraction, Figure 5, Figure 6, we can see their decrease with increasing concentration for all the size categories of nanoparticles.

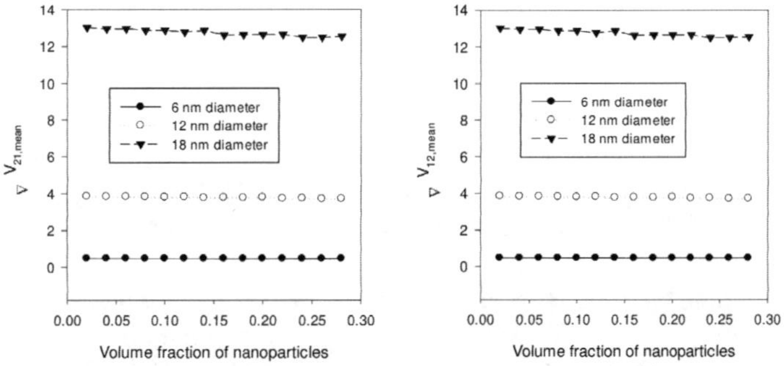

a) b)

Figure 5. Average normalized energy barriers versus volume fraction of nanoparticles

Analysing the Figures 3-5, we can conclude that, in case of nanoparticles without size dispersions or with small dispersions, for medium sizes, the decrease of energy barriers with increasing concentration influences significantly the decrease of relaxation time with increasing concentration. In this case, the dependence of the attempt frequency factors on the local magnetic field, in the relation (15), has no major influence. In case of very small and quite large nanoparticles, the values of these frequency factors change the shape of the relaxation time dependence on the nanoparticle concentration. So, their approximation through constant values, as appearing in most models in the literature, is not very appropriate.

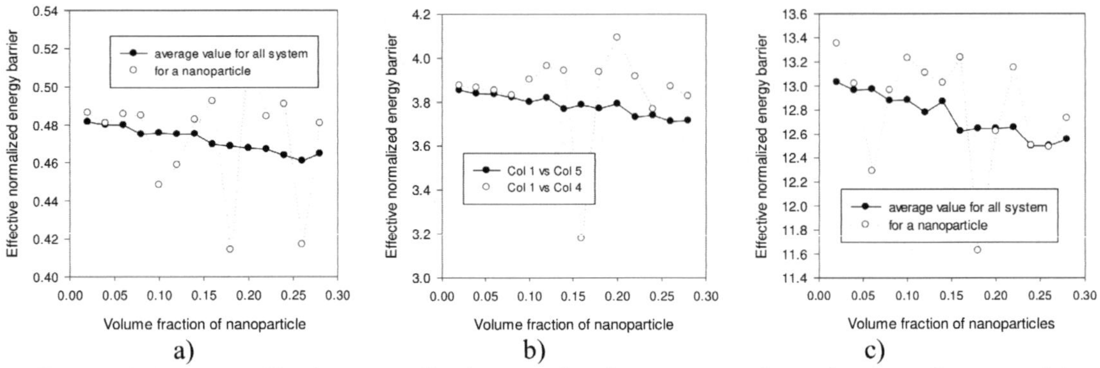

Figure 6. Average effective normalized energy barriers versus volume fraction of nanoparticles a) small-sized; b) medium-sized; c) high-sized nanoparticles

Moreover, we keep the nanoparticle volume fraction at a constant value and assess the Néel relaxation time according to the standard deviation parameter of nanoparticle diameters – Figure 7.

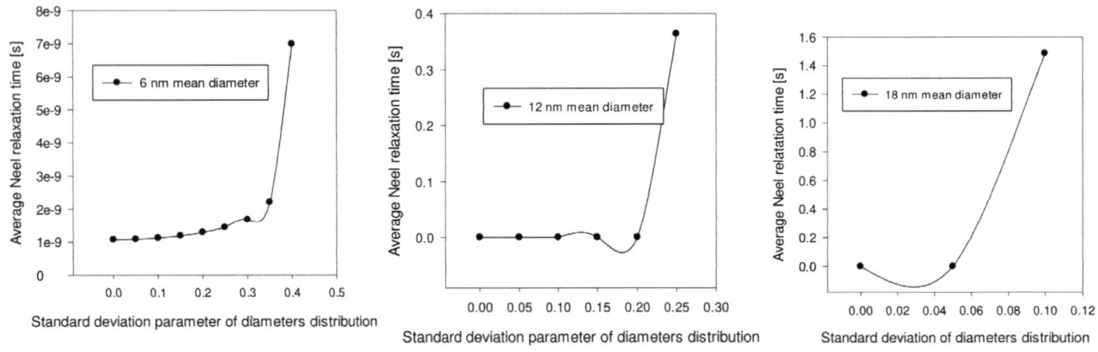

Figure 7. Average Néel relaxation time versus volume fraction of nanoparticles a) small-sized; b) medium-sized; c) high-sized nanoparticles

We consider lognormal distribution of the diameters, with the standard deviation $v \cdot dm$, where v is the standard deviation parameter. If the nanoparticle sizes are dispersed, the Néel relaxation time increases sharply with increasing standard deviation parameter of the nanoparticle diameters - Figure 7. We keep the standard deviation for size distribution at a constant value, for example $0.1 \cdot d_m$, and assess the dependence of the Néel relaxation time on the nanoparticle volume fraction. It is noted that the size dispersion does not affect the type of Néel relaxation time dependence on the nanoparticle concentration, but only the values – Figure 8.

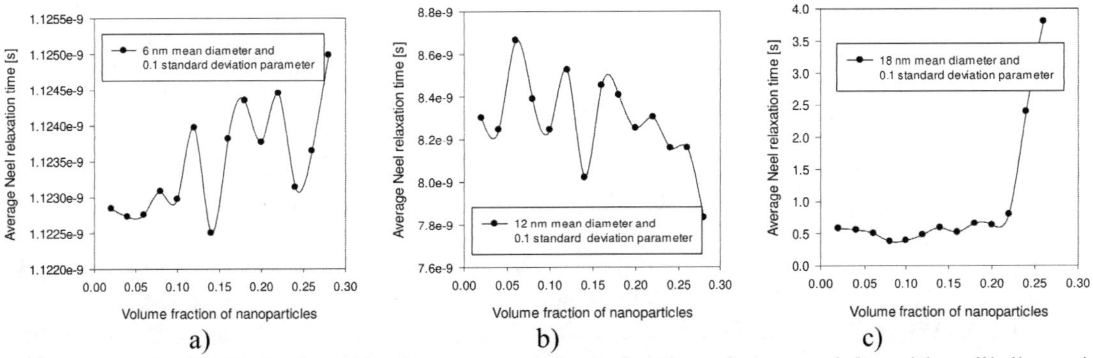

Figure 8. Average Néel relaxation time versus volume fraction of nanoparticles with radii dispersion a) small-sized; b) medium-sized; c) high-sized nanoparticles

5. Conclusions

In this paper, we tried to clarify the effect of magnetic dipolar interactions on the Néel relaxation time for nanoparticle systems with medical applications, i.e. not very concentrated and not subjected to very intense external magnetic fields. The Néel relaxation time has been assessed by using the Coffey solution in oblique magnetic field, discretized and adapted to local magnetic field on the nanoparticle. The 3D random distribution of the nanoparticle positions and the random distribution of the easy axes anisotropy orientations were obtained by numerical simulation.

Considering the nanoparticle sizes without dispersion or with small dispersion, we found the fact that, in case of small-sized (≤ 6 nm) and large-sized (≥ 18 nm) nanoparticles, the average Néel relaxation time increases with increasing concentration, whereas, in case of medium nanoparticle sizes, the average Néel relaxation time decreases with increasing concentration. The increase of Néel relaxation time with increasing nanoparticle concentration is sensitive at small sizes, so it can be considered approximately constant, but this increase should be taken into account at large sizes of the nanoparticles. In case of medium sizes, the influence of the energy barrier variation with the concentration prevails, so the attempt frequency factors have no major influence. In case of small and large sizes, the attempt frequency factors dependence on concentration (through the local magnetic field on particle) and on nanoparticle sizes, can overturn the concentration dependence of the Néel relaxation time, being significant at large nanoparticle sizes.

If the nanoparticle sizes show dispersion, the Néel relaxation time increases strongly with increasing standard deviation parameter of the nanoparticle diameters, the more strongly the larger are the nanoparticle sizes. Also, we found that the size distribution does not affect the type of Néel relaxation time dependence on the nanoparticle concentration, but only the values.

The study presents a theoretical tool to control the parameters influencing the dynamics of magnetic moments of the nanoparticle systems, with biomedical applications, and not only.

References

[1] Hansen M F and Mørup S 1998 Models for the dynamics of interacting magnetic nanoparticles, *Journal of Magnetism and Magnetic Materials* **184** 262-274

[2] Prené P, Tronc E, Jolivet J P, Livage J, Cherkaoui R, Noguès M, Dormann J L and Fiorani D 1993 Magnetic properties of isolated γ-Fe2O3 particles, *IEEE Transactions on Magnetics* **29** 2658-2660

[3] Mørup S and Tronc E 1994 Superparamagnetic relaxation of weakly interacting particles, *Physical Review Letters* **72** 3278- 3281

[4] Urtizberea A, Natividad E, Arizaga A, Castro M and Mediano A 2010 Specific Absorption Rates and Magnetic Properties of Ferrofluids with Interaction Effects at Low Concentrations, *Journal of Physical Chemistry C* **114** (11) 4916–4922

[5] Dormann J L, Bessais L and Fiorani D 1988 A dynamic study of small interacting particles: superparamagnetic model and spin-glass laws, *J. Phys. C: Solid State Phys*ics **21** 2015- 2034
[6] Dormann J L, Spinu L, Tronc E, Jolivet J P, Lucari F, D'Orazio F and Fiorani D 1998 Effect of interparticle interactions on the dynamical properties of γ-Fe2O3 nanoparticles, *Journal of Magnetism and Magnetic Materials* **183** 255-260
[7] Brown W F Jr 1963 Thermal Fluctuations of a Single-Domain Particle, *Physical Review* **130** 1677-1686
[8] Shtrikman S and Wohlfarth E P 1981 The theory of the Vogel-Fulcher law of spin glasses, *Phys. Lett. A* **85** 467-470
[9] Kleemann W, Petracic O, Binek C, Kakazei G N, Pogorelov Y G, Sousa J B, Cardoso S and P P Freitas 2001 Interacting ferromagnetic nanoparticles in discontinuous $Co_{80}Fe_{20}/Al_2O_3$ multilayers: From superspin glass to reentrant superferromagnetism, *Phys. Rev. B* **63** 134423-1-134423-5
[10] Petracic O, Kleemann W, Binek C, Kakazei G N, Pogorelov Y G, Sousa J B, Cardoso S and Freitas P P 2002 Superspin glass behavior of interacting ferromagnetic nanoparticles in discontinuous magnetic multilayers, *Phase Transitions* **75** 73-81
[11] Dormann J L, Fiorani D and Tronc E 1997 Magnetic Relaxation in Fine-Particle Systems, *Adv. Chem. Phys.* **98** 283-494
[12] Dormann J L, Fiorani D and Tronc E 1999 On the models for interparticle interactions in nanoparticle assemblies: comparison with experimental results, *Journal of Magnetism and Magnetic Materials* **202** (1) 251–267
[13] Coffey T, Crothers D S F, Dormann J L, Geoghegan L J, Kalmykov Y P, Waldron J T and Wickstead A W 1995 Effect of an oblique magnetic field on the superparamagnetic relaxation time, *Physical Review B* **52** (22) 15951-15965
[14] Haase C and Nowak U 2012 Role of dipole-dipole interactions for hyperthermia heating of magnetic nanoparticles ensamble, *Physical Review B* **85**(4) 045435-1-045435-5
[15] Coffey W T and Kalmykov Y P 2012 Thermal fluctuations of magnetic nanoparticles: Fifty years after Brown, *Applied Physics Reviews Section of the Journal of Applied Physics* **112** 121301-1-121301-47
[16] Coffey W T and Cregg P J 1993 On the Theory of Debye and Néel Relaxation of Single Domain Ferromagnetic Particles, *Advances in Chemical Physics* **83** 263-464
[17] Fannin P C, Malaescu I and Marin C N 2005 The effective anisotropy constant of particles within magnetic fluids as measured by magnetic resonance, *Journal of Magnetism and Magnetic Materials* **289** 162–164
[18] Coffey W T, Crothers D S F, Dormann J L, Kalmykov Y P, Kennedy E C and Wernsdorfer W 1998 Thermally Activated Relaxation Time of a Single Domain Ferromagnetic Particle Subjected to a Uniform Field at an Oblique Angle to the Easy Axis: Comparison with Experimental Observations, *Pys. Rev. Lett.* **80** 5655-5658
[19] Wang Z, Holm C 2001 Estimate of the cutoff errors in the Ewald summation for dipolar systems, *J. Chem. Phys.* **115** 6277-6798

IOP Conf. Series: Materials Science and Engineering **106** (2016) 012005 doi:10.1088/1757-899X/106/1/012005

Vector control structure of an asynchronous motor at maximum torque

C P Chioncel[1], G O Tirian[2], N Gillich[1] and E Raduca[1]

[1]Eftimie Murgu University of Resita, Electrical Engineering and Computer Science Department, P-ta Tr. Vuia, no. 1-4, 320085 Resita, Romania
[2]Politehnica University of Timisoara, Department of Electrical Engineering and Industrial Informatics, 5 Revolution Street, 331136 Hunedoara, Romania

E-mail: c.chioncel@uem.ro

Abstract. Vector control methods offer the possibility to gain high performance, being widely used. Certain applications require an optimum control in limit operating conditions, as, at maximum torque, that is not always satisfied. The paper presents how the voltage and the frequency for an asynchronous machine (ASM) operating at variable speed are determinate, with an accent on the method that keeps the rotor flux constant. The simulation analyses consider three load types: variable torque and speed, variable torque and constant speed, constant torque and variable speed. The final values of frequency and voltage are obtained through the proposed control schemes with one controller using the simulation language based on the Maple module. The dynamic analysis of the system is done for the case with P and PI controller and allows conclusions on the proposed method, which can have different applications, as the ASM in wind turbines.

1. Introduction

Currently, the speed control on synchronous and asynchronous machine is performed by changing the frequency of the power supply, based on the fundamental relationship:

$$n_1 = \frac{f_1}{p_1} \tag{1}$$

where: f_1 – the frequency of the stator voltage; n_1 - rotating speed magnetic field of the stator and p_1 – represents the number of pole pairs [1].

As the number of pole pairs - p_1 - is given by construction, results that the speed - n_1 - can be only changed by modifying the frequency of the power supply. This frequency change can be achieved through various techniques [2] that use power electronics based on more efficient thyristors.

Different, complex, control methods of the asynchronous machine are proposed and analysed in different publications, with a difficult practical implementation [3], [4], [5]. The proposed method is simpler, presenting a high efficiency, as shown by the obtained results. All those control strategies require a real time monitoring system to provide a correct overview of the systems behaviour [8].

When only the frequency is changed, the voltage remaining unchanged, in the electric circuit with ferromagnetic coil, occurs the phenomenon of saturation of the ferromagnetic core, higher harmonic in current are reached and therefore the operation poses problems, both, to the supply system as to the machine that operates at variable speed [5].

2. Vector control scheme adjusting the voltage, frequency aiming the voltage variation

The control algorithm for ASM implements the speed [rpm] control by three different strategies: constant stator flux Ψs, constant rotor flux Ψr and constant useful flux Ψu.

For the proposed control strategies, two methods can be applied: first, the voltage at the controllers output is proportional to the speed error and the frequency is calculated from the condition of constant rotor flux and second, at the controller output we obtain a pulsation (i.e. stator frequency) proportional with the speed error and the voltage is computed, as in the first case, from the condition to maintain the rotor flux constant. Further, we present near the 1st method and the obtained results. From the general equation of the asynchronous machine, we obtain the structural scheme, Figure 1, control structure through voltage and frequency follows the voltage variations.

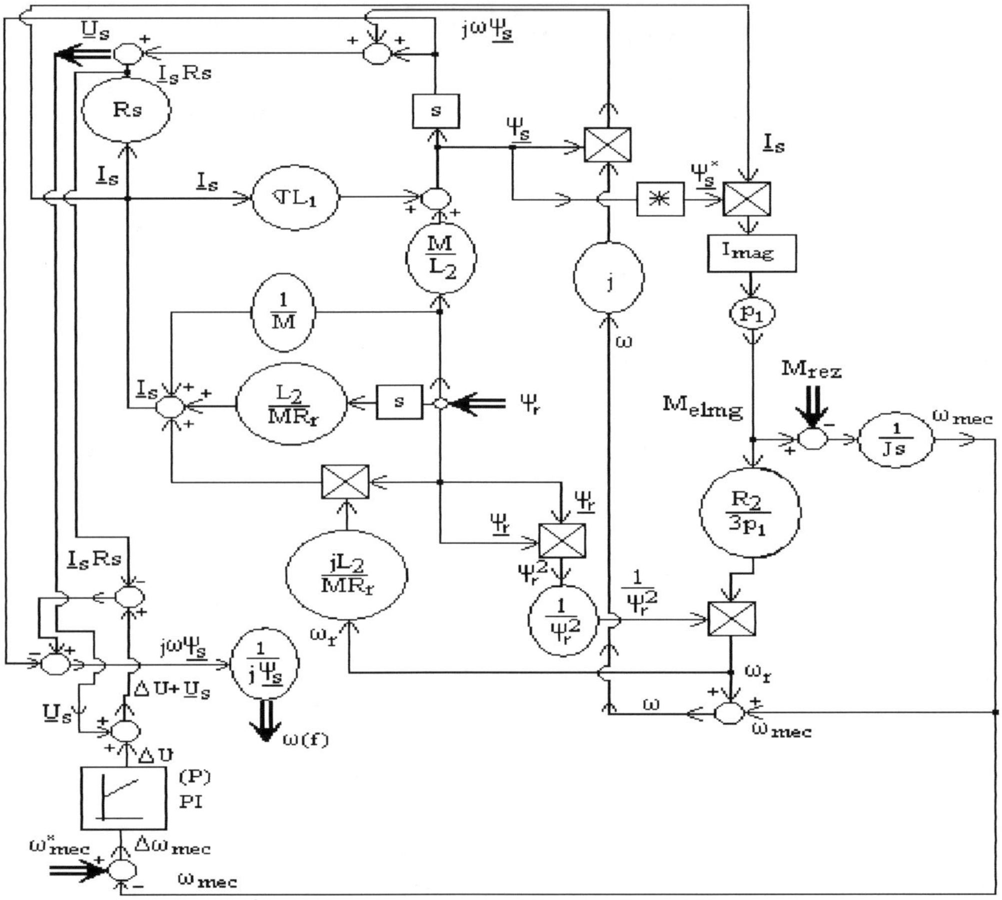

Figure 1. Control structure through voltage and frequency follows the voltage variations

The primary parameter is the stator voltage and the second parameter is the stator frequency. Through the controller, P or PI, the stator voltage is continuously changed proportionally to the error of the mechanical speed: $\Delta\omega_{mec} = \omega^*_{mec} - \omega_{mec}$. The condition of constant rotor flux is achieved by changing by changing the stator frequency until reaching the prescribed speed, ω^*_{mec}, voltage and frequency are variable.

At the controllers (P or PI) output, we obtain a scalar value: α, sub-unitary, dependent on the value $\Delta\omega_{mec} = \omega^*_{mec} - \omega_{mec}$. If the speed leap [rpm] is higher ($\Delta n = n^* - n$), α will be higher too.

The speed control through by voltage, the scalar component α is multiplied by the voltage Us, so that the stator voltage increases his value to $U_s(1+\alpha)$.

The stator frequency f (i.e. pulsation ω) is obtained from the stator voltage equation:

$$U \cdot (1+a) - I_S R_S - \frac{d\psi_S}{dt} = j\omega\psi_S, \tag{2}$$

as it can be seen in the structural scheme given above.

The disturbing size – resistant torque M_{rez}, and the electromagnetic torque – M_{elmg}, based on the movement equation

$$J \cdot \frac{d\omega_{mec}}{dt} = M_{elmg} - M_{rez}, \tag{3}$$

will generate the mechanical angular velocity ω_{mec} [6]. The current value of ω_{mec} is compared with the prescribed value ω^*_{mec} resulting the error

$$\Delta\omega_{mec} = \omega^*_{mec} - \omega_{mec}. \tag{4}$$

This value (4) represents the input in the P or PI controller.

The rotor flux is given as value (Ψ_{rN}), and, therefore, it determines the stator flux Ψs, stator current Is and then the electromagnetic torque [7]:

$$M_{elmg} = p_1 I_{mag}\left[Is \cdot \psi^*_S\right] \tag{5}$$

The proposed, simplified, control diagram block for adjusting the voltage, using a P or PI controller, is shown in Figure 2.

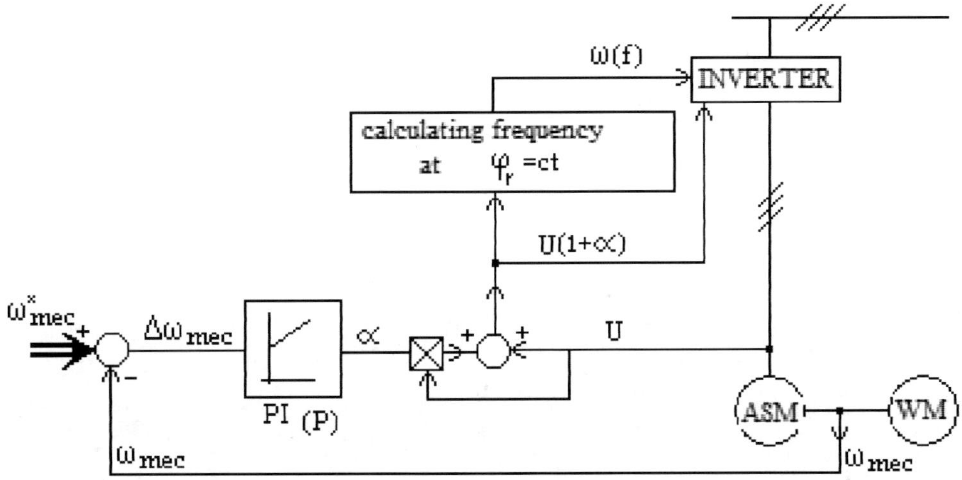

Figure 2. Control structure through voltage and frequency follows the voltage variations

3. Control structure with the P-Type regulator

The parameters of the asynchronous machine used in the implementing the simulation analysis, are: $L_1 = 0.1$ H, $L_2 = 0.1$ H, $M = L_u = 0.08$ H, $R_1 = 5\Omega$, $R_2 = 5\Omega$, $p_1 = 1$, $\sigma = 0.36$, $L_{1\sigma} = L_{2\sigma} = 0.02$ H, $1 - \sigma = 0.64$, $U_N = 380$ V, $\Psi_{sN} = 1.2$ Wb. The initial data, imposed by the technological process:

- stator flux Ψ_s (from the construction data of the motor): $\Psi_s = 1.2$ Wb,
- initial speed $n(0) = 2866.2$ rpm or $\omega_m(0) = 300$ rad/s,
- initial frequency $f(0) = 50$ Hz,
- rotor flux ψ_r is obtained from the stator flux ψ_s:

$$\psi_r = \psi_S \frac{L_u}{L_1} \cdot \frac{R_2}{\sqrt{(R_2)^2 + (\omega_r L_2 \sigma)^2}} = 0.96[Wb]. \tag{6}$$

The final parameters of the system, are:

- final speed $n_{\text{final}} = 2960.4$ rpm,
- final mechanical angular velocity = prescribed angular mechanical speed,
- from $\omega = \omega_r + \omega_{\text{mec}} = 324$ rad/s,
- stator frequency $f = 51.568$ Hz,
- final stator pulsation $\omega_{\text{final}} = 324$ rad/s,
- rotor flux $\Psi_{\text{rfinal}} = 0.96$ Wb,
- final stator voltage $U_{\text{final}} = 406.13$ V,
- rotor angular frequency $\omega_{\text{rfinal}} = \omega - \omega_{\text{mec}} = 14$ rad/s,
- final electromagnetic torque equal with the initial, 7.74 Nm.

The dynamic of the system analyzed by numerical simulations, assumes the evolution of the system from the initial voltage U and pulse ω: $U(0) = 394.2$ V, $\omega(0) = 314$ rad / s, $f(0) = 50$ Hz to the final of values: $U(\infty) = 406.13$ V, $\omega(\infty) = 324$ rad / s, $f(\infty) = 51\ 568$ Hz.

The control parameter based on the proposed control structure, Figure 2, is:

$$\alpha = K \cdot \left(\omega^*_{mec} - \omega_{mec}\right) \tag{6}$$

INITIAL: $\omega = 314$, $\omega_r = 14$, $\omega_{\text{mec}} = 300$

FINAL: $\omega = 324$, $\omega_r = 14$, $\omega^*_{\text{mec}} = 310$.

In order that the control system, using a P-type regulator, has to reach the prescribed speed in a control period of $T = 0.4$[s], this will require a certain number of adjustments brought to voltage and frequency values.

For $K = 1$, the firts adjustment range 0 – 0.4s; initial: $\omega_{\text{mec}} = W = 300$ rad/s; voltage = 394.2 + 1(310 - W) = 404.2V.

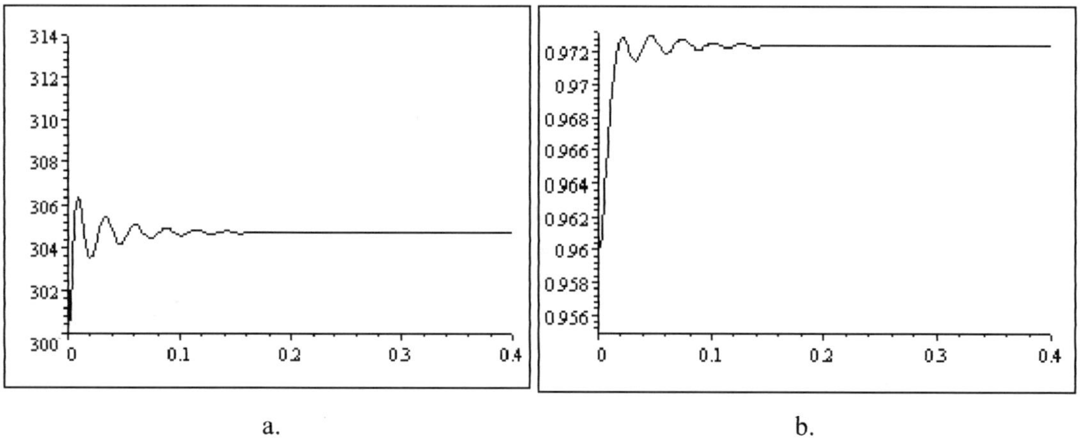

Figure 3. Evolution in time of a) mechanical angular velocity and b) rotor flux

It is noted that for the first interval, the angular mechanical speed oscillates and reach a stationary value after 0.15s and the rotor flux also stabilizes in the same time, exceeding the nominal value. The final values, which consist the baseline for the next adjustment period, are: $U(0.4) = -0.25881$, $Z(0.4) = -4.5883$, $Y(0.4) = -20.696$, $X(0.4) = 6.9209$, $\omega_{\text{mec}} = 304.73$[rad/s] $= W(0.4)$.

For a permissible error $\Delta\omega_{\text{mec}} = 0.3$[rad/s] the process is concluded, because $\omega_{\text{mec}} - \omega_{\text{mec}} < 0.3$, Figure 4. The adjustment period assumes seven intervals, a time of 2.8s.

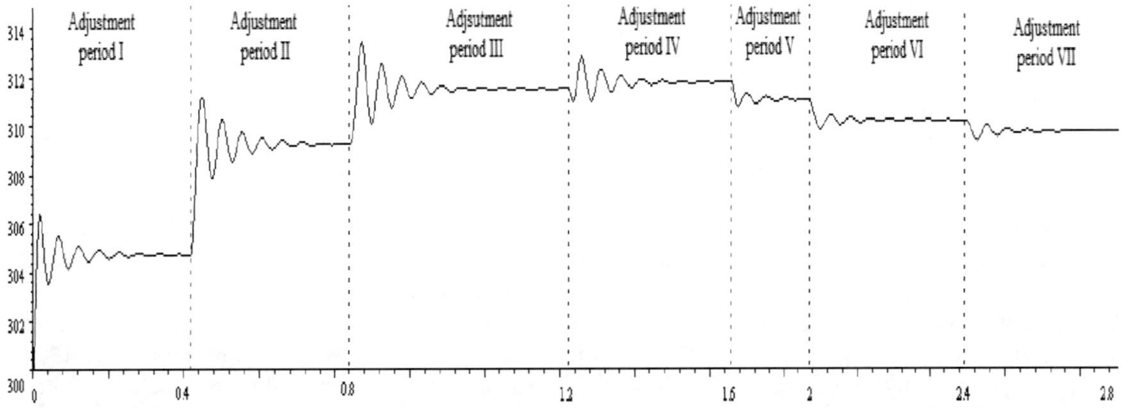

Figure 4. General evolution of the angular mechanical speed in the adjusting period, for $K = 1$

Another studied case is that for the proportionality constant $K = 0.5$. After going through the control steps, as shown in the case $K = 1$, we obtain the general evolution of the pulse, in five adjustment periods, in a time of 2s, Figure 5.

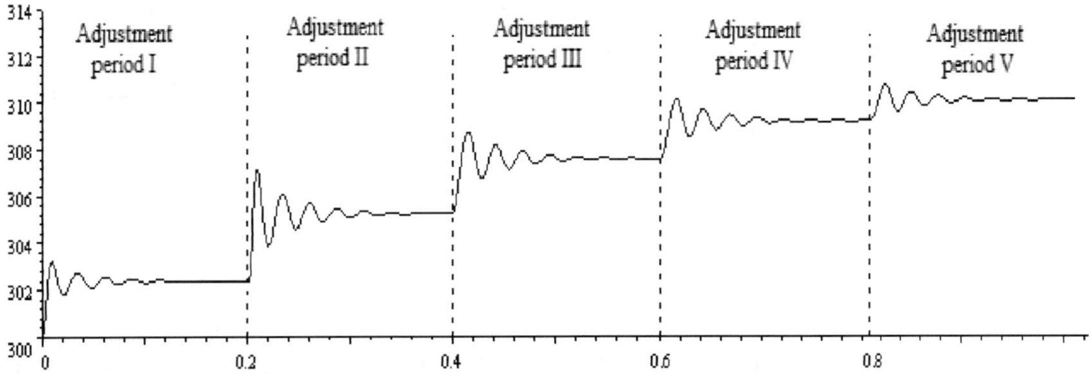

Figure 5. General evolution of the angular mechanical speed in the adjusting period, for $K = 0.5$

Concerning the proportional controller P, of the equation $\alpha = K \cdot \Delta\omega_{mec}$, following conclusion can be drawn: the proportionality constant K of the controller influence significantly the systems stability.

- at $K = 10$ and the inertial moment $J = 0.001$[kgm2], the system becomes unstable,
- at $K = 1$ and the same inertial moment $J = 0.001$[kgm2], the system is stable and the duration of the transient process is about 2.8s,
- at $K = 0.5$, the duration of the transitory process, at same J, is approximately 2s.

In order to achieve the prescribed speed, they are two possibilities:

- the control system provides periodic the voltage and frequency values and K remains fixed,
- K changes with the values of speed and resistant torque.

4. Conclusion

The proposed control scheme for the speed of the asynchronous machine (ASM) represents a modern solution for the automatic control for many electric drive systems, where a fast answer to the appeared perturbations is required.

The prescribed speed is achieved in an asymptotic way, by voltage adjusting, the frequency following the voltage variation, using a proportional controller. The proposed control scheme can be also checked in other configurations, i.e. PI controller. Until the prescribed speed is reached, the rotor flux exceeds the nominal value that brings the motor function in the saturation regime.

The proposed control algorithm can be adapted in other control structure, as controlling the frequency and the voltage will follow the frequency variations. For future researches, we propose to compare the accuracy of different other control methods, applied in various domains such as wind power plants, and bring our contribution to improve their outcome.

References

[1] Popescu M 2000 *Induction Motor Modelling for Vector Control Purpose*, Helsinki University of Technology, Laboratory of Electromechanics, Report, Espoo, pp 1-144

[2] Chioncel C P, Chioncel P and Gillich N 2008 *Scalar control structure of an asyncronous motor at maximum torque*, Annals of DAAAM and Proceedings of the 19-th International DAAAM Symposium "Intelligent Manufacturing & Automation: Focus on Next Generation of Intelligent Systems and Solutions", Trnava, Slovakia, October 22-25, pp 233-234

[3] Deepthi P and Padmaja S M 2012 Adjustable speed system of induction motor using vector controlled, *International Journal of Engineering Research and Technology* **1**(6)

[4] Chioncel C P, Erdodi G M, Petrescu D I, Spunei E and Gillich N 2015 *Control of wind power systems imposing the current in the intermediate circuit of the converter at variable wind speed*, The 9th International Symposium on Advanced Topics in Electrical Engineering, Bucarest, Romania, May 7-9, pp 923 – 928

[5] Tutulea L N, Deaconu S I and Popa G N 2015 Dual stator winding variable speed asynchronous generator: optimal design and experiments, *IOP Conf. Ser.: Mater. Sci. Eng.* **85** 012010

[6] Chioncel C P, Babescu M, Chioncel P, Gillich N and Gillich G R 2007 *Speed control method for asynchronous motor*, Annals of DAAAM and Proceedings of the the 18th International DAAAM Symposium "Intelligent Manufacturing & Automation: Focus on Creativity, Responsibility and Ethics of Engineers", Zadar, Croatia, November 21-24, pp 137-138

[7] Boraci R, Babescu M, Budisan N and Boraci A R 2009 *Mathematical model of the two orthogonal three-phase stator generator*, 5th International Symposium on Applied Computational Intelligence and Informatics SACI'09, Timisoara, Romania, May 28-29, pp 447-451

[8] Chioncel C P, Gillich N, Chioncel P and Gillich G R 2007 *CMS Solutions in monitoring and real time data transfer of photovoltaic plants*, XVth International Symposium on Electrical Apparatus and Technologies, SIELA, Plovdiv, Bulgaria, May 31-June 1-st, pp 14-17

IOP Conf. Series: Materials Science and Engineering **106** (2016) 012006 doi:10.1088/1757-899X/106/1/012006

Study of thermal regimes of an electric resistance furnace using AT-503 controller

A Iagăr[1], G N Popa[1] and C M Diniş[2]

[1]Politehnica University of Timisoara, Department of Electrical Engineering and Industrial Informatics, 5 Revolution Street, 331128 Hunedoara, Romania

E-mail: angela.iagar@fih.upt.ro

Abstract. In this paper were studied some thermal regimes of a low-temperature electric resistance furnace, with indirect heating. To determine the heat losses was used a thermal imager, which achieved a global image of the thermal field inside the furnace and upon the carcass. Thermal images of the furnace carcass revealed the following areas of heat losses greater than admissible in steady-state: furnace door, temperature measuring point and bonding areas of furnace metal carcass. To determine dead time and time constants were analyzed heating and cooling of the furnace until reaching the steady-state. In the experiments has been used AT-503 temperature controller, equipped with a K type thermocouple. Were analyzed following control methods of furnace temperature: on/off with hysteresis of 5°C and autotunig.

1. Introduction

Electric resistance furnaces (ERF) translate electric energy into the thermal energy. These furnaces are widely used in metallurgy, machinery, building materials and semiconductor industries. Some applications include: age hardening, temper hardening, annealing, heating before forging, stretching, pressing, glazing and baking of ceramics [1], [2].

The heating process is very complicated, and depends on the type of furnace, the materials used in its construction, its geometry, being a mixture of several nonlinear characteristics. In the case of ERF a problem of slow dynamics may occur, which represents a great disadvantage [3]. An important characteristic of ERF is the possibility of adjusting the working temperature very accurate, an essential condition for obtaining quality pieces [1-10].

The analyzed furnace has a rectangular chamber with external volume of 58.464 dm^3. The masonry is made of firebrick, with thickness of 45 mm; the carcass that consolidates chamber is made of steel, with thickness of 5 mm.

Two heating elements (resistors) made of Cr20-Ni80 alloy, placed near the furnace hearth (bottom), in parallel connection, are used to supply heat to the furnace chamber. Resistors are coiled on ceramic tube and their terminals are electrically insulated by ceramic beads. Active power generated by heating elements is 1kW/230Vc.a.

2. Heat losses computation in steady-state

The correct construction of the masonry and its careful execution at mounting will influence: heat losses, power consumption, heating time, quality of thermal processing, operation time, weight, overall size and cost of ERF.

Total power dissipated by heating elements is transferred to the processed material (charge) and toward the furnace masonry (Figure 1.a).

IOP Conf. Series: Materials Science and Engineering **106** (2016) 012006 doi:10.1088/1757-899X/106/1/012006

The power received by the charge is the output power (P_u, corresponding to useful heat flow rate Φ_u), and the heat losses of ERF (P_p, corresponding to losses heat flow rate Φ_p) include:

- heat losses accumulated in the masonry of furnace and other auxiliary materials Q_a;
- heat losses transmitted through the walls of the furnace (by conduction) Q_z;
- radiation losses through the furnace door, measuring points, etc. Q_r.

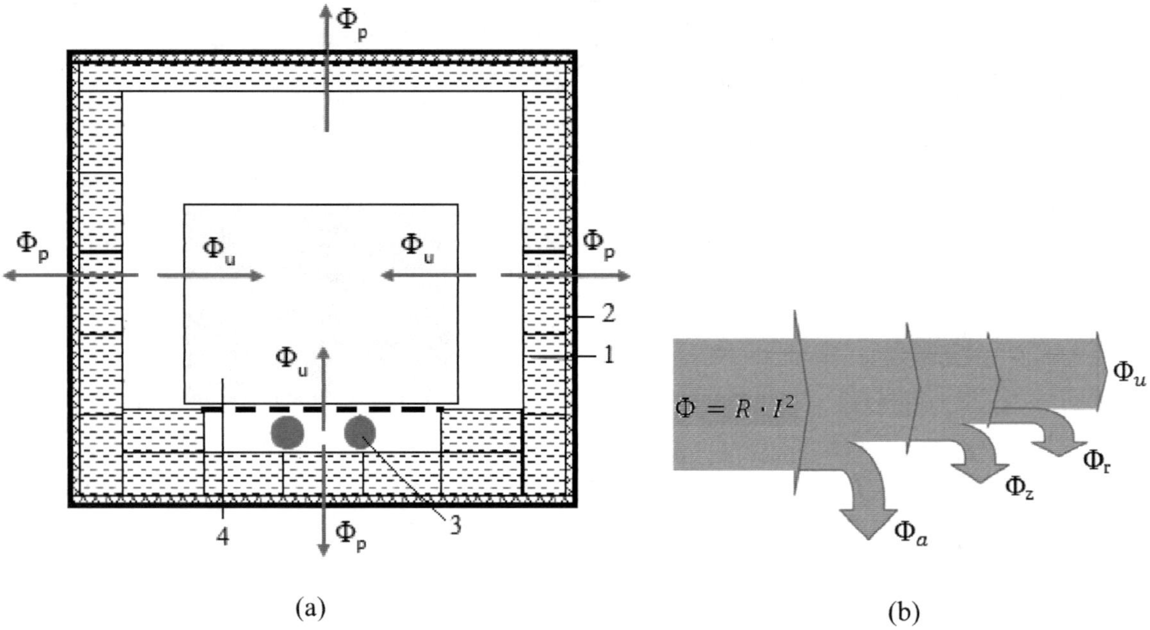

Figure 1. a - Schematic diagram of heat transfer in an ERF with indirect heating; b - distribution of heat flow: 1 – firebrick; 2 – steel carcass; 3 – heating elements; 4 – processed material; Φ_u – useful heat flow rate; Φ_p – heat flow rate of losses

In order to achieve a better efficiency of electric furnace is necessary to minimize the heat losses. Masonry of electric furnace must be designed in such a way the following condition will be satisfied [2]:

$$\frac{Q_a}{t_r} + \Phi_p = \text{minimum}, \quad (1)$$

where: Q_a is the heat stored in the furnace masonry until steady-state; Φ_p are the heat losses through the walls of the furnace to environment, in steady-state; t_r is the time of furnace operation.

Dimensioning of electric furnace masonry (thicknesses of layers, number of layers and their characteristics) is satisfactory if allows to obtain a temperature on the exterior surface $\theta_e = 40\ldots80$°C (if the temperature inside the furnace, θ_i, is less than 1000°C), or $\theta_e = 150\ldots400$°C (if $\theta_i \geq 1000$°C), for heat losses at most 750 W/m^2 [1].

Figure 2 shows the thermal images of the furnace carcass in steady-state, obtained by thermal imager [11]. The temperature on the exterior surface, in steady-state, is below 80°C, except furnace door, measuring point of temperature in the center of furnace chamber, and bonding areas of metal carcass. So in these areas heat losses are higher than admissible.

The maximum temperature on the furnace door is 124.8°C because it is not thermally insulated. The maximum temperature at the measuring point of temperature inside furnace is 78.8°C, and the maximum temperature in bonding areas of metal carcass is 86.6°C.

IOP Conf. Series: Materials Science and Engineering **106** (2016) 012006 doi:10.1088/1757-899X/106/1/012006

Figure 2. Thermal images of furnace carcass in steady-state

Thermic balance of an ERF with indirect heating is done for steady-state, when the absorbed energy from the network is entirely transferred to environment as heat losses.

Heat losses through the masonry of low-temperature electric furnaces are mainly due to conduction and convection.

Mathematical formulae used to compute the thermal resistances in the processes of conduction in masonry and convection on the inner and outer surfaces of furnace chamber are expressed by equations (2-5).

$$R_s = \frac{g_s}{\lambda_s \cdot A_{med1}}, \text{ [°/W]} \quad (2)$$

$$R_c = \frac{g_c}{\lambda_c \cdot A_{med2}}, \text{ [°/W]} \quad (3)$$

$$R_{\alpha i} = \frac{1}{\alpha_i \cdot A_i}, \text{ [°/W]} \quad (4)$$

$$R_{\alpha e} = \frac{1}{\alpha_e \cdot A_e}, \text{ [°/W]} \quad (5)$$

In the previous relations: R_s represents the thermal resistance in the process of conduction in firebrick layer; g_s is the thickness of firebrick; $\lambda_s\, [W/(m\cdot°C)]$ is the thermal conductivity of firebrick; A_{med1} is the average area of firebrick layer; R_c represents the thermal resistance in the process of conduction in steel carcass; g_c is the thickness of carcass; $\lambda_c\ [W/(m\cdot°C)]$ is the thermal conductivity of steel; A_{med2} is the average area of steel carcass; $R_{\alpha i}$ and $R_{\alpha e}$ represent the thermal resistances in the process of convection on the inner and outer surfaces of furnace chamber; $\alpha_i\ [W/(m^2\cdot°C)]$ and $\alpha_e\ [W/(m^2\cdot°C)]$ are the heat transfer coefficient of air inside and outside furnace, and A_i and A_e are the areas inside and outside of furnace chamber.

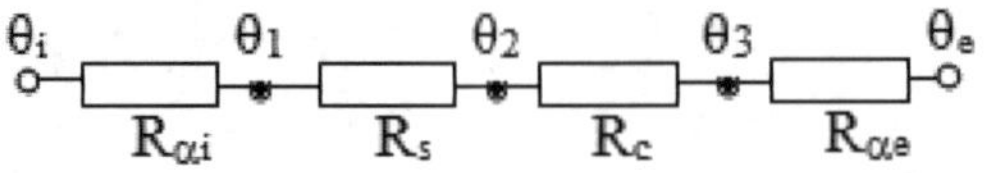

Figure 3. Equivalent electric diagram corresponding to heat transfer through the masonry

The inner surface of the chamber of furnace is A_i = 69.26 dm^2, surface between firebrick and metal carcass is A_c = 87.105 dm^2, and the outer surface of the chamber of furnace is A_e = 91.26 dm^2. Furnace door has area A_u = 1.82 dm^2.

The air temperature in the vicinity of the inner wall of the furnace is $\theta_i = 247°C$, so $\alpha_i = 23.5$ W/(m^2·°C), and the air temperature in the vicinity of the metal carcass is $\theta_e = 40°C$, and $\alpha_e = 10.5$ W/(m^2·°C).

From the relations (4) and (5) were determined thermal resistances in the process of convection on the inner and outer surfaces of furnace chamber: $R_{\alpha i} = 61.439779\cdot10^{-3}$ °C/W, $R_{\alpha e} = 104.359\cdot10^{-3}$ °C/W.

Were calculated the average areas: A_{med1} = 77.2725 dm^2 (for firebrick layer) and A_{med2} = 90.0925 dm^2 (for steel carcass).

The thickness of firebrick and carcass are g_s = 45 mm, respectively g_c = 5 mm, and the thermal conductivity of steel carcass is $\lambda_c = 50.93$ W/(m·°C). Was calculated the thermal resistance in the process of conduction in metal carcass, $R_c = 0.11\cdot10^{-3}$ °C/W, using equation (3).

To calculate the thermal resistance in the process of conduction in the firebricks was considered a linear variation with temperature of firebrick thermal conductivity:

$$\lambda_s = 0.093 + 0.16\cdot10^{-3}\cdot\theta_{smed} \quad \text{W/(m·°C)} \tag{6}$$

where $\theta_{smed} = 188$°C represents the average temperature of the chamotte layer (according to the thermal images from Figure 4).

Thermal resistance in the process of conduction in the layer of firebrick was determined by relation (2): R_s=473.1513276·10^{-3} °C/W.

Heat flow rate can be computed taking into account the equivalent electric diagram corresponding to the thermal transfer through the resistance furnace masonry (from Figure 3) with the relation:

$$\Phi = \frac{\theta_i - \theta_e}{R_{\alpha i} + R_s + R_c + R_{\alpha e}} = 323.913\text{W} \tag{7}$$

Heat losses through the masonry represent 32.39% from the installed power of analyzed furnace, according to relation (7).

The temperature field in the furnace chamber is not uniform in steady-state (from Figures 4 and 5). The temperature in the front part (near the door) is lower than that in the back; also, the temperature in the top of chamber is lower than that in the bottom one.

IOP Conf. Series: Materials Science and Engineering **106** (2016) 012006 doi:10.1088/1757-899X/106/1/012006

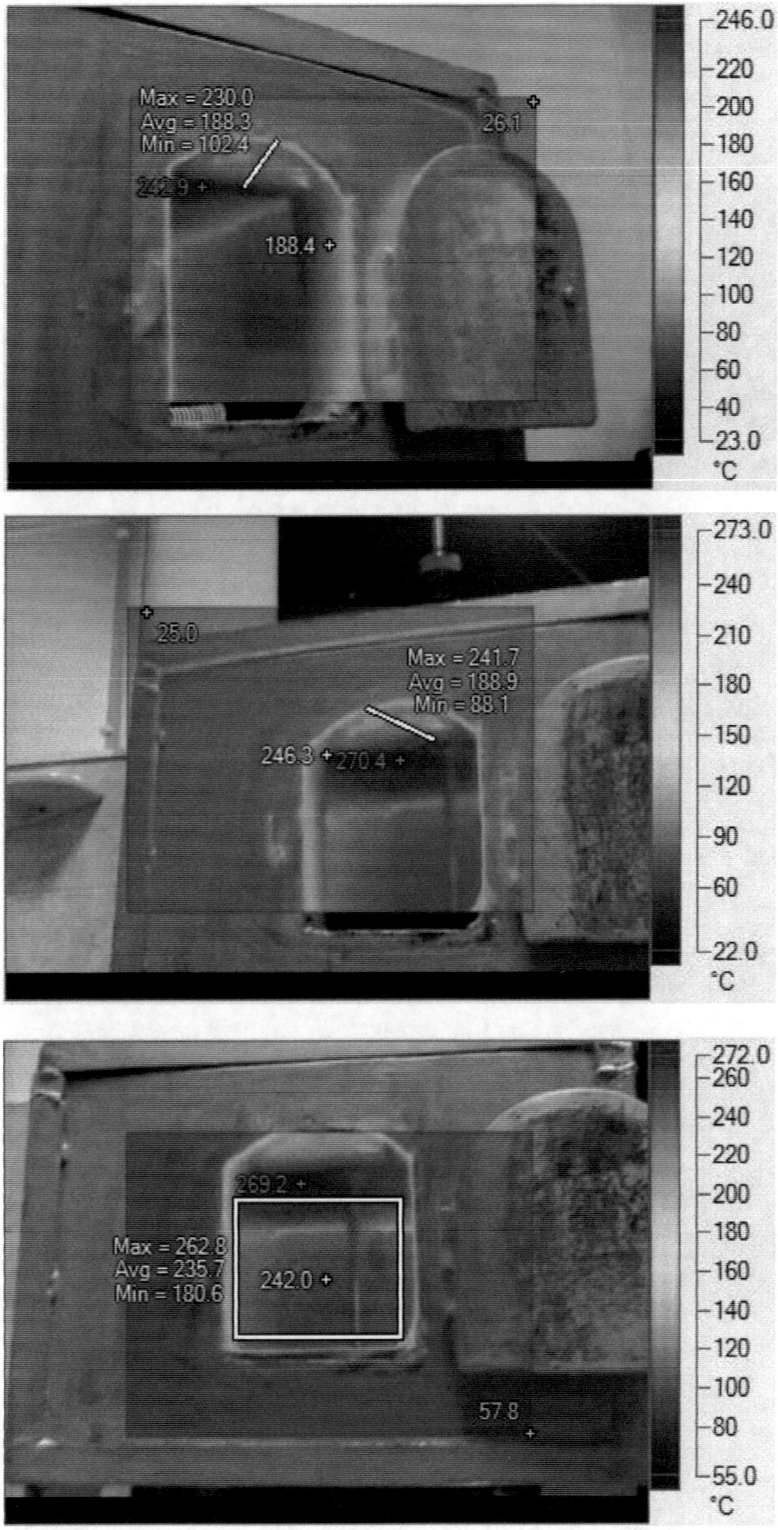

Figure 4. Details regarding furnace thermal field in steady-state

IOP Conf. Series: Materials Science and Engineering **106** (2016) 012006 doi:10.1088/1757-899X/106/1/012006

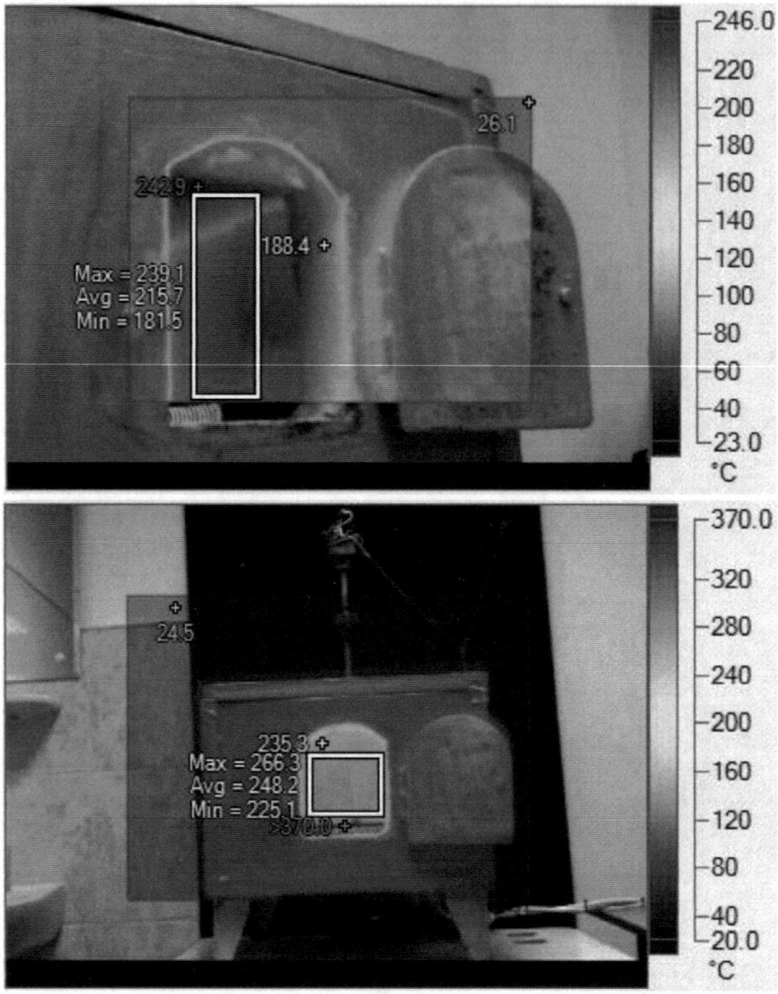

Figure 5. Details regarding furnace thermal field in steady-state

3. Determination of dead time and time constants of the furnace

It is very important to determine the optimal settings of temperature controller to achieve good control accuracy. This requires a good knowledge of the furnace parameters.

To determine dead time and time constants were analyzed heating and cooling of the furnace until reaching the steady-state (Figures 6 and 7).

The heating time constant of electric resistance furnace are experimentally determined from the heating equation:

$$\theta = \theta_{max}\left(1 - e^{-\frac{t}{T_i}}\right) + \theta_a \cdot e^{-\frac{t}{T_i}} \tag{8}$$

where: θ_{max} [°C] is the maximum temperature of the furnace corresponding steady-state; θ_a [°C] is the ambient temperature; $T_i = m \cdot c/(\alpha \cdot A)$, [s] is heating time constant; m - the batch weight [kg]; c - specific heat of processed material, [J /(kg·°C)]; α - heat transfer coefficient, [W/(m^2·°C)]; A - heat exchange surface furnace-environment, [m^2].

Knowing the maximum temperature of furnace θ_{max} = 321°C, ambient temperature θ_a = 26°C, heating time t_i = 1600 s and the final temperature θ_I = 216°C, was determined the heating time constant of the furnace:

$$T_i = \frac{t_i}{ln(\theta_{max} - \theta_a) - ln(\theta_{max} - \theta_1)} = 1549 \text{ s} \tag{9}$$

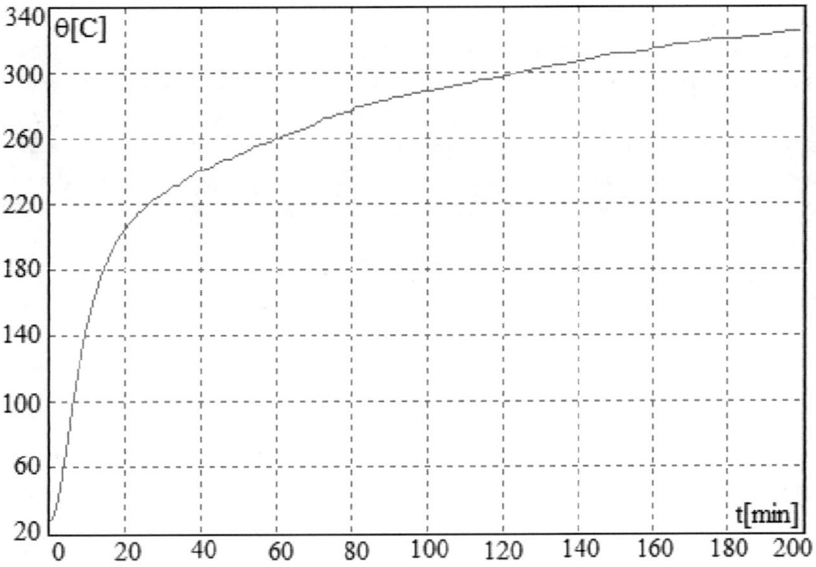

Figure 6. Time-variation of the temperature inside the furnace during the heating regime

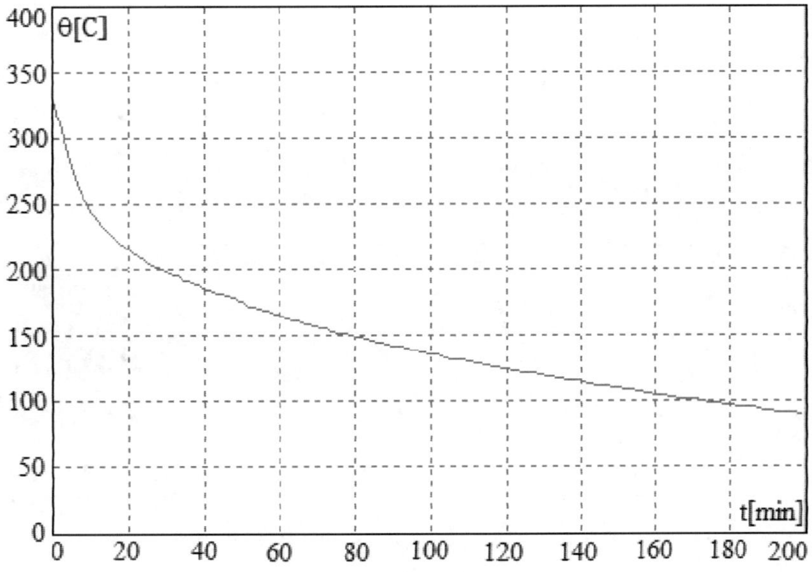

Figure 7. Time-variation of the temperature inside the furnace during the cooling regime

The time constant for cooling process was experimentally determined using the cooling equation. For this purpose has been made the cooling of furnace starting from θ_{max} to the final temperature θ_2 = 27°C, within the time period t_r = 12000 s.

From furnace cooling equation:

$$\theta = (\theta_{max} - \theta_a) \cdot e^{-\frac{t}{T_r}} + \theta_a, \tag{10}$$

result:

$$T_r = \frac{t_r}{ln(\theta_{max} - \theta_a) - ln(\theta_2 - \theta_a)} = 2110 \text{ s.} \tag{11}$$

It is noted that the cooling time constant of ERF is higher than the heating time constant. Thus, the dynamic characteristics of furnace in the heating process differ from its dynamic characteristics in the cooling process.

From the heating curve (Figure 6) was graphically determined the dead time of electric furnace, L=37.24 s. These parameters will be used later to determine the optimal settings of AT-503 temperature controller in the control methods P, PI and PID, to achieve good control accuracy.

4. Study of ERF temperature control using AT-503 controller

Experiments were performed using AT-503 temperature controller [12]. This controller can have as inputs the following types of temperature sensors: thermocouples (K, J, T, R, E, S, B, N type), or RTD (Pt100 or JPt100). To measure the temperature at the center of the furnace chamber was used a K type thermocouple. Output controls of AT-503 include relay, SSR, linear voltage, linear current and signal retransmission.

Figure 8 shows the schematic diagram of experimental installation. The command for connecting/disconnecting the furnace resistors was done through an electromechanical contactor (Un = 230V AC, In = 10 A) using the relay output of temperature controller. Link to the computer was achieved by RS-232 communication module.

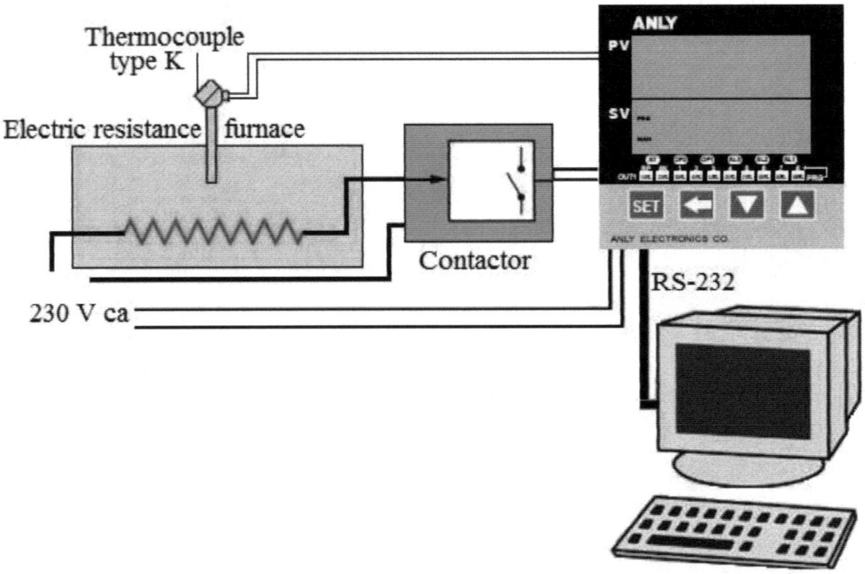

Figure 8. Schematic diagram of experimental installation

Control methods that can be implemented with AT-503 controller are: PID, PI, P, on/off and dead band. Further are presented experimental results in the case of autotuning and on/off control with hysteresis of 5°C. The set value (SV) of temperature was 100°C.

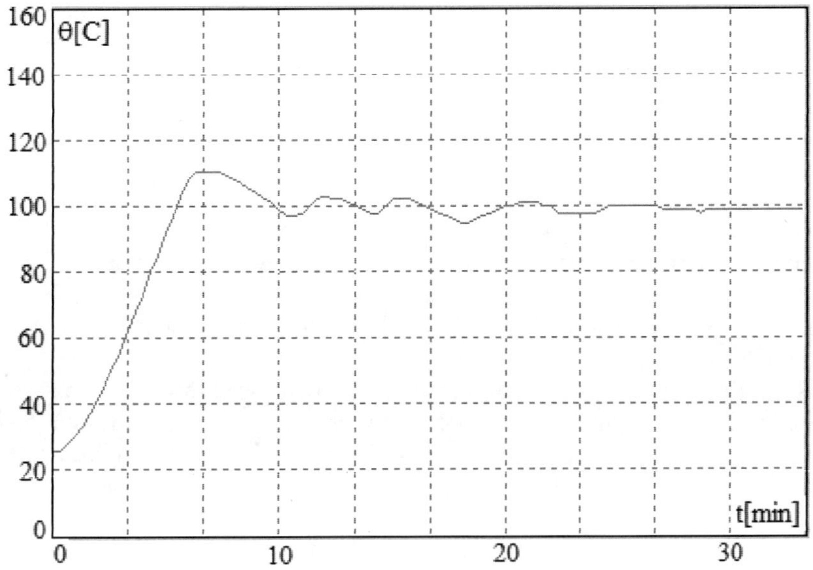

Figure 9. Time-variation of temperature inside the electric furnace in the case of autotuning

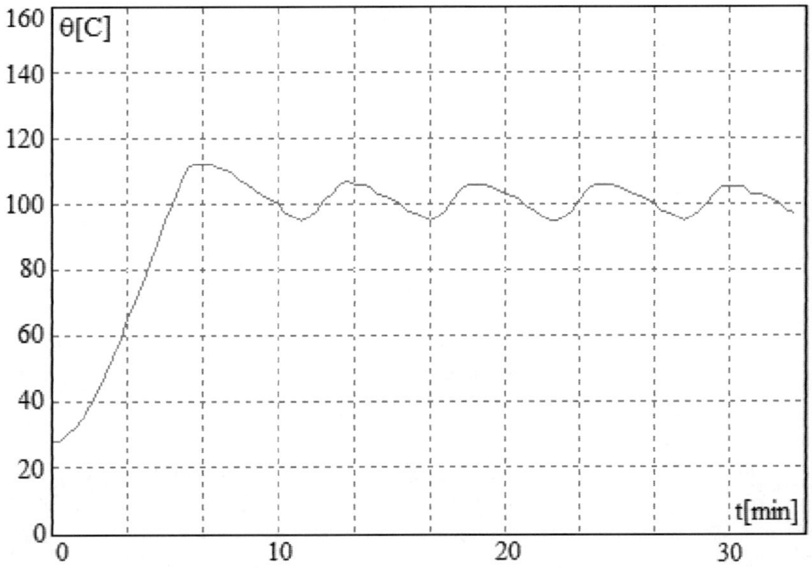

Figure 10. Time-variation of temperature inside the electric furnace at on/off control with hysteresis of 5°C

Figure 9 shows the time-variation of the temperature inside the electric furnace at autotuning. Rise time (time necessary for the temperature to rise beyond 90% of the set value SV for the first time) is about 5 min.

It is noted an overshoot of 10°C, which occurs after 7 minutes from the heating start. Temperature in the center of furnace chamber is stabilized after 24 minutes from the beginning of heating; this is settling time. The difference between set value of temperature and measured temperature at the end of recording is 1°C (steady-state error).

Figure 10 shows the time-variation of the temperature inside the electric furnace in the case of on/off control with hysteresis of 5°C. This is the simplest and cheapest of all temperature control methods. Using electromechanical contactor, by alternative connecting and disconnecting of heating elements from the network, the active power is modified in two steps: rated power Pn and 0 in this case. However control on/off may be oscillatory and leads to uneven distribution of the furnace temperature.

The dead band is of 10ºC in the case of on/off control with hysteresis of 5°C. Dead band is around the set value (desired temperature) in this case, but may be outside.

5. Conclusions

The main advantages of chamber-type furnaces consist of simple construction and the possibility of achieving of various thermal regimes.

Experiments indicate that analyzed ERF is a non-linear element, with a high inertia and hysteresis and time-varying parameters.

Non-linearity of ERF is mainly due to heat transfer by convection and radiation. Heat transfer coefficient of the air is not constant, but increases with temperature. For this reason heating dynamics are faster than cooling dynamics, the cooling time constant of the furnace being greater than heating time constant.

In steady-state, thermal images of the furnace carcass, obtained with a thermal imager, revealed the following areas of heat losses greater than admissible: furnace door, temperature measuring point and bonding areas of furnace carcass. Heat losses through the masonry represent 32.39% from the installed power of analyzed furnace.

To improve the efficiency of ERF heat losses must be minimized, by insulating the door, and PI or PID fine control must be used. These control methods can reduce the overshoot and settling time, and eliminate the steady-state error.

References

[1] Golovanov N, Sora I, Cantemir L et al. 1997 *Electrothermal conversion and electrotechnologies* (in Romanian), Technical Publishing Bucharest, Romania

[2] Metaxas A C 1996 *Foundations of Electroheat, A unified approach,* John Wiley&Sons, Chichester, England

[3] Chook K C and Tan A H 2007 Identification of an Electric Resistance Furnace, *IEEE Transactions on Instrumentation and Measurement* **56**(6) 2262-2270

[4] Chook K C and Tan A H 2008 *Comparison of Three Proportional-Integral-Derivative Based Controllers on a Bilinear Electric Resistance Furnace*, IEEE International Instrumentation and Measurement Technology Conference, Victoria, Vancouver Island, Canada, May 12-15, pp 190-195

[5] Dimitrov B and Yatchev I 2012 *Study of High-Temperature Chamber-Type Electric Resistance Furnace and Attainment of Uniform Temperature Field*, ICATE - 2012 International Conference on Applied and Theoretical Electricity, Craiova, Romania, October 25-27, pp 1-5

[6] Cojuhari I, Izvoreanu B, Vicol N and Melnic R 2014 *Sistem Automat de Reglare a Temperaturii în Cuptor Electric* (in Romanian), 8th International Conference on Microelectronics and Computer Science, Chisinau, Republic of Moldova, October 22-25, pp 217-220

[7] Sankowski D, Kucharski J and Lobodzinski W 1997 Autotuning temperature control using identification by multifrequency binary sequences, *Control Theory and Applications, IEE Proceedings* **144**(3) 233–240

[8] Laciak M, Fazekas P and Kacur J 2011 *Monitoring and control of temperatures in electric furnace in PROMOTIC system,* 12th International Carpathian Control Conference (ICCC), Velke Karlovice, Czech Republic, May 25-28, pp 239-242

[9] Panoiu C and Panoiu M 2010 *Using Adaptive Filters in Controlling of Electrical Resistance Furnace Temperature Based on a Real Time Identification Method*, Modelling Simulation and Optimization, INTECH, Edited by Gregorio Romero Rey and Luisa Martinez Muneta,

Rijeka, Croatia
[10] Wang J and Spanos C J 2002 Real-Time Furnace Modeling and Diagnostics, *IEEE Transactions on Semiconductor Manufacturing* **15**(4) 393-403
[11] ***Fluke Ti25 Thermal imager User's Manual
[12] ***AT03 PID Temperature Controller User's Manual

IOP Conf. Series: Materials Science and Engineering **106** (2016) 012007 doi:10.1088/1757-899X/106/1/012007

Interoperability framework for communication between processes running on different mobile operating systems

A Gal[1], I Filip[1] and F Dragan[1]

[1]University Politehnica Timisoara, Automation and Applied Informatics Department, Vasile Parvan str., no. 2, 300223 Timisoara, Romania

E-mail: andrei.g88@gmail.com

Abstract. As we live in an era where mobile communication is everywhere around us, the necessity to communicate between the variety of the devices we have available becomes even more of an urge. The major impediment to be able to achieve communication between the available devices is the incompatibility between the operating systems running on these devices. In the present paper we propose a framework that will make possible the ability to inter-operate between processes running on different mobile operating systems. The interoperability process will make use of any communication environment which is made available by the mobile devices where the processes are installed. The communication environment is chosen so as the process is optimal in terms of transferring the data between the mobile devices. The paper defines the architecture of the framework, expanding the functionality and interrelation between modules that make up the framework. For the proof of concept, we propose to use three different mobile operating systems installed on three different types of mobile devices. Depending on the various factors related to the structure of the mobile devices and the data type to be transferred, the framework will establish a data transfer protocol that will be used. The framework automates the interoperability process, user intervention being limited to a simple selection from the options that the framework suggests based on the full analysis of structural and functional elements of the mobile devices used in the process.

1. Introduction

A mobile device is a computational device having a hardware structure similar to that of a personal computer (PC) and also a processing power compared to it. Mobile devices are conceived to be moved very easily from a place to another by one person. Their dimensions are basically compared to that of a human hand. Virtually any device with a hardware structure similar to that of a computer, but much smaller size is considered a mobile device.

In the field of mobile devices manufacturers around the world have noted several manufacturers of such devices such as *Nokia*, *Samsung*, *Apple*, *Motorola*, *BlackBerry*. Each of these companies have managed to get noticed by producing mobile devices as sophisticated as they could, meaning as technologically advanced, using both hardware and software equipments to produce next-generation devices.

Very fast technological development has led to the possibility of creating some very technologically advanced mobile devices, giving birth to a new concept, the smart device (smartphone in particular). If from hardware's point of view every company wanted to rapidly improve the structure of the devices they produced, reaching a point where the structure of the devices could be compare to the one of a PC, leading to a "hardware uniformization", from a software point of view things were slightly different. If initially each company produced small pieces of software dedicated especially for the use of phone features, with the development of hardware, companies began to produce complex software for use of those hardware elements, reaching a point where a mobile device had its own operating system (OS) installed. The operating systems differed depending on the companies which produced them, and their use opened the path to the development of OS based applications, thing that lead to an operational incompatibility between the applications run on different operating systems. This incompatibility between operating systems was further deepened by alternating dominance of these systems appeared on the market. Since the development of an application for each operating system in part involves a complex process that requires time, each company (even ordinary developers) thought solutions that do not involve communication / interoperability / synchronization between these applications running on different operating systems, though the applications propose same functionalities on each operating system.

Thus, with all the applications available on the mobile devices we use every day, it feels the need to create such a framework that offers the possibility of communicate/ interoperate / synchronize between applications installed on different mobile operating systems. Though we speak about same application installed on different operating systems or same application installed on the same operating system, no matter what the version of the operating system is, the solution is the same. The framework could handle even different applications noting that the applications interchange data relevant to each of the applications involved in the process of switching data.

On the market there are currently 3 large dominant mobile operating systems that are installed on nearly 90% of mobile devices available. Having this in consideration the proof of concept we present further will be implemented using these 3 large operating systems along with the devices where they are installed which are: *iOS* installed on an *iPhone 5S*, *Android OS* installed on a *Samsung Galaxy S4* and *Windows Phone OS* installed on a *Nokia Lumia 520* [1].

iOS is an operating system developed by Apple Inc. company and it's the second more used mobile operating system on mobile devices in USA as in the whole world. Apple Inc. being the only company which distributes iOS on the mobile devices they produce, the company occupies the first place in a top of most sold mobile devices both in USA as in the whole world [1]. In particular the mobile device we will use to implement the proof of concept is iPhone 5S which uses iOS 8 as its operating system. iPhone 5S is a mobile device belonging to the smartphone category which has the following technical characteristics we are looking for: A7 chip with 64 bit architecture, Wi-Fi module which can use protocol 802.11 a/b/g/n, Bluetooth 4.0 wireless technology, Assisted GPS and GLONASS (Global Navigation Satellite System), accelerometer, gyroscope [2].

Android OS is an operating system based on the Linux Kernel and it's developed by Google. Its user interface, much as like iOS, is using direct manipulation to provide interaction between the user and the operating system. Taking this in consideration, Android operating system was designed only to be used by devices such as touchscreens, smartphones, tablets or smart TV's, as this type of devices offer the possibility to handle the objects provided by the operating system directly [3]. Android OS is the most used mobile operating system in the world, and being an open source operating system it's distributed by a lot of companies on the mobile devices they produce [1]. For the proof of concept further presented, the chosen mobile device is Samsung I9500 Galaxy S4 which uses Android OS version 4.2.2 and has the following technical specifications: Cortex-A7 chip, Wi-Fi module which can use protocol 802.11 a/b/g/n/ac, Bluetooth 4.0 wireless technology, Assisted GPS and GLONASS, NFC (Near Field Communication), accelerometer, gyroscope.

Windows Phone OS is an operating system developed by Microsoft exclusively for smartphones. It is an operating system based on Windows Kernel. Windows Phone OS replaced the first mobile operating system developed by Microsoft, called Windows Mobile and all the devices that had Windows Mobile installed are now compatible with Windows Phone. Like both Android and iOS, Windows Phone is using direct manipulation to let the users interact with the objects it provides. It's the third used mobile operating system in the whole world after the first two presented above. Though it's not an open source mobile operating system Windows Phone unlike iOS it's distributed by many smartphone manufacturers on their devices. We chose to use for implementing the proof of concept the smartphone Nokia Lumia 520 which uses Windows Phone 8 OS and technical specifications like: Snapdragon S4 chip (similar to Cortex-A8 chip) having a dual-core 1.2 GHz, Wi-Fi module which uses protocol 802.11 b/g/n, Bluetooth 4.0 wireless technology, GPS and GLONASS, accelerometer, gyroscope [4].

The technical characteristics of the mobile device on where the operating system is installed present a lot of interest as the framework will make use of them as communication channels with which data is transmitted between applications, thus we have to know the technological limits they have.

2. Framework architecture

The framework is intended for working on the three different mobile operating systems presented above by being installed on each of the mobile operating system. The structure of the framework is represented in the figure below (Figure 1). The core of the framework is represented by the 6 modules that altogether work for offering the solution to communicate easily and optimal between the applications which use it. The architecture of the framework may vary a little bit depending on the mobile operating system where it is installed. The changes in the architectural structure will be made so that the framework is able to use properly the hardware available on the device using that operating system. Any additional modules that may be needed in the implementation could be added as new modules communicating with the core of the framework [5].

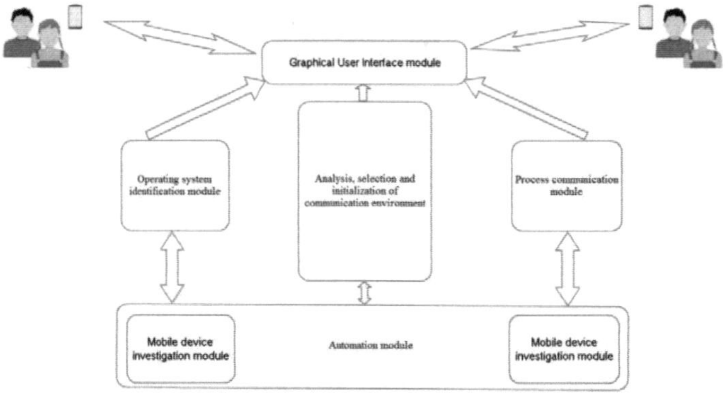

Figure 1. Framework architecture

2.1. Graphical User Interface module

Given that the framework offers the possibility to the users to communicate/synchronize between processes installed on the mobile device they use, the Graphical User Interface module is the module responsible for making available all the necessary information to the users in terms of selection of processes that will start the communication, the data chosen to be synchronize/switched between the processes, the way in which the communication/synchronization is done (selection of communication environment). The module is situated at the top of the framework's architecture and it communicates either directly or indirectly with all the others module below it.

2.2. Operating system identification module

As the main aim of the framework is to make possible the communication between processes running on different mobile operating systems, a first step in making this possible is the identification of the operating system with which the process has to communicate. In order to identify this, the communicating processes have to switch information that will point to the type of the operating system where they are running, any information related to the version of that operating system in order to have a better understanding in how the communication will be performed being necessary [5]. The whole process of switching operating system information will be done by the operating system identification module (Figure 2). In the identification process the module will initiate a handshake type communication that will basically negotiate the rules under which the data will be exchanged between processes involved. The set of rules that the framework will establish will be encapsulated in a protocol for manipulating the data switched between the communicating processes, according to the standards set by the operating systems on which the processes are running. The information gathered by the module will also be used to implement the communication strategy between the processes involved in exchanging the data [6].

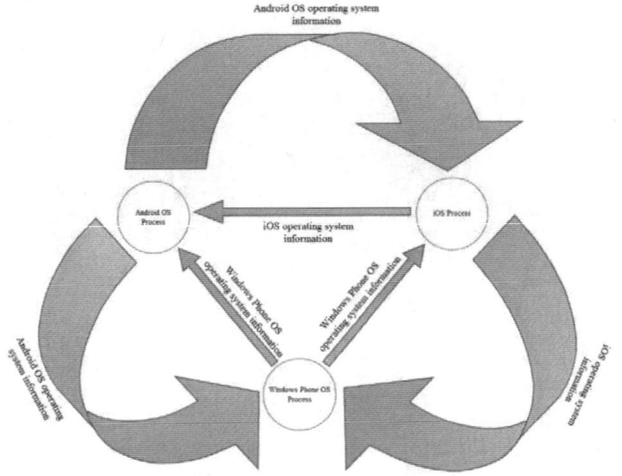

Figure 2. Operating system identification module

2.3. *Analysis, selection and initialization of communication environment module*

As every mobile device used for implementing the proof of concept provides several hardware capabilities that can be used as communication environments, the selection of the optimal environment in each case is a problem that has to be solved in order to have an efficient exchange data process. The framework module responsible for this is the *analysis, selection and initialization of communication environment module* (Figure 3). This module will search hardware communication capabilities of the mobile device to find possible communication environment to be used by processes that use the framework. Each possible communication environment will be examined and analyzed and as a result of the analysis there will be selected the optimal communication environment which will be used by the processes during the data exchange [7].

The analysis will take into account several parameters such as the communication environment data transfer speed, accuracy transfer, the possibility of losing data, the maximum amount of data that can be transferred, possibility of transferring blocks of data that can be sent through this medium of communication. An important thing to mention is that the result of the analysis will provide the optimal communication environment as being the main communication environment between those processes, but this does not mean that other communication environments cannot be used in the exchange process. Even more than that, through this module the framework will select, if needed, based on the analysis of all the processes involved in exchanging the data, a communication environment for each type of data the processes commits to switch between them [8], [9].

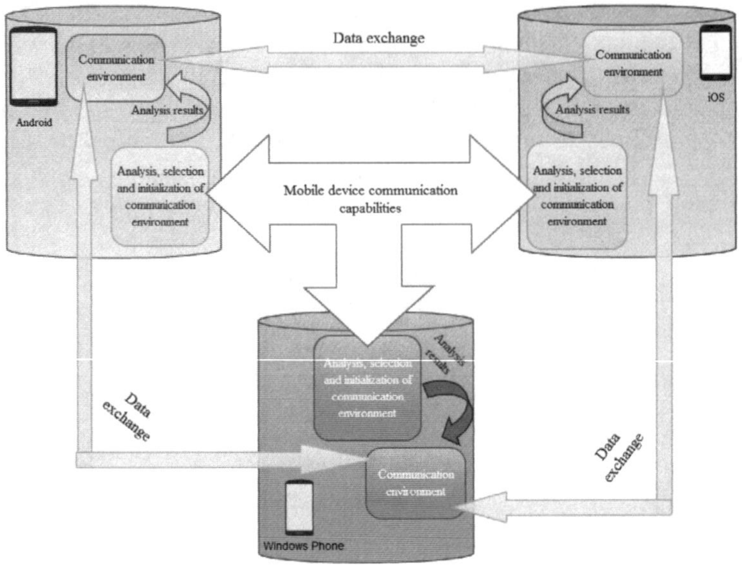

Figure 3. Analysis, selection and initialization of communication environment module

2.4. Process communication module

The module is responsible for initializing and performing the communication/data exchange between processes which will use the framework. When initializing the communication process, a new handshake type communication is initiated, and communication rules between processes will be negotiated (Figure 4). The set of rules established will control the format of the data transferred, the communication protocol the processes will use, timeout periods, limitation of bandwidth or limitation of data packages length sent along the communication channels. This module will effectively make the transfer of data from one process to another. An important thing to mention about the module is that it will maintain a live link with the *Analysis, selection and initialization of communication environment module* (Figure 2) as per having a fast switch from a communication environment to another in case of failure of sending data or reaching an impasse. If the module cannot make the switch to another communication environment, the communication is interrupted without altering the already exchanged data. Maintaining the data integrity has high priority when the communication environment is changed not only when the communication is interrupted [9], [10].

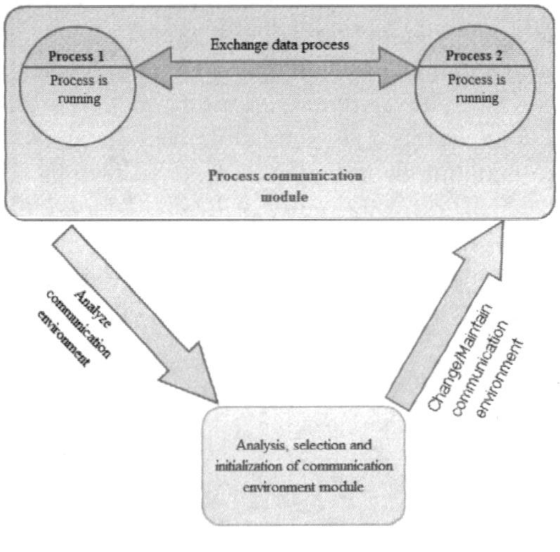

Figure 4. Process communication module

2.5. Automation module

The most important part of the data flow that will be manipulated by the framework will not be manipulated directly by the users from within the processes they are using, as the framework will have an automation module that will automatically manipulate the data so that a user's implication is reduced to accepting/choosing protocols that will be used by the processes when exchanging the data. The automation module will act as a wrapper over the data flow coordinating it from one process to another. The automation part of the framework can be integrated in a separated module of the framework, or it could be added separately in each of the existing modules in the framework, in this particular case creating the automation at the module level. In each case the result will be the same: data flow will be handled automatically by the framework with a minimum of user input. As depicted in Figure 1 the automation module is an independent module of the framework communicating directly with all the modules that deal with data manipulation [9].

2.6. Mobile device investigation module

The idea being that of a framework that offer users the opportunity to switch data between processes they run on different mobile devices, it has to be taken in consideration a way in which the process ran on a mobile device is able to find another device which runs a process with which it can switch relevant data for each of the process. In fact the process itself does not find the mobile device, but the mobile device finds another mobile device. The framework provides a module that handles this type of investigation, a module that will automatically search for all available devices that are able to communicate with it, using the framework and any of the communication environments it has available. The module will both get and provide data relevant for starting to communicate between processes run on mobile devices using the framework [8], [9].

3. Conclusion

In this paper we have presented the proof of concept of a framework that provides the ability to inter-operate between processes running on different mobile operating systems installed on smartphones (smart devices) that are different in terms of technical communication capabilities that they offer. Each of the mobile operating systems being different in terms of kernel structure, the framework will have to be installed on each mobile operating system, so that the process running on those operating systems could have access to it. Processes that will use the framework will initiate any communication by using the mobile device investigation module in order to gather information about other available processes.

The framework provides information about the type of the operating system where the partner process is running is by initiating a handshake type communication using the operating system identification module. All the information acquired by this module is passed to all the other module so that a proper data transfer protocol can be established. The data transfer protocol will be composed not only of operating system information but will also be taken in consideration communication environment used, format of the data to be transferred, and technical limits that the devices involved in the data transfer have. As manipulation of data could reach a high level of complexity depending on the processes involved, the framework will use an automation module that will take care of data packages switched between processes so that the user's intervention is reduced to only apply or select from options made available by the framework.

So far the framework is in a proof on concept state but in the future we want to consider a real implementation on each of the three mobile operating systems presented. Of course the proof on concept might suffer structural changes when reaching the implementation level as there might be limitations imposed by the operating systems, and more particular by the companies which developed the operating systems.

References

[1] ***Worldwide Smartphone OS Market Share. *http://www.idc.com/prodserv/smartphone-os-market-share.jsp* (accessed July 2015)

[2] ***Apple Inc. iOS. *http://web.archive.org/web/20071006005308/http://www.apple.com/iphone/features/index.html* (accessed July 2015)

[3] ***Google. Android OS. *http://www.android.com/history/* (accessed July 2015)

[4] ***Nokia. Windows Phone OS. *http://www.windowsphone.com/en-us/phones* (accessed July 2015)

[5] Xuan X, Wang Y and Shanping L 2014 *Privacy requirements patterns for mobile operating systems,* IEEE 4th International Workshop on Requirements Patterns, Karlskrona, Sweden, August 26, pp 39-42

[6] Enck W, Gilbert P, Chun B, Cox L P, Jung J, McDaniel P and Sheth A N 2010 *TaintDroid: An information-flow tracking system for realtime privacy monitoring on smartphones*, Proceedings of the USENIX Symposium on Operating Systems Design and Implementation, Vancouver, Canada, October 4-6, pp 99-106

[7] Ahmad M S, Musa N E, Nadarajah R, Hassan R and Othman N E 2013 *Comparison between Android and iOS Operating System in terms of security*, IEEE 8th International Conference on Information Technology, Kota Samarahan, Malaysia, July 1-4, pp 1-4

[8] Burns J 2008 *Developing secure mobile application for Android*, ISEC Partners

[9] Lashkari A H and Moradhaseli M 2011, *Mobile Operating Systems and Programming: Mobile Communications,* VDM Verlag

[10] Enck W, Ongtang M and McDaniel P 2009 *On lightweight mobile phone application certification*, Proceedings of the 16th ACM Conference on Computer and Communications Security, Chicago, USA, November 9-13, pp 235-245

IOP Conf. Series: Materials Science and Engineering **106** (2016) 012008 doi:10.1088/1757-899X/106/1/012008

Methods for cost estimation in software project management

C V Briciu[1], I Filip[1] and I I Indries[2]

[1]University Politehnica Timisoara, Automation and Applied Informatics Department, Vasile Parvan str., no. 2, 300223 Timisoara, Romania
[2] University Politehnica Timisoara, Computer Science Department, Vasile Parvan str., no. 2, 300223 Timisoara, Romania

E-mail: briciucatalin@yahoo.com

Abstract. The speed in which the processes used in software development field have changed makes it very difficult the task of forecasting the overall costs for a software project. By many researchers, this task has been considered unachievable, but there is a group of scientist for which this task can be solved using the already known mathematical methods (e.g. multiple linear regressions) and the new techniques as genetic programming and neural networks. The paper presents a solution for building a model for the cost estimation models in the software project management using genetic algorithms starting from the PROMISE datasets related COCOMO 81 model. In the first part of the paper, a summary of the major achievements in the research area of finding a model for estimating the overall project costs is presented together with the description of the existing software development process models. In the last part, a basic proposal of a mathematical model of a genetic programming is proposed including here the description of the chosen fitness function and chromosome representation. The perspective of model described it linked with the current reality of the software development considering as basis the software product life cycle and the current challenges and innovations in the software development area. Based on the author's experiences and the analysis of the existing models and product lifecycle it was concluded that estimation models should be adapted with the new technologies and emerging systems and they depend largely by the chosen software development method.

1. Introduction

Nowadays, the main concern of the producers (e.g. carmakers) is reducing the costs of production for different products but in the same time the increasing of the quality of them. At first seen, this sounds not to be very plausible since in the real life it is known the fact that with less money there cannot be bought products with a better quality. What if the producers find the method to do this? In mathematical terms, this can be described as an optimization problem of the quality triangle (Figure 1) from the costs point of view.

As a mathematical description, this can be considered as *max(F(u))* where u can be considered the entry considered for all disciplines of the project (software, hardware, mechanical, financial, human resources). In this relation, *F* is the objective function and in the same manner as u is a vertex quantity. A detailed approach above this function can be found later in the article.

Even at the start of the article, all branches of the product development has been taken into consideration, in the next parts of the paper the focus will be set to software development, even maybe there are products in which the software component is the less time/money consuming part, but in these days, a higher degree of products consists from software parts. (e.g. websites, smart houses, security services). Starting from the beginning of this century, due to higher demands on the software components from requirements point of view, the costs had been increased also and it comes to the situation where most of the projects were been underestimated from time and costs point of view. This fact changed the mindset of the software producers to have a good process to determine the costs and the finishing time even before the quotation time. Also, it is known the fact that defining a development process is not a task to be defined and finished in a half a year or one year. One issue that appears in this situation is related to the startup companies that do not have very experienced people. For example, an estimation of the project from the costs points of view in automotive field can be not a so hard operation if the project team has at least two or three experienced people. They can estimate, of course, with some errors the project costs and the project duration based on their previous experience. But what happens if this experienced people are missing from the quotation phase of the project. For a small company, this can be maybe instead of the "start" phase, the "stop" phase. For this, each company defines its process development rules for achieving the desired level of quality. But this process development rules during the product lifecycle cannot guarantee if they are fulfilled then the projects are successfully. It is estimated that 1/3 of a projects are considered unsuccessfully [1].

Figure 1. Quality triangle

Starting from the beginning of the 1980', software managers and quality assurance employers tried to develop methods to forecasting the evolution of the projects though resolutions regarding products life cycle, software development methods and even methods for estimating the projects costs. They were intuitive and very easy to understand, but in our times it seems that they don't have applicability on the current projects. In the second part of the paper, this will be presented in detail. Also, a tendency that can be seen currently is to replace the old development software processes (they are still in use) with new ones that are more closed to the developers' daily work. Also, it is known the fact that a single process cannot be applied in all fields of software activity because in one sector activity the quality is first (maybe in terms of safety and security), but in others functionality is in top (web services, PC desktop application). Also, an old C++ application that have a graphical user interface it can be done now very fast because the IDE (Integrated Development Environment) it has now implemented a drag-and-drop functionality. This was also anticipated by the Philip G. Amour in [2], which considers the software process as "a race for knowledge and new": "Like a climber planning a route over glaciers and up a mountain face, the destination is clear, but getting a safe route to the mountain and back down again takes a mixture of careful planning and an adaptive approach to events. The climbers may find impassable chasms, dangerous overhangs or unpredictable changes in weather. The plan will probably change".

This research paper is the first paper from a research study of three years where the final scope is the respond to the following questions and acts as an introduction for understanding the input parameters of the research:

a) Is there a software development process applicable to all software disciplines in which the obtained results are the same or near the same? If yes, is feasible the development of a new process model that can be generalized to all software process disciplines?

b) Are the software estimation models available in the literature accurate to all projects types? If yes, what is the optimum existing model for cost estimation, otherwise, can be the cost estimation model generalized for all projects? In this point, it is necessary to be proven that the existing models estimate truly also real projects and not only the projects in which they were developed.

c) Is feasible and realizable an estimation model where the human component to be considered not as an abstract resource, but as a state variable of the mathematical model?

Before continuing, there are necessary some definitions for the main terms and topics described further in the paper.

The second section of the paper presents the existing most known software development processes. For a better understanding, they will be presented in a comparative way, in which the applicability of them in the current times will be pointedly. In the third section of the article, some of the most important and know cost estimations models are briefly described. The fourth section presents a short description of the existing models for estimations. Next section will be reserved for the presentation of the idea in which the study will be continued for achieving the answers to those three questions.

2. Software Development Process Models

Before introducing the definition for software development model, it is necessary to start with the definition of the Software Development Life Cycle (SDLC). SDLC can be considered as a set of techniques and methods that describes the phases needed for the development of the product from order point of view and deliverables. The SDLC suppose that phase X supposed to be performed after Y phase and as inputs for the phase Y (outputs from the phase X) with the resulting deliverables. Some examples can be considered the requirements are translated into design at certain level, design can be considered as input for the code, and the code is an input for the verification and validation phase. The main phases of the SDLC are:

a) Quotation and Concept Refinement: this phase is started with the innovation & roadmap where an opportunity is identified. After the identification of the opportunity there is a needed for the system concept development. As a deliverable, it should be considered that system boundaries are clearly defined, and the documents for cost benefits, risk management and assessment plans are defined.

b) Development: this phase is started with the planning of the activities that acts as a basic for acquiring the resources for achieving the refine concept. As first deliverables documents of this phase can be considered the project management development plan and all other planning documents. Then, this phase is continued with the requirements gathering and analysis whose deliverable is the Functional Requirements Documents (System Requirements Specification - SRS). Based on this document, the requirements are translated into design documents at system level (Software Architecture Document - SWAD) and then at component level (CSDD - Component Software Detailed Design). In this phase, the design is frozen and approved. The approved design is considered as input for the Implementation phase where coding, review protocol documents and all activities regarding the SW integration and SW validation documents are performed. After the implementation phase is end, the integration and validation test cases are performed for assuring the bug-free of the project. As inputs for this phase, it should be taken into consideration the System Requirements Specification when the system testing is performed and component Requirements Specification when the validation tests are done.

c) Industrialization - Production Ramp-up and the series production. This phase starts with the deploying of the application in the client side and describes tasks to operate and maintain information systems in a production environment. It is often use also as Post-Implementation phase means implementation of new features.

d) Disposition - after series, where the delivery obligation is ended and also describes the end-of-system activities.

According to Walt Scacchi [3] the software development life cycle is either a descriptive and prescriptive way to describe how the system is or should be developed. The descriptive component presents the way in which a previous software system or component was developed while the prescriptive model suggests the order of operations to be done for achieving the initial scope of the project. In facts, it acts as a guidelines or framework of activities that should be performed.

A software process model can be characterized as "networked sequence of activities, objects, transformations, and events that embody strategies for accomplishing software evolution" [3]. From the beginning of the software development, there was a tendency to learn from previous projects and to organize the lesson learnt into models that can be reused later as improvements. These sequences can be seen as non-linear subsystems which can be done and done until the final product achieves the desired level of quality. It is hard to define for each sequence the exact input parameters without omission of at least one of them due to the fact that maybe these sequences can be executed in parallel. Also, only based on the successful/unsuccessful of a sequence or a part of the sequences it cannot be guaranteed the success of the project. From qualitative point of view, these parameters can be defined with sufficient prevision, but from quantitative point of view, this operation can be often considered subjective depending by the person in charge with these tasks. It is important to know is the fact that planning of the testing sessions will depend by the chosen development process. If a process supposes the developing of the system in linear steps then the testing will be performed through the end of the project. If it is done in increments or repetitive phases then preparation of test cases scripts and specification will be started in the same time with the development of the executables. As a representation of the sequences as a system using system theory notation this can be seen in the picture from below (Figure 2).

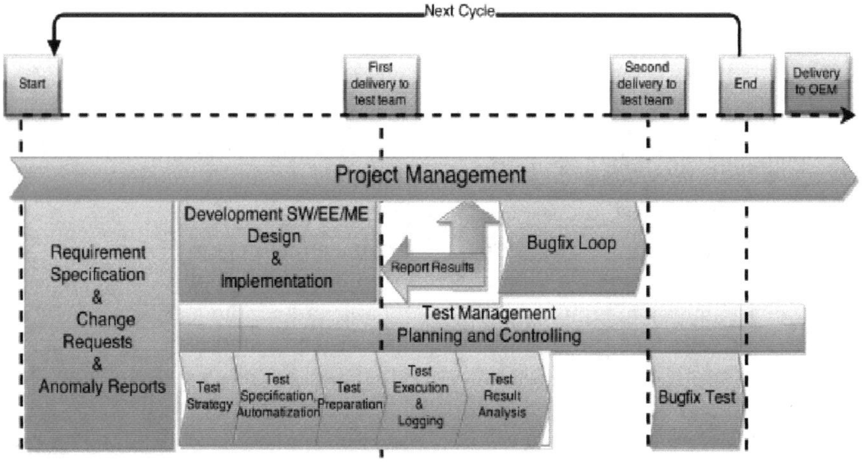

Figure 2. Software product life cycle

The main known software process models are the following:

a) **Waterfall model:** this model is known for its linearity because a phase is not started until the previous phase is not ended successfully. The phases of the project are well-defined from input and outputs point of view. A good definition of phases also leads for project managers to an easily way to define the needed activities. What is interesting to follow here is the fact that testing phase is done near the end of the project. This cannot be accepted in today world because customer wants to see at least some prototypes or releases for continuing work. Also, due to complexity of the systems, many times the customers change the requirements during lifecycle and without executables it is hard for them to check if the client understood very well the requirements. Using this model for the software development this cannot be achieved and acts as a structural defect of the model. Also, this structural defect is the main argument of the objectors who said that this is not applicable anymore. This implies also a high risk of uncertainty because any defect from the design phase cannot be corrected. In fact, this method is supposed to be used for low complexity projects where the requirements are defined and clarified.

b) **V-cycle model:** it is one of the most used software process models used in the embedded systems industry and it seems to be feasible also for long and very complex projects. In the same way as waterfall model, a phase cannot be started if the previous one is not finished. In practice, due to elasticity of the model, some phases can be started in the same time. As an advantage of the model, it is very simple to be understood and the verification and validation is done in the same time with the corresponding phase. In fact, on both wings of the V-shape there is a correspondence between the phases that can be seen in Figure 3.

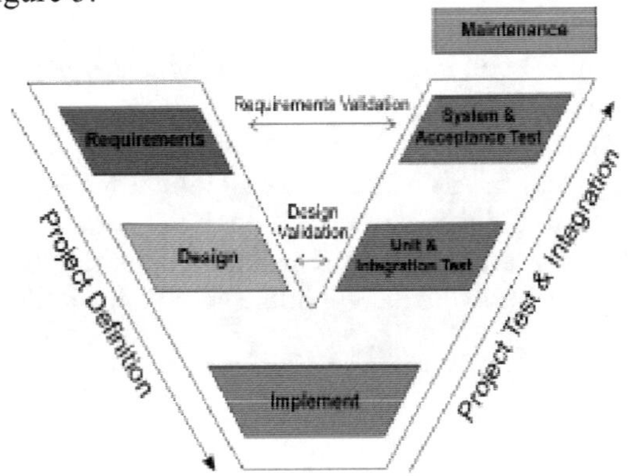

Figure 3. V-cycle model

In the same manner as required for choosing the waterfall model, it can be used for the projects where requirement are clearly and fixed and the customer has a higher degree of confidence in the client. The phases meaning are described briefly in the [4].

c) **Incremental model**: using this type of software model, the application is developed in small increments until the final application is ended from the requirement point of view. To have an example we can consider a 3x3 panel which represents the final project and each delivery is similar with the installation of one panel. For each increment, all phases of the SDLC are passed. As a comparison, we can considered it a multi-waterfall method since for all increments all phases should be passed (software is done step by step). What is important to remember from here is the fact that after each increment working software can be delivered to the customer. Also, this means that requirements should not be know all from the beginning and the client can "respond" to the working software after each build, but there is a disadvantage because in this way costs risks management introduces more uncertainties. Comparative with the waterfall method, the costs can be higher.

d) **RAD model** - (Rapid Application Model) - is a simple model that allows complex systems to be developed as independent systems. This requires very good knowledge of the project and of the requirements from the systems engineers. "The generic" solution is the most used software paradigm where components are developed as independent modules and then built together at the integration. Nowadays, this model is very used since the AutoSAR suppose that software components for building a software is not necessary to be developed by the same client. In facts, in [5], authors describe the methodology for building such types of modules.

e) **Agile model** - The agile model starts from the Manifesto Agile, a reference of software development methods. It is based on an approach based on development and deliveries in very small increments. The main idea of the model is based on the continuous integration, continuous improvement of the code and pair programming. What is different related to the models presented before is the fact that requirements are classified in Story Cards and based on this the stories and team velocities are calculated and the releases are planned. From development point of view, it is almost the same with the incremental model because functionality is added after each release. Refactoring and simple design are others paradigm that are intensively used during the development using this type of model. Nowadays, this model seems to be the most used in the web-development and enterprises areas, except the embedded area, but the tendency is to start the usage of it also in that area.

f) **Iterative model** - the iterative model supposed from beginning that requirements are not clarified and it will be clarified after reach release based on the review on the current software. It acts like a genetic algorithm whose behavior improves after each population and at the end of the maximum generation of populations is gives the best solution.

g) **Spiral model** - the spiral model is different from the first fourth presented models because it is split in other phases where the known SDCL phases are performed. These four phases are: "Planning", "Risk assessment", "Engineering", "Evaluation". The deliverables for each phase are:

- Planning - SRS document
- Risk assessment and evaluation: risk evaluation and risk prevention countermeasures
- Engineering: the software builds executables and the test reports of validation test cases
- Evaluation: feedback from the customer and approval for the next spiral.

Due to evaluation from the client after each spiral round results that the spiral method is applicable for the projects where the risks should be taken into consideration and evaluated because not all requirements are clear from the beginning.

3. Software Development Process Models

A main component of the software models from the quantitative perspective is the cost of the project. Depending by this factor, a software development company can be a considered as a good example from financial point of view or the opposite. Theoretically, the company tries to maximize the quality of the delivered software, but practically on most cases, the balance at the end of the project is considered as maximization function. Taking into consideration the quality triangle presented in the first part the costs represents a quality indicator for the final estimation of the project. The software estimation suppose to accurate forecasting the size of the software that should be developed, the time of the software until it can be released and project schedules. From the beginning of the software engineering, there were researches for finding a mathematical model of the cost estimation, but the rhythm of new programming languages and CASE tools did the researcher job very ineffective because the new programming languages were based on high level languages, therefore the number of ELOC (Effective Line Of Codes) decreased. In first part of this chapter a brief description of the empirical parametric estimation cost functions will be presented and in the second part new techniques that are applicable for the current development sandboxes (COCOMO II model, estimation through function points).

There are several methods to forecasting the costs of a project based on some product values measure that can be the number of code lines in the application or other measured values to quantify some characteristics of the software application.

a) **Empirical mathematical estimation models**: first model is a linear model that was developed at the beginning of the software development in a structured way and it has the mathematical formula Effort = LOCb, where effort is expressed in the man-month and the LOC is effective lines of code [6]. Regarding the constant, some researches quotes' that it should be above par, others states that should be under par. Continuing with the enumeration of the researched solutions, it was stated that a linear model cannot expressed entirely an exact model, so a non-linear approach was introduced according to the following: $Effort = a + bLOC + cLOC^2$ [7]. The research based on the datasets project found that the value of the b is almost 1 [8]. Nowadays, growing complexity and CASE tools make the cost estimations models not applicable in the "field". Therefore, a new type of estimation that not takes into consideration the number of lines of code was introduced. This is based on the so called function points where the $Effort = a + bFP$ [9]. A different approach was done by the Basili and Panilip-Yap in 1985 where they suggest that the cost of the project can be estimated based on the number of pages of documentation instead of the code. Therefore they built a linear mathematical description of this according to $Effort = a + b * NumberOfPagesOfDocumentation$, where the number of pages did not include sections of the codes. Regarding the mathematical expression, it was demonstrated later that the model is not usable because the number of pages of documentation vary from the chosen software process, from individual to individual; it depends by the chosen programming language and at least by the company where it was applied. Apparition of the object oriented objects determines the researchers to take into consideration the number of objects, packets, methods and services based on the mathematical description $Effort = a + b*NumberOfObjects + c*NumberOfServices$ [10]. After 2000, the models are not based exclusively on the LOC because the LOC is not relevant anymore if is not taken into consideration with others characteristics of the software. Some examples are the models of the Jeffery from 2000 according to $Productivity = aLOC^{b}MaximalTeamMember^{-c}$ where the productivity is the productivity is expressed in the man-month effort. Also a good example is the Putnam model [11] where suggests a model related to the lines of code based on the time needed to deliver a system and total effort of lifecycle and size of the system according to the mathematical model: $LOC = C_k K^{1/3} t_d^{4/3}$, where the K if the total effort related to the chosen lifecycle, t_d is the time until delivery and the C_k is a project related coefficient for expressing the project features. As an assumption for this model, the effort invested by the team follows the Rayleigh curve.

One of the most known estimation model is the COCOMO (Constructive Cost Model) model. It was developed in two phases and the models are known under the name of COCOMO I and COCOMO II. According to Dr. Barry Boehm, the researcher that creates the mathematical description for the estimation model, the model is built "to help people reason about the cost and schedule implications of their software decision". The first version of COCOMO was stable for the waterfall model, but evolution of software to object oriented paradigm and RAD approach make the model ineffective. Therefore, COCOMO II was introduced for supporting also the new models which capabilities are KLOC, Function Points and Object Points [12]. Also, it has three sub-models whom preconditions are the type of the software process used [13]:

- **Post Architecture model**: this model estimates the actual development process and the maintenance of the actual project. It is known the fact that not all calibration parameters for all COCOMO II are fixed, but they can be used in an experimental way.
- **Early Design model**: according to [14] involves exploration of alternatives software/system architecture and concepts of operation.
- **Application Compositions model**: this model estimates also the prototyping phase and the effort indicated by the risk assessment procedure and the technology maturity.

As a conclusion about the COCOMO models, they can be used for setting the project budgets according to the estimated cost, deciding which parts of the projects can be developed from scratch, reused from an old part or just bought as a COTS (components off-the-shelf) component and also allows companies to set up their development process and product life cycle for balancing the "triangle of the quality".

b) **Estimation models based on the expert's experience**: this fundamental of the approach is based on the fact that a number of experts in domain are grouped for estimating the costs. For this, some steps should be followed:

- Every expert will received a set of the requirements of the estimated system and will have an amount of time in which all requirements will have to be understood at all.
- Project manager will give the experts some forms to be filled with the estimation for each component. These forms will be filled individually.
- In a meeting, the results are presented and the differences will be discussed and reasoned.
- The steps 2), 3) will repeat until there will not be differences higher than a predefined threshold.

c) **Estimation models based on comparison with other existing models**: in the automotive industry it is preferred a "bottom-up" approach because each component is estimated then the costs for each component is summed and this is the total cost. Now, also these estimations are not very cleared because the estimated cost frequently is not taken into account the interruptions or the changes in the requirements in a release starting from the middle of the project. Now, since the AutoSAR is "entering" faster in the automotive industry, the forecasting of the costs will be easier because the basic software modules can be build from others suppliers(tiers) and based on the requested prices a estimation can be done. The costs for them can be requested because the components are already built.

4. A proposal for a cost estimation model using genetic algorithms

Evolution of the software procedures and technologies makes very difficult to estimate the work packages for a developer and directly this in involving also in the estimation of the entire project. As we can see in the picture from below that show the dependency between the productivity and the effort estimation based on the current tasks of the developer. In a developer life, in the beginning, the productivity (or development velocity) is at a lower value, but due increasing in the experience growth this value. Also, after reaches to know the current tools, the functionality and language, the developer will be able to develop generic component that is something new for him and resulting another decrease in the velocity. New programming languages and the development of the COTS acts as obstruction in increasing of velocity.

The model proposed to be analyzed considered a person as the following triplet $P = \{, S, A, W\}$, where $S = \{ s_x | s_x \subset SKILLS, 1 < x < MAX\_SKILLS\}$, A is the allocation for the person for the current task, and W represent a coefficient for a people that is willing to learn; $0 < W < 1$. A project is consists from several task that can be defined as a cost means the task duration and the necessary skills to be executed. As a mathematical description $T = \{C, S\}$ where C means the task duration and it will be expressed in man days *(MD)* and the required skills can be one or several skills from the set of skills. Also as an assumption for calculation of the costs for the overall project is the fact that a *MD* represents a fixed rate because even if a person has a higher wage than other the differences can be calculated as a medium value. A project also comprises a plurality of tasks until if finished, but also this task has dependency between them because some of them cannot be performed until others are finished. There are constraints of types "no earlier than", "as soon as possible". It is clear that the set of tasks can be grouped into a matrix as the adjacency. Now because all of these have been considered, my proposal is to define a model for cost estimation using as inputs the resources available at the beginning of the project and the multitude of the tasks that can be performed in terms of man days. For calculation of these there are several methods, but the preferred one is using genetic algorithms in the first phase. Thus the mathematical model is formulated as follows:

a) **Input parameters:** T_i - means the totality of the tasks of the project. The relation between two tasks is as follows: it $T_i - T_j = 0$ means that the tasks can be started in the same time, $T_i - T_j = X > 0$ means that the T_j can be started after T_i after X days. Also as inputs parameters will be taken into consideration the person available at the beginning of the proposed model and for each task a cost expressed in the man days is taken into consideration. Taking into consideration all described above, a project will be represented as an oriented graph with weights on arcs.

b) **Genetic representation:** a chromosome will be represented by a binary array where in each position is coded the ID of the person which is needed to perform the task. Number of position in the array it will be the same with number of tasks.

c) **Fitness function**: the reason for the fitness function is to select the "fittest" chromosome (gene) from the current genome for reproduction in later stages. The risk for the selection of a "not-fit" fitness function is to have a local maximum instead of a global one as it can be seen in the picture from below. As fitness function the following algorithm is proposed: calculate for each task the effectiveness factor considering the allocation factor and the current skills. If the person assigned has not the skills for performing the task, then the cost should be increased with the **cost * willing to learn coefficient**. It is understandable that at the total cost of the task will be added also the **allocation factor * total cost**. For all tasks, the effectiveness factor shall be calculated in the same way. The function to minimize is the total cost of the project optimizing in the same time also the resource allocation.

d) **Selection of the fittest chromosomes for reproduction** will be done using the roulette wheel method where the probability to extract an individual will be proportional with the sum of the fitness functions for all chromosomes from the current genome. The extraction will be repeated as many times as the number of individuals, but in the end the same number of chromosomes into the population we should have, so it might need for regenerating the of the population. Next step is to recombine the existent chromosomes through crossover operation with a chosen probability of 0.6. No special operators will be used for the first variant of the genetic algorithm, but it will be taken into consideration for the next variant the definition of some operators for not allowing the generation of not allowed chromosome. What is important to know is the fact that the aging is not taken into consideration in the current model.

As additional improvements of the algorithms can be used the model using the adaptive evolutionary algorithms for travelling salesman problem with precedence constraints as authors described in [15]. The results will be compared and based on this the tuning of the parameter will be done. Also an alternative solution for estimation of the costs for the proposed model is using neural networks. The results obtained using the neural network can be compared with the results obtained using genetic algorithms. For all steps above, the datasets will be used from the COCOMO 81 datasets provided by the PROMISE organization. The reason for comparing the results is to have a unified estimation model for different software process model.

As additional improvements of the algorithms can be used the model using the adaptive evolutionary algorithms for traveling salesman problem with precedence constraints as authors described in [15]. The results will be compared and based on this the tuning of the parameter will be done. Also an alternative solution for estimation of the costs for the proposed model is using neural networks. The results obtained using the neural network can be compared with the results obtained using genetic algorithms. For all steps above, the datasets will be used from the COCOMO 81 datasets provided by the PROMISE organization. The reason for comparing the results is to have a unified estimation model for different software process model.

5. Conclusion

As a conclusion, it can be seen that the estimation models are adapting in the same time with the new technologies and emerging systems and they depend largely by the chosen software development process. An Agile based project for sure cannot be compared with a V-cycle based software. Finding a mathematical model for forecasting the costs of a project it is very likely to not to be developed, but a mathematical description that can be applied on one or two development process with a high precision can be found. Now, taking into consideration the things stated at the related work section, these models even they can predict now precisely the cost estimation, it can happen that starting with the third generation of the product in the same conditions as was considered for the previous generations the used model to not allow forecasting with the precision of the costs. This can be solved by doing also continuous improvements on the models in the same way that for software development the continuous integration and improvements methods are used at a large scale. Mathematics is also extending from algorithms point of view (e.g. genetic algorithms, neural networks) so these new methods together with the extending of databases for projects can be the solution of counteract time.

References

[1] Ambler S V 2014 The Non-Existent Software Crisis: Debunking the chaos report, *Dr. Dobb's Journal* (Online: *http://www.drdobbs.com/architecture-and-design/the-non-existent-software-crisis-debunki/240165910*)

[2] Amour P G 2003 *The laws of software process: A New Model for the Production and Management of Software*, First Edition, Auerbach Publication, Boca Raton, FL USA

[3] Scacchi W 2002 *Process Models in Software Engineering, Encyclopedia of Software Engineering*, John Wiley and Sons, Hoboken, NJ USA

[4] Balaji S and Sundararajan M M 2012 Waterfall Vs V-Model Vs Agile: A comparative study on SDLC, *International Journal of Information Technology and Business Management* **2**(1) 26-29

[5] Briciu C V, Filip I and Heininger F 2013 *A new trend in automotive software: AUTOSAR concept*, 8th IEEE International Symposium on Applied Computational Intelligence and Informatics SACI 2013, Timisoara, Romania, May 23-25, pp 251 – 256

[6] Walston C E and Felix C P 1977 A method of programming measurement and estimation, *IBM*

Systems Journal **16**(1) 54-73

[7] Banker R D, Chang H and Kemmerer C F 1994 Evidence on Economies of Scale in Software-Development, *Information and Software Technology* **36**(5) 275-282

[8] Zivadinovic J, Medic Z, Maksinovic D, Damnjanovic A and Vujici S 2011 *Methods of effort estimation in software estimation*, International Symposium Engineering Management and Competivness EMC 2011, Zrenjanin, Serbia, June 24-25, pp 417-422

[9] Kemerer C F 1993 Reliability of function points measurement: a field experiment, *Communications of the ACM* **36**(2) 85-97

[10] Brownlow L S 1994 *Early Estimating in the Object-Oriented Analysis and Design Environment,* Proceedings of the European Software Cost Modelling Conference, Ivrea, Italy

[11] Putnam L H and Myers W 2003 *Five core metrics: the intelligence behind successful software management*, Dorset House Publishing, New York, NY USA

[12] Arnuphaptrairong T 2013 Early Stage Software Effort Estimation using Function Point Analysis: An Empirical Validation, *International Journal of Design, Analysis and Tools for Intergrated Circuits and Systems* **4**(1) 15-21

[13] Boehm B W 2006 *The COCOMO II Suite of Software Estimation Models*, COCOMO/SSCM Forum 21 Tutorial, University of Southern California, Center for Software Engineering, Las Vegas, USA

[14] Selby R W 2007 *Software Engineering: Barry W. Boehm's Lifetime Contributions to Software Development, Management and Research*, Wiley-IEEE Computer Society Press, Hoboken, NJ USA

[15] Sung J and Jeong B 2014 An adaptive evolutionary algorithm for traveling salesman problem with precedence constraint, *The Scientific World Journal* **2014**(4) 1-11

Study on road sign recognition in LabVIEW

M Panoiu[1], C L Rat[1] and C Panoiu[1]

[1]Politehnica University of Timisoara, Department of Electrical Engineering and Industrial Informatics, 5 Revolution Street, 331128 Hunedoara, Romania

E-mail: manuela.panoiu@fih.upt.ro

Abstract. Road and traffic sign identification is a field of study that can be used to aid the development of in-car advisory systems. It uses computer vision and artificial intelligence to extract the road signs from outdoor images acquired by a camera in uncontrolled lighting conditions where they may be occluded by other objects, or may suffer from problems such as color fading, disorientation, variations in shape and size, etc. An automatic means of identifying traffic signs, in these conditions, can make a significant contribution to develop an Intelligent Transport Systems (ITS) that continuously monitors the driver, the vehicle, and the road. Road and traffic signs are characterized by a number of features which make them recognizable from the environment. Road signs are located in standard positions and have standard shapes, standard colors, and known pictograms. These characteristics make them suitable for image identification. Traffic sign identification covers two problems: traffic sign detection and traffic sign recognition. Traffic sign detection is meant for the accurate localization of traffic signs in the image space, while traffic sign recognition handles the labeling of such detections into specific traffic sign types or subcategories [1].

1. Introduction

Road and traffic signs, traffic lights and other traffic devices are used to regulate, warn, guide or inform road users. Traffic signs help achieve an acceptable level of road traffic quality and increase safety with orderly and predictable movement of traffic and pedestrian, by helping drivers anticipate the road ahead in addition to any associated problems and hazardous situations.

Looking at the task of road and traffic sign recognition, it appears to be a well-defined, seemingly simple task. Road signs are designed, manufactured and installed according to strict regulations in order to make them readily distinguishable from their surroundings (Figure 1). However, looking at road sign recognition in its full scale reveals a series of problems that affect the performance of the detection system [1], [3].

These signs are designed to be easily recognized by drivers, but considering the object recognition and interpretation abilities of humans, it is a hard task to try to develop a computer based system which should be able to do what people do in everyday life.

It can be concluded that it is extremely important for the algorithms to be developed for the detection and recognition of road and traffic signs to have high robustness of color segmentation, high insensitivity to noise and brightness variations, and should be invariant to geometrical effects such as translation, in-plane and out-plane rotations and scaling changes in the image [1], [4].

Our system is based on taking images through a camera from a moving vehicle and analyzing it using the recognition program. It uses traffic sign features as filters in a step by step algorithm designed to eliminate false positives with every step. It provides correct results on a wide variety of image distortion factors, such as the weather and the environment in which the image was taken, its position in relation to the camera, etc. The program, due to the fact that it works with thresholds and probability, is robust and efficient.

Because the efficiency and speed of detection are important factors in this study, algorithms have been developed to reduce the search space and indicate potential only the potential traffic signs.

Such a system could help improve traffic flow and safety, could avoid hazardous driving conditions, such as collisions, but it does require real-time image processing [1], [2].

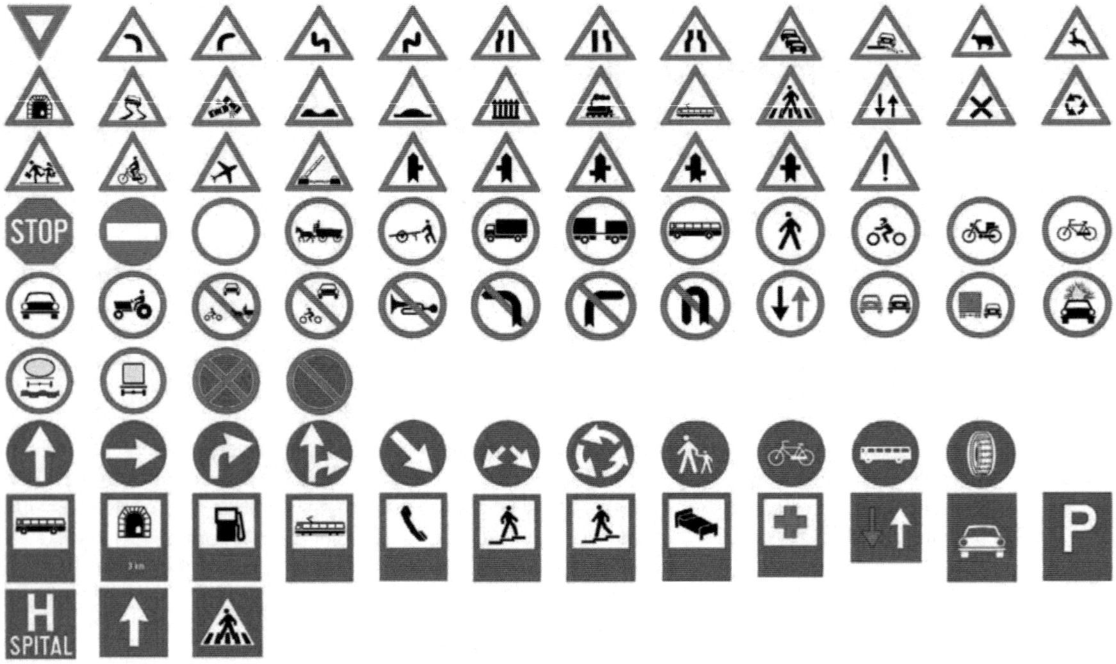

Figure 1. All the traffic signs recognized in this study

2. Identification Process

The software we have developed in LABVIEW works in two basic stages; namely detection and recognition (Figure 2). In the detection stage, the sign is detected according to the sign's color information, and then according to the sign's shape. Using these features the sign is extracted and sent through a classifier to decide what type of sign it is, in accordance with the pictogram [5].

The color of the road sign together its shape determines its category. Detection and recognition can be achieved even if one component, color or shape, is missing [1], [2], but if the two features are available, combining color and shape gives better results.

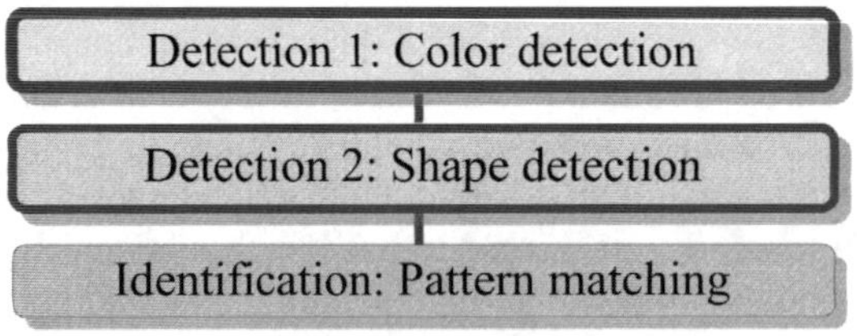

Figure 2. General road sign recognition diagram

In order to program a computer to recognize the signs in the scene, the main strategy is to find the right combination of colors in the scene, so that one color is located inside the contour of another color in combination with the shapes of the 2D objects defined by the individual colors. If a candidate possessing the right combination of shapes and colors is found, the system tries to classify the object according to the rim-pictogram combination and give the result of this classification.

The system performs a parallel search through color threshold for 2 main colors: blue and red. The program looks for the rims of potential signs. The output is a segmented binary image containing potential regions which could be recognized as possible road signs. Each of the candidates is tested against a certain set of features (a pattern) to decide whether it is in the group of road signs or not, and then according to these features they are classified into different groups. These characteristics are chosen to highlight the differences between classes of traffic signs. The shape of the sign plays a central role at this stage. It may be a triangle, a circle, an octagon, etc. Pictogram analysis is another classification step. By analyzing the pictogram, the number of patterns to test can be reduced (testing a pattern is, in itself, a time and resource-consuming task) and only the most likely models will be chosen to be tested on the sign in question [5].

2.1. Color Detection

The system performs a parallel search for 2 of the 4 main colors, namely blue and red. Regarding the other 2 colors, there are relatively few signs (2-3) that contain them and are governed by the same rules, therefore these are considered negligible. Including these colors can be a further development.

The programming environment we chose to use is LabVIEW, due to the fact that it contains a large number of color processing functions, known as IMAQ functions. Colors are represented in images through color spaces. A color space uses a three-dimensional coordinate system, each color being represented by a point. IMAQ uses the RGB color space to display images, and the HSL color space to process them. The RGB color space is an additive color system. The three primary color components can combine to reproduce almost all possible colors. The RGB space reproduces an image as the driver would expect to see it.

The HSL (hue, saturation and luminance) color space provides more accurate color information than the RGB space. The HSL space de-couples of the intensity component from the color information and the close relationship between chromacity (hue and saturation) and human perception of color. This feature leads to a more robust color representation independent of light intensity variation. It is used in image processing where the lighting conditions can vary considerably [6], [4].

At this stage, a color threshold is applied to the image, transforming it into a binary image. Binary images pixel have only values of 0 or 1 and, therefore require less computing time with greater precision in processing. The threshold is defined so as to separate a specific color range found in the rims of traffic signs from other colors in the image. This process sets (logic 1) all the pixels that belong to that range (interval), and resets (logic 0) the pixels outside the range. This is used to isolate objects of interest (the rims of road signs) in an image: pixels inside the threshold interval that form particles (contiguous regions of nonzero pixels). Transforming the image to binary does present some other advantages, among which the possibility of working with IMAQ functions specific to object identification.

The particles obtained are passed through morphological functions to remove the small shape defects they may have, in order to enhance their precision for the recognition stage. Morphological functions have also been used to retouch the shapes of the particles and, correct the faulty selections that may have occurred by applying the threshold. These functions change the shapes of pixel regions by expanding or reducing particles, smoothing the borders of the preparation of particles of quantitative analysis, etc. These operations change overall size and shape of particles in the image. The operation used in this case is an expansion function: dilation removes small holes within particles and extends the particle contour according to the model defined by the structural element.

The resulting binary image is then passed through a particle filter. This has the purpose of finding particles whose spatial characteristics satisfy certain criteria and eliminating the particles that are of no interest. In real-time applications, particle filtering is extremely important in directing the processing tasks and focusing the analysis only on the relevant particles [6], [7]. In this application, the length, width and diameter of the template particle are used as criteria for removing all particles that do not match them within some tolerance. They are also filtered by the number of holes they contain holes, this being the only sign that a particle might represent the rim of a road sign. A more refined search is then performed on the remaining particles.

The object or objects derived from morphological operations are modified to form masks for the inside the road sign. These operations are used for expanding or reducing particles, smoothing the borders of objects, filling the holes they contain, preparing particles for quantitative analysis, finding the external and internal boundaries of particles, observing the geometry of regions, locating particular configurations of pixels and extracting them for modeling and identification purposes [6], [8].

The resulting image is used as a mask on the original image in order to extract regions of the image that represent potential signs for the shape processing stage. A mask is a binary 8-bit image that isolates the parts of an image for processing. The pixels of the mask determine if the inspected corresponding pixels in the image are processed. If a pixel of the mask has a non-zero value, the corresponding pixel in the image is selected. If a pixel of the mask is 0, the corresponding pixel in the image is converted to a background pixel (a black pixel). Therefore, the pixels in the source image pixels are processed if the mask contains values other than zero [6], [7]. If at the end of this algorithm there is at least particle in the mask, then the analysis module is activated.

Since there can be more than one road sign in an image, there can be more than one particle in the mask. Hence, the mask must be divided in single particle masks. To this purpose, particle analysis is used. This consists of a series of processing operations and analysis functions that produce information about particles in an image, including statistical information, such as the particle size or the number, location, and the presence of particles. A typical particle analysis process scans through an entire image, detects all the particles in the image, and builds a detailed report on each particle. This can be used to uniquely identify an object (a particle) in an image [6], [8]. The positions and dimensions of the mask are inserted into a vector, further processing being applied for the generation of each individual mask.

The resulting binary images (single particle masks) are passed through a shape detection function (Figure 3).

2.2. Shape Detection

Two major shape detection algorithms have been used in this project depending on the characteristics of the sign category, namely Shape Matching and Shape Detection. Shape Detection (Figure 4.a) is more accurate, but it can only detect rectangles and circles. Shape Matching (Figure 4.b) may seek a shape after a given model, but it cannot find regular shapes with accuracy with which Shape Detection can. Therefore circles and rectangles are sought for with Shape Detection and triangles are sought for with Shape Matching. Both functions are able to detect shapes regardless of size and orientation. The output of these functions is a report detailing the location of each part, the centroid of the part, and a score that indicates the degree of match between the template and the part. This helps focus the search on the regions of the image that are likely to be traffic signs, rather than scanning the entire image for possible results. if the following stage is based on a different characteristic of a road sign than the stage that produced the mask, it also becomes less likely to yield false positives [9].

Depending on the results of the shape function, the single-particle mask then is used to extract the sign candidate from the original image. Before the original image is masked, color equalization is applied in order to enhance it and to improve contrast and brightness. The extracted symbol is rotated in accordance with the detection results of the shape matching step, and then resized to 105x105 pixels in size. Since IMAQ functions use template-based recognition, they only work with objects (signs) the same size as the template. Hence resizing the image is a key point in the algorithm. A black 10-pixel border is also applied to the sign, all to ensure recognition

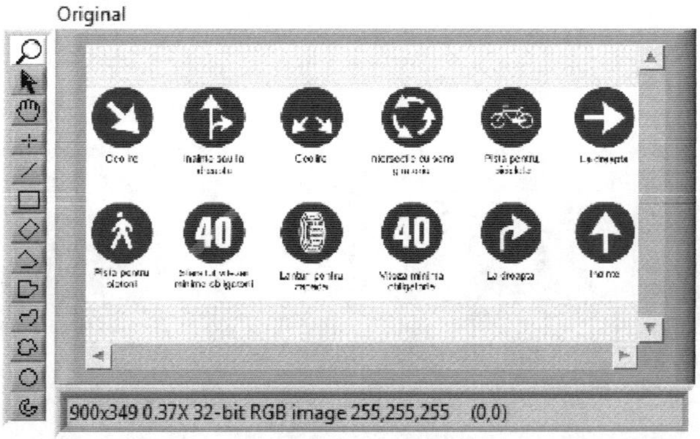

a) Original image

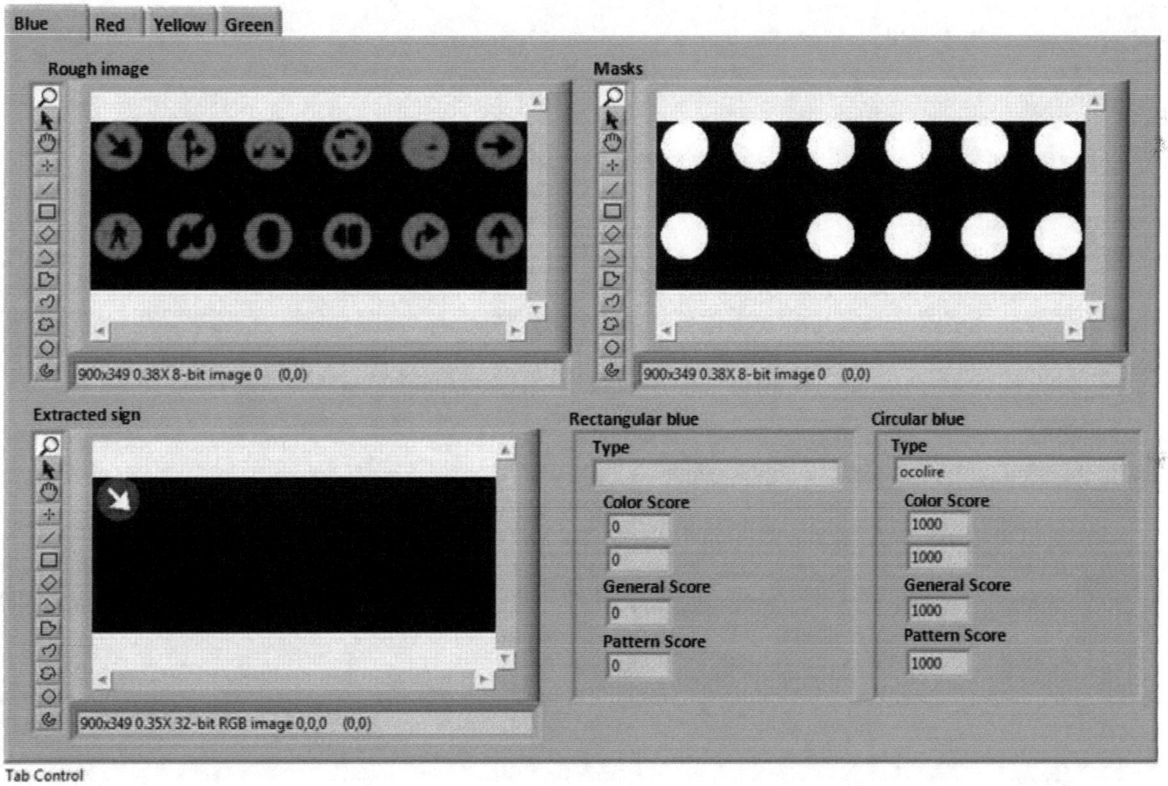

b) Images from the processing steps and the final result

Figure 3. Front Panel of the recognition program

2.3. Recognition

The obtained image is then passed through three different recognition techniques, grouped in a recognition module. These consist of pattern matching, color matching and color-pattern matching.

Pattern matching locates regions of a grayscale image that match a predetermined template, a reference pattern which constitutes an idealized representation of a feature. Pattern matching works reliably under various conditions including poor lighting (only under uniform changes in the lighting across the image), blur, and noise, shifting of the template or rotation of the template. The machine vision application searches for the model in each acquired image, calculating a score for each match, that relates how closely the model matches the pattern found. If the measurement falls within a tolerance range, the part is considered good. If it falls outside the tolerance range, the component is rejected. The result is a fast and accurate pattern matching [6], [8].

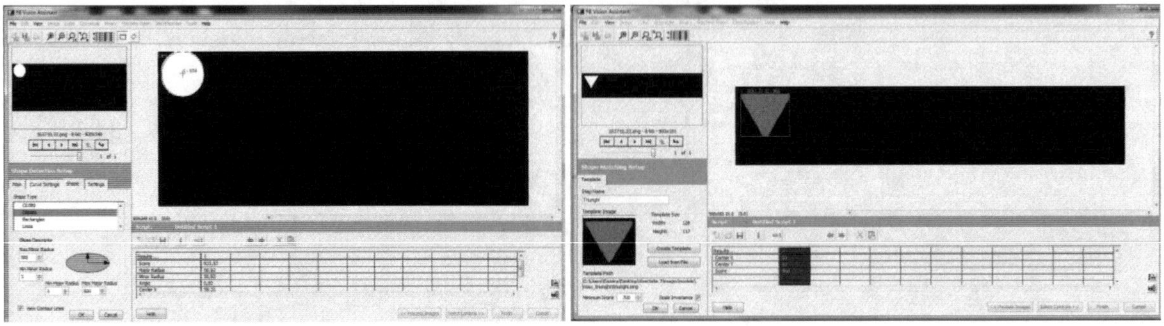

a) Shape Detection b) Shape Matching

Figure 4. Shape functions

Color matching is used to select regions within an image that contain the color information used as a reference. The color information in the image may consist of one or more colors. The machine vision software then learns the three-dimensional color information in the image and represents it as a one-dimensional color spectrum, comparing the color information in the entire image or regions in the image to the learned color spectrum, calculating a score for each region. The match score, ranging from 0 to 1000, relates how closely the color information in the image region matches the information represented by the color spectrum, defining the similarity between the color spectrums. A score of zero represents no similarity between the color spectrums, whereas a score of 1000 represents a perfect match.

Color pattern matching is a unique approach that combines color and spatial information to quickly find color patterns in an image. It uses a combination of color location and grayscale pattern matching to search for the template. Color pattern matching uses color matching first to locate the objects, and then pattern matching to refine the locations, providing more accurate results. The addition of color helps to improve the accuracy and reliability of the pattern matching tool, making it less sensitive to factors such as orientation of the part, scale changes, and lighting changes and maintaining its ability to locate the reference patterns and gives accurate results despite these changes [6], [7].

The machine vision application searches for a model or template that represents the object, in each acquired image, calculating a score for each match. The score indicates how closely the model matches the color pattern found. It also provides the application with information about the number of instances and location of the template within an image.

A sign is recognized (identified) if it obtains a score higher than a threshold level on each of these algorithms. An object is passed through the recognition module of a particular indicator if there is a chance that the object might represent the sign in question. The possibility is computed by analyzing the score with which the rim shape was detected. This is not necessarily an accurate result, and if it were left only to this, a sign would usually have to be passed through all the templates for that particular sign category. Considering the fact that the execution time for a sign category that contains only one sign (the stop sign) is 0.12 seconds, the duration for a category containing 34 signs, as is the triangular red category, would be 4.08 seconds. This is obviously too long for real-time sign recognition. Considering that the Stop sign is the only sign in its category (red hexagon), its duration is minimal possible in this configuration (Figure 5), but for other sign categories further processing is requires in order to reduce the number of templates tested and, implicitly, the execution duration.

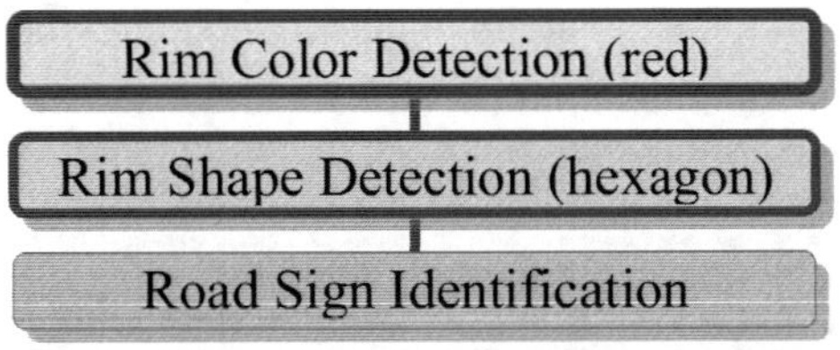

Figure 5. Stop sign recognition diagram

2.4. Comparison between processing algorithms

Several attempts have been made in order to reduce the duration of this algorithm. First, we used the color and shape detection functions applying them, this time, to the background of the shape. The image is then taken through another color threshold in order to detect the background color (usually white but can also be blue, depending on the sign type) or the interior color of the sign. Therefore, the lack of the specific background color in the candidate constitutes a possible sign rejection criterion. The resulting binary image is taken through the shape matching step, this time for the shape of the background. Using Shape matching instead of Shape detection is mandatory in this case, since the shape of the background can form the pictogram of the sign and does not constitute a regular shape [6].

Shape in road signs is particularly difficult to assess; it is, generally speaking, irregular and can be easily influenced by objects that obstruct the image. From this point of view it is necessary that the shape detection be as robust as possible, making it, therefore, less accurate. This facilitates confusion between signs of similar shape. Therefore, shape detection is used to classify rather than to recognize the individual road sign. Pattern matching is used to detect the individual signs. This automatically means that the pattern matching function will have to search through several models and choose the best fit for the sign sought after. Shape analysis helps sort the number of models which should be tested and focus the search on the most likely model. This step has the purpose of activating only the recognition module based on the most likely matches for the sign in question. The search result represents the sign with the maximum found score, score that exceeds a certain threshold. If the maximal score is below the threshold, the candidate either does not represent a sign or does not represent a sign belonging to that category. One example of such an algorithm, which can be seen in Figure 6, is used on the circular blue signs. It presents an average execution time of 0.66 seconds to 11 signs.

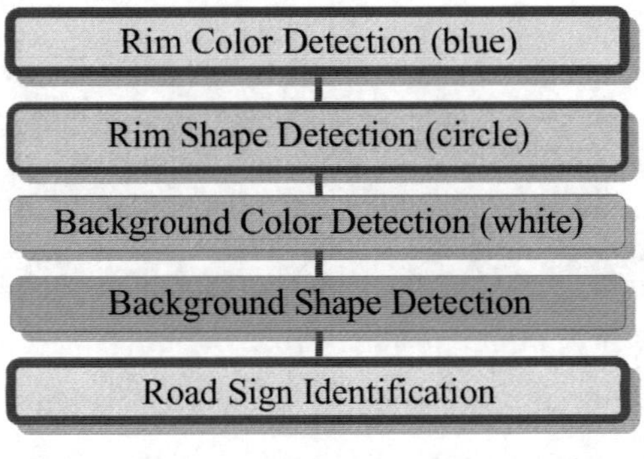

Figure 6. Circular blue signs recognition diagram

Using this algorithm, for a number of 34 signs, the duration would increase to around 2 seconds, which is a significant improvement from the performance of the stop algorithm. Another improvement would be the use of a more complex shape matching function that searches for a correspondence between the background shape and the rim shape. The program basically looks for an irregular shape formed from the exterior and the interior margins of the contour of the sign (Figure 7). This verifies the existence of a correspondence between the size and orientation of the contour and the background. This operation has not shown significant improvements in duration (15 signs in 0.87 seconds), but it increases the accuracy of the algorithm.

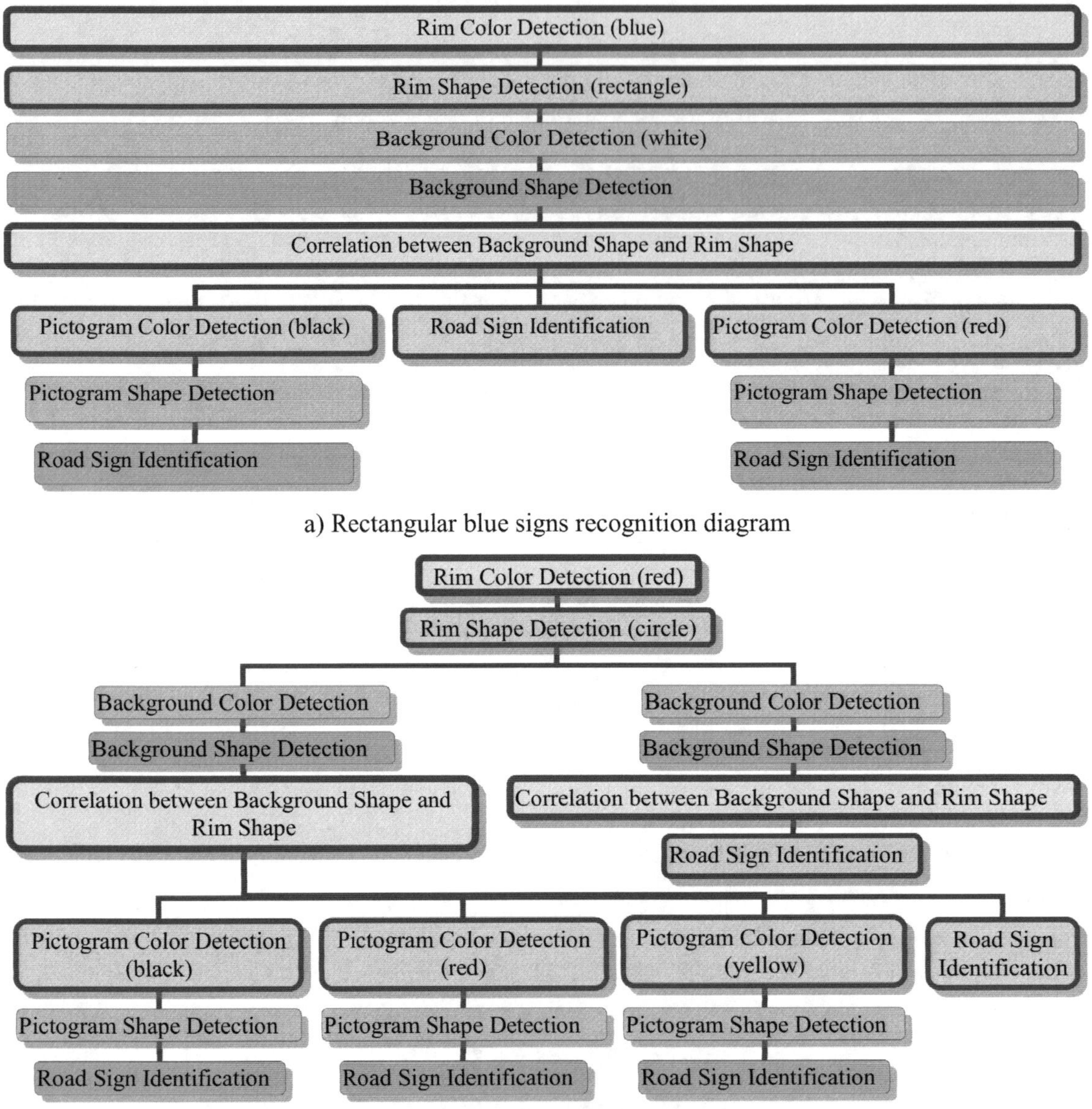

a) Rectangular blue signs recognition diagram

b) Circular red signs recognition

Figure 7. Diagrams of subprograms that use pictogram analysis

Another option is to search according to the color/colors of the pictogram. There are road signs where the pictogram is the background, but many have both a background and a pictogram, different from one another. Therefore a secondary search based on the color of the pictogram could improve processing time. This operation was used on the rectangular blue signs, these having different colored pictograms. The resulting algorithm can be seen in Figure 7.a.

A more complex version of this algorithm was used for the red circular sign category. There is significant improvement in this case (27 signs in 1.2 seconds). The diagram can be seen in Figure 7.b.

A variation of the latter is the triangular red sign recognition algorithm. It can check 34 patterns in 0.63 seconds. The improvement consists of the use of pattern matching for determining the shape of the pictograms. This seems counterintuitive, since pattern matching functions are those that give the duration of the process, being the most time consuming of the all the function used in this program. The logical motivation would be the fact that the execution time of these functions depends on size of the template and on the image to be checked. Since the pictogram of the sign is significantly smaller than the sign itself, it takes much less to check its presence in the image than of the entire pictogram. It is also much more precise than a shape matching function, leaving less pictogram patterns to check in comparison to shape matching. This algorithm can be seen in figure 8. These methods could be improved in order to reduce the execution time, and so increase the execution speed. Execution duration is influenced by the number of different types of signs simultaneously detected in the image, the results of the intermediate functions of shape and color, the number of matching modules simultaneously active, the time it takes to receive data from the module, etc. Therefore, the execution time can vary within a fairly wide range (± 100%). However, the average duration of an algorithm still represents a possibility to compare algorithms between them.

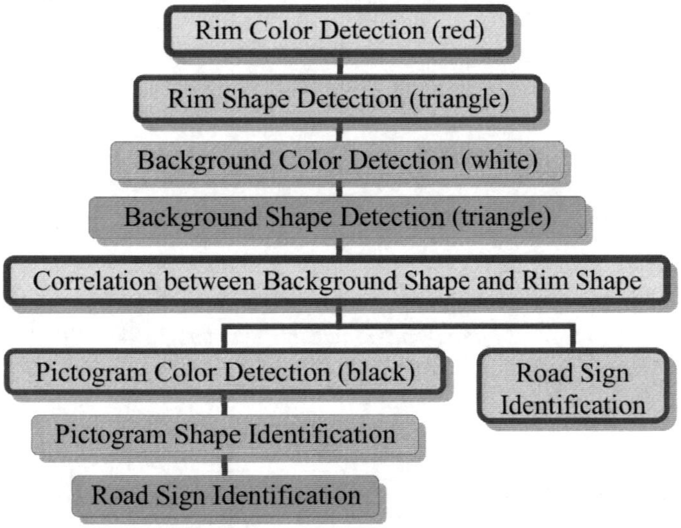

Figure 8. Triangular red signs recognition diagram

3. Conclusions

Identifying road signs is a simple task for human beings, but it appears to be quite the challenge for computers. If a person is asked to point out the traffic sign in an image, they can do this easily. However, from the point of view of computer vision, there are a number of difficulties, such as lighting, weather condition, incorrect positioning, obstructions, etc. To deal with them, the road sign recognition system is provided with large amounts of training data to allow the system to respond correctly when a traffic sign is encountered.

The program, because it works with thresholds and probability, is robust and effective. The chance of false alarms has been reduced by filtering the results gradually through successive stages. Each potential sign is tested with a set of features to decide if it can represent a road sign or not, and then, depending on these characteristics, potential signs are classified into different groups. Different levels of detection work with different procedures and are based on different principles of seeking different attributes of road signs, eliminating false signs.

Road signs can be either pictogram-based or text-based. The road and traffic signs considered in this paper are those that use a visual/symbolic language that can be interpreted by drivers. The signs containing text are not included, as these are subject to further developments based on optical character recognition. Supplementary signs are also not included, because their meaning is dependent on the associated road sign. In this regard, an algorithm capable of combining sign meanings based on proximity is needed. This is also an opportunity for further improvement.

Road signs may appear rotated, resulting in perspective errors. Perspective errors are caused by the incorrect positioning of the sign itself, which is not perpendicular to the axis of the road, and therefore not perpendicular to the axis of the camera. The sign should be directed towards the road to which it corresponds, but this is not always the case, making the sign difficult to identify even for experienced drivers. Our system works well regardless of lighting conditions, faded colors, obstructions, etc. but it has limited capabilities regarding perspective errors. A further improvement to the program would be, therefore, to make it invariant to perspective errors.

The algorithm presented in this paper is based on the detection of the contour color of road signs, but there are signs that do not have a continuous contour, for example the End of compulsory minimum speed sign. Generally speaking, these are known signs crossed with a red stripe. Theoretically, by using a HSI threshold composed of both colors, this sign could be detected with this system. The problem is that the signs that have a contour consisting of only one color would also be detected in this category and would complicate the identification of the sign category. This could be solved through a color matching function that would determine the category to which the sign belongs, based on its color spectrum.

References

[1] Fleyeh H 2008 *Traffic and Road Sign Recognition,* Napier University, UK, Doctoral Thesis

[2] Fleyeh H and Dougherty M 2005 *Road and traffic sign detection and recognition*, Proceedings of the 16th Mini-EURO Conference and 10th Meeting of EWGT, Poznan, Poland, September 13-16, pp 644-653

[3] Greenhalgh J and Mirmehdi M 2012 Real-time detection and recognition of road traffic signs, *IEEE Transactions on Intelligent Transportation Systems* **13**(4) 1498-1506

[4] Mathias M, Timofte R, Benenson R and Van Gool L 2013 *Traffic Sign Recognition - How far are we from the solution?*, Proceedings of IEEE International Joint Conference on Neural Networks (IJCNN 2013), Dallas, USA, August 4-9, pp 1-8

[5] Torresen J, Bakke J W and Sekanina L 2004 *Efficient recognition of speed limit signs,* Proceedings of IEEE Conference on Intelligent Transportation Systems (ITS), Washington DC, USA, October 3-6, pp 652-656

[6] ***National Instruments 2003 *IMAQ Vision Concepts Manual*, June 2003 Edition, 322916B-01

[7] ***National Instruments 2004 *IMAQ Vision for LabVIEW*, August 2004 Edition, 371007A-01

[8] ***National Instruments 2005 *NI Vision for LabVIEW User Manual*, November 2005, 371007B-01

[9] ***National Instruments 2008 *NI Vision Assistant Tutorial*, June 2008, 372228K-01

IOP Conf. Series: Materials Science and Engineering **106** (2016) 012010 doi:10.1088/1757-899X/106/1/012010

Research on models of biological systems that can be integrated into mechatronic systems

P P Pop[1], A Pop-Vădean[1], C Barz[2], T Latinovic[3] and O Chiver[2]

[1]Technical University of Cluj Napoca, Department of Mechatronics and Machine Dynamics, Muncii str., no. 103-105, 400641 Cluj Napoca, Romania
[2]Technical University of Cluj-Napoca, North University Center of Baia Mare
[3]University of Banja Luka, Faculty of Mechanical Engineering, Republic of Srpska, Bosnia and Herzegovina

E-mail: poppaulpetrica@yahoo.com

Abstract. The models of biological systems that we find on Earth can be the subject of research to develop a few mechatronic systems. Such models are offered by bees, ants, crows, cranes, etc.

Article aims to investigate these models and their manifestations. Imitating this behavior and studied him offer ideas for develop models that can be integrated into mechatronic systems. They can be integrated into mechatronic system as algorithms for finding local optimum, to search, to detect an optimal way travel on a network, to find best decision, etc.

1. Introduction

It's easy to build a network of sensors that covers a certain area but it is very difficult to adapt to the environmental conditions for maximum efficiency operation. Problems related to: validate the measurement points, power supply, signal transmission, the degree of automation of the system, the influence of environmental factors in quality measurement, system maintenance, safety equipment, etc. Two models are the best candidates to be followed in the composition of the monitoring system: flocks of crows and central nervous system of the human body. Specifically the combination of these two models provides design, construction and successful operation of the system. Crows for their ability to orient themselves in their travels by detecting the location and trajectory adjustment according to information provided by other crows from the edge of the flock. But flock behavior according to information that is constantly updated during the trip. The central nervous system model is copied and adapted following functions: command and control, real time transfer of information and interfacing. The interface is based on the brain's ability to create virtual images of objects placed in crisis situations. The aim is to better manage information and eliminate redundant signals. It is a form of understanding of reality and communication with the environment.

For the most part of our existence we can make fairly accurate and precise forecasts about weather events on geographical areas. Weather radars, satellites and ground stations are helpful tools in collecting and processing field data needed in order to make the best prognosis. A variety of information is available via the Internet. Network interconnection ensures real-time data exchange and the correct information that lead to making the best decisions.

But unfortunately there are short time phenomena that occur on small areas. They cannot be predicted due to weather effects generated moments before its occurrence. An example of such effect is the thermal inversion effect or the massive floods that happen in a short period of time. How do we protect the assets and the population from such phenomena impossible to predict?

2. Flooding caused by short downpours on a small area

Such a phenomenon is the downpour, where enormous quantities of water fall over a short period of time on a less extensive surface. The application aims to monitor the water level of a river basin. Sensors are installed on the main river and on the main tributaries. They transmit data using GSM network (Figure 1) to a central server. This data is retrieved and compared with the data from the location where the sensor is located.

This method can provide information on sensor functionality. The data is then stored in an online database stored on the server (Figure 3). This model allows access to data history transmitted in a given time and space. This information can also be used to generate future models regarding the amount of water reaching the rivers from a certain surface, resulted from normal rains. Such data can be used in various further research.

Figure 1. GSM antenna

3. The database structure of the application

Database tables are interconnected so as to ensure optimal management of information [1-3]. They store information about the technical specifications of ground sensors: size measured, measuring interval, operating mode, etc. [4], [5] (Figure 2). The locations of the sensors are important to making the correlation between the size measured and the river location where the measurement is being performed. If data from a sensor repeatedly fails to be received, the information will not be filled in the database and the faulty sensor will be signaled in another section. The program will thus alert the administrator about the fault occurred as well as the location of the deffective sensor. These locations are stored and transmitted through the GPS coordinates system: latitude and longitude. For greater accuracy, along with these coordinates the database also stores the measurements of the riverbed at the location of the sensor.

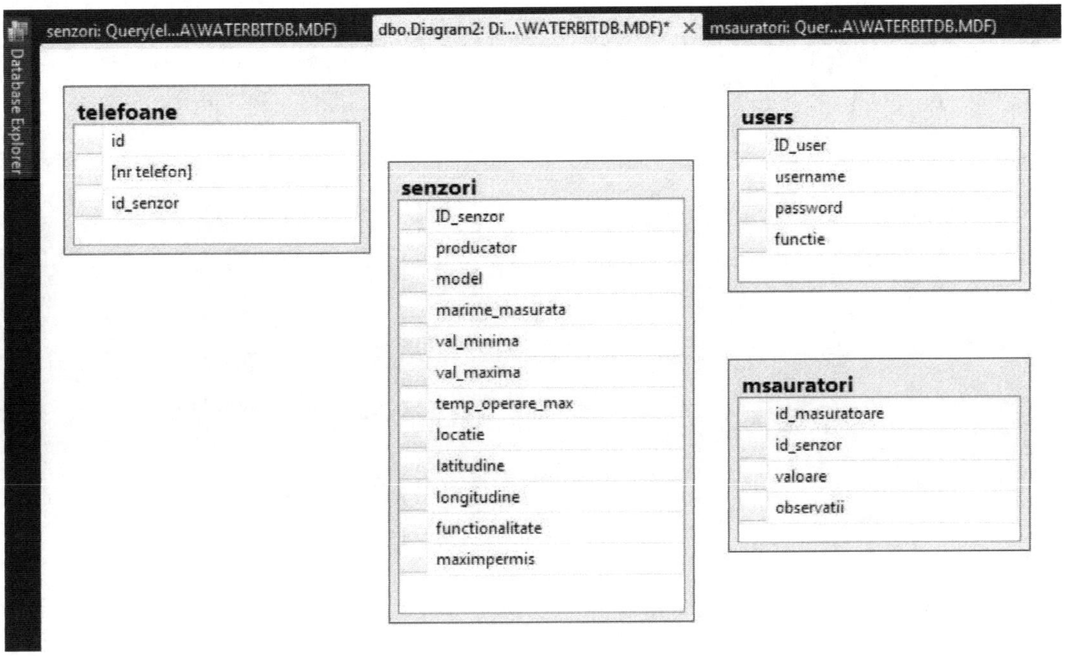

Figure 2. A part of diagram database apllication

4. Running the application

The program, which is installed on a PC, manages the information in the system, processes it and takes decisions. It is implemented using C # (Sharp) programming technology with an interactive graphical interface [6-9]. It benefits from access zones well defined between administrator, user and system operator (Figure 4). It enables the management and the usage of the integrated system, as well as the access to database information.

The moment the sensors register that the level of water is exceeded, the program launches the hazard alert. Alerting is done by using the GSM network for mobile telephony (Figure 1). The SMS (Texting) services of GSM network application send alert messages both to the local authorities in the endangered areas and to the population.

The application uses a database in which the contact information of local authorities and emergency intervention teams are stored. Alerting the population located in the risk area can be done in two ways. The alert message can be sent to the phone numbers stored in the database (Figure 2), or use the GSM antennas (Figure 1) in the endangered area to send the alert message to all the GSM users connected to the network in the range of the respective antenna (Figure 1).

5. Monitoring water chemistry

The application can also monitor the pollution of a river basin. The sensors placed in rivers may refer certain chemicals present in the water. They can send measured values to the database server using the telephone network. The program processes field data and decides the level of alert. Then it alerts the competent authorities and the population living along the area where the pollution will spread. This way one can precisely know the point of origin of the pollution, may establish its magnitude and also how it can be mitigated. It's important that the measures taken in order to reduce pollution impact on the environment be taken in the shortest time possible from the onset of pollution. Such a real time monitoring system is very useful.

id_masuratoare	id_senzor	valoare	observatii
1	1	51	Niciun pericol i...
2	2	52	Niciun pericol i...
3	1	51	Niciun pericol i...
4	2	52	Niciun pericol i...
5	1	300	Sansa de inund...
6	2	400	Niciun pericol i...
7	1	300	Sansa de inund...
8	2	400	Niciun pericol i...
9	1	300	Sansa de inund...
10	2	400	Niciun pericol i...
11	1	300	Sansa de inund...
12	2	400	Niciun pericol i...
13	1	300	Sansa de inund...
14	2	400	Niciun pericol i...
15	1	300	Sansa de inund...
16	2	400	Niciun pericol i...
17	1	300	Sansa de inund...
18	2	400	Sansa de inund...
19	1	300	Sansa de inund...
20	2	400	Sansa de inund...
21	1	300	Sansa de inund...
22	2	400	Sansa de inund...
23	1	480	Niciun pericol i...

1 of 76 | Cell is Read Only.

Figure 3. Sensor measurements recorded in the database

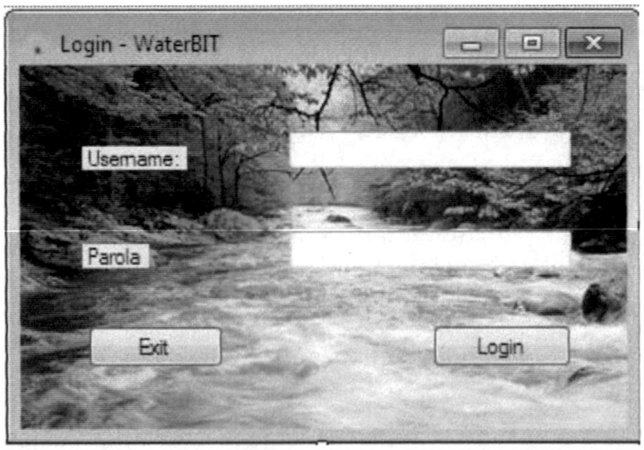

Figure 4. Login application window

6. The automation of systems for defense against natural disasters

This program can be used to automate the flood defense systems in case of flooding and pollution in rivers. These systems can be used for any hidrographic system which has water reservoirs in order to reduce pollution levels or the level of water in case of flooding. The system can control the discharge or intake of considerable quantities of water from the river.

Knowing the storage capacity of reservoirs in real time, the application can decide the regulation of the flow of water from the river through automatic handling of locks. This can be operated by long distance actuators or through its own network. The signals can also be sent through the GSM network.

7. Conclusion

This article presents an application that proposes ideas and solutions for a sustainable development of the knowledge-based society. One can correlate data regarding the annual average rainfall with the amounts of water that flow into rivers or with the influence of environment on water.

- Research possibilities on the measured data are endless;
- Monitoring pollution is also an added value that can be integrated into the system;
- Planning rational water consumption while maintaining its quality parameters is another option that this application can offer.

References

[1] ***www.programming.de/programming_cs.php
[2] ***www.dreamincode.net
[3] ***www.codeproject.com
[4] ***www.infoap.utcluj.ro/vcsharp
[5] ***www.functionx.com
[6] Donis M 2006 *Programming Microsoft Visual C# 2005: The Language,* Microsoft Press, Seatle, USA
[7] Jhon S 2010 *Microsoft Visual C# 2010. Step by Ste,* Microsoft Press, Redmond, Washington, USA
[8] Paul K 2002 *Advanced C# Programming,* McGraw Hill, Osborne, Australia
[9] ***https://msdn.microsoft.com/enus/library/dd831853 (v=vs.100).aspx

Modelling of some parameters from thermoelectric power plants

G N Popa[1], C M Diniş[1], S I Deaconu[1], Şt Maksay[1] and I Popa[1]

[1]Politehnica University of Timisoara, Department of Electrical Engineering and Industrial Informatics, 5 Revolution Street, 331128 Hunedoara, Romania

E-mail: gabriel.popa@fih.upt.ro

Abstract. Paper proposing new mathematical models for the main electrical parameters (active power P, reactive power Q of power supplies) and technological (mass flow rate of steam M from boiler and dust emission E from the output of precipitator) from a thermoelectric power plants using industrial plate-type electrostatic precipitators with three sections used in electrical power plants. The mathematical models were used experimental results taken from industrial facility, from boiler and plate-type electrostatic precipitators with three sections, and has used the least squares method for their determination. The modelling has been used equations of degree 1, 2 and 3. The equations were determined between dust emission depending on active power of power supplies and mass flow rate of steam from boiler, and, also, depending on reactive power of power supplies and mass flow rate of steam from boiler. These equations can be used to control the process from electrostatic precipitators.

1. Introduction

Plate-type electrostatic precipitators are most commonly used in de-dusting equipment from industry. They can be used in the treatment of very large gas flows (thousands of m^3/s) at different temperatures and gas pressures. It have applications in producing electricity (thermoelectric power plants), metallurgical, cement and glass industry [1].

The steam thermoelectric power plant is a chemical energy transformer in electric energy of the greatest proportions. Energy transformation can't be done directly, but through transformation of the combustion process in the heat, the heat is restored water to other the state of aggregation: liquid turns to steam with high pressure. Steam turbine drive produces such energy, mechanic, and electric generator – mechanical coupled with the turbine - transform this energy into electrical energy [2].

Fossil fuel combustion is burned in boiler absorbing oxygen from the air. Non-combustible elements from fuel remain in the form of ashes and slag (can reach 40-50% of the total fuel) boiler's stoker, and a part will be discharged into the atmosphere. Gas with dust can't be discharged directly through the boiler due to excessive pollution which would create. For this reason, between the boiler and the funnel is installed, usually plate-type electrostatic precipitators, which collects most of the dust particles (over 95%), before being discharged to the funnel. In principle, a higher pressure steam will cause an increase in turbine speed. However, the turbine speed is kept constant, because it determines the voltage frequency made by synchronous electric generator. For maintaining constant turbine speed is used automatic adjustment installation [3], [4].

Automatic tuning system has as main task to preserve equality between power developed by turbine and generator load. The size of the informant's best of imbalance between them is the turbine speed. Speed controller used as the size of the execution mass flow rate of the steam in the turbine. The turbine power is directly proportional to the mass flow rate of steam and with the adiabatic drop in the turbine [5].

Mathematics modelling of parameters in the industry can be achieved with the least squares method [6-9].

Automating processes from thermoelectric power plants is important, so electricity production improvements, through various optimization processes in a power plant. Starting and stopping the filtration installations (or part of them) must be made in such way to save time, not to exceed the allowable pollution limits, and also not to be affect lifetime of the facility. It must at the same time, to improve the operation of the installations, to ensure a low electricity consumption (in terms of collecting the required dust emission) and to ensure safe operation through automation [10], [11].

Precipitators with three sections are often used in industry. Each section is supplied with power from a separate voltage power source (silicon-controlled rectifier type) [3], [4]. This power source provides a high voltage in section (usually up to 60 kV) in order to create the Corona effect to produce substantial charge carriers that will attach to the dust particles from the gas. Dust particles will have the electrical charge and will collect on the surface of the collecting surface of precipitator [12].

It shows, for the case of precipitators with three sections, mathematical modelling, using the least squares method, dust emissions of precipitators (dependent parameter) depending on the total power active, reactive, and apparent power sources, and mass flow rate of steam from boiler.

2. Electrical Powers and Parameters Measurements on Precipitator Power Supplies

To measure the electrical parameters of precipitator power supplies is necessary to acquisition the voltages and currents [13].

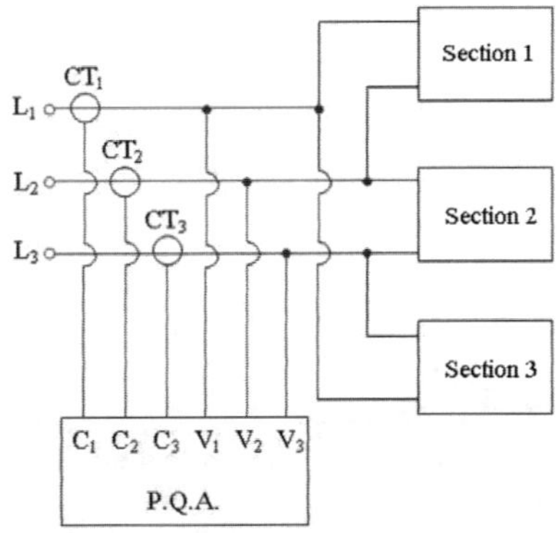

Figure 1. The sensors' and probes' connection at power quality analyser CA 8334 B at the power supplies of precipitator with three sections

It is analysed a precipitator with three sections, every power supply, for each section, is connected at two phases to balance the network (Figure 1).

It is used power quality analyser (PQA), CA 8334 B (Chauvin Arnoux, France) [14]. Using PQA can be done fast Fourier transform and can be identify voltages and currents harmonics, the waveforms of voltages and currents, the total harmonic distortion, the powers (active, reactive, and apparent), and power factor [15] , [16].

The high voltage in precipitator sections is generated by silicon-controlled rectifier that is connected at two phases (400 V, 50 Hz).

PQA was connected at inputs currents C_1, C_2, C_3, and inputs voltages V_1, V_2, V_3, for each power supply using one current sensor (*CT*) AmpFLEX (current measuring up to 3000 A), and one two line low voltage probes (L_1, L_2 or L_2, L_3 or L_3, L_1) [15]. Because precipitator has three sections the PQA was connected like in Figure 1 using three current sensors (CT_1, CT_2, CT_3) and three line low voltage probes (L_1, L_2, L_3). The current sensors are connected on the aluminium bus bars, in front of fuses and circuit breaker, to measure the electrical parameters.

Some specific electrical measurements for deforming state are given below [6], [13], [17]. In reality, each of the periodic signals (currents, voltages) has an infinite number of harmonics. From the practical point of view, to be able to accomplish calculations by PQA, the maximum number of harmonics to be taken into account is, in most cases, 40. In relationships (1) to (4), *k* is degree harmonics and with *0* is indicated continuous component.

The root-mean square (*RMS*) of current is:

$$I = \sqrt{I_0^2 + \sum_{k=1}^{40} I_k^2}\,, \tag{1}$$

and, the root-mean square of voltage is:

$$U = \sqrt{U_0^2 + \sum_{k=1}^{40} U_k^2}\,. \tag{2}$$

If it makes φ the phase displacement between voltage and current, for each of harmonics, the active power in un-sinusoidal is defined as:

$$P = U_0 \cdot I_0 + \sum_{k=1}^{40} U_k \cdot I_k \cdot \cos\varphi_k\,, \tag{3}$$

and, the reactive power is:

$$Q = \sum_{k=1}^{40} U_k \cdot I_k \cdot \sin\varphi_k\,. \tag{4}$$

Apparent power can be determined according to the root-mean square of the voltage and current:

$$S = U \cdot I\,. \tag{5}$$

In deforming state, it is used another specific power, namely the deform power:

$$D = \sqrt{S^2 - \left(P^2 + Q^2\right)}\,. \tag{6}$$

In Figure 2 is the Fresnel diagram, for voltages and currents fundamentals (at 50 Hz), in the case of the three sources of the sections. With the indexes *1,2* and *3* it is noted the source's phases (like in Figure 1). From the diagram it can be seen the symmetrical and balanced voltage, as well as the inductive nature of loads (the current is after the voltage).

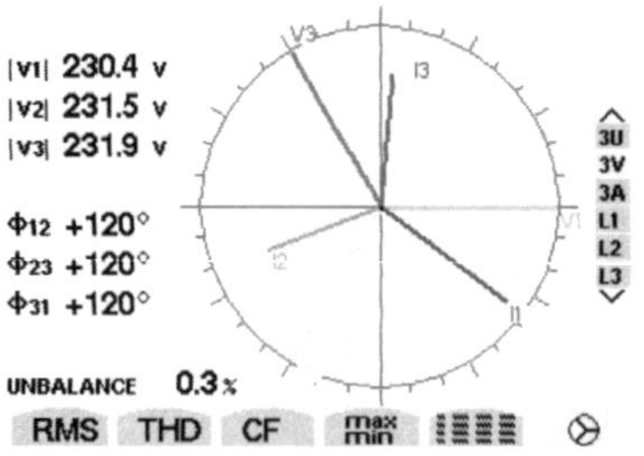

Figure 2. Fresnel diagram for fundamental currents and voltages for power supplies

The power sources have the two thyristors connected in anti-parallel, which are control by firing angle. A higher value of the firing angle dues a lower root mean square of voltage. In this way, it can change the voltage in the rise-up transformer, and in the secondary winding where is connected the precipitator's section.

In Figure 3, for a power supply (behaviour is the same for the other two power sources) are present the voltage and the current absorbed by the network. Due to the principle of the operating of the silicon-controlled rectifier power sources, the supply voltage is sine (measured frequency, f = 50.02 Hz), but the current is not the sine shape and will introduce a deforming state. The deforming state of the power sources can be highlighted in Fig. 4 where it shows the analysis of harmonic currents in the three phases (L_1, L_2, L_3).

Thanks to automatic voltage adjustment in sections (due to the events into sections), the *RMS* voltage in the primary winding transformer changing every time, and also, the *RMS* value and the shape of the current absorbed by the network.

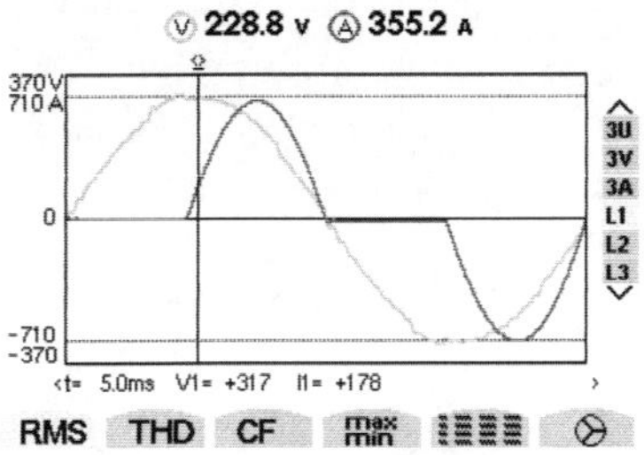

Figure 3. Current and voltage on phase L_1 for input section power supply for one period of time (20 ms)

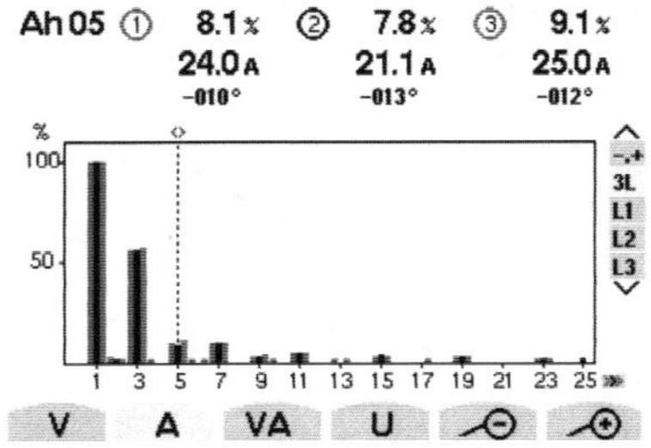

Figure 4. Spectrum analysis of currents through phases L_1, L_2, and L_3 at one moment of time

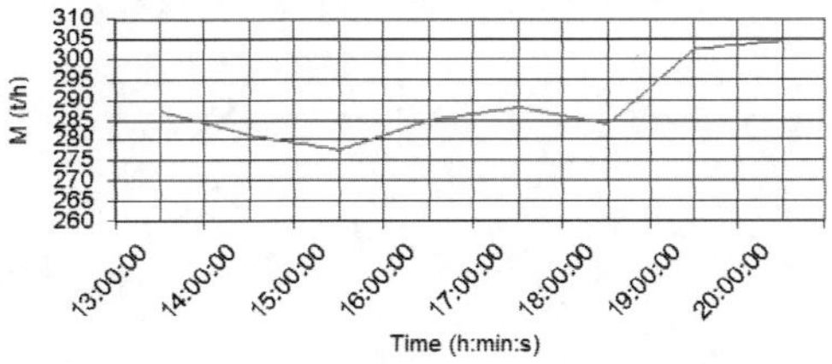

Figure 5. Mass flow rate of steam *M* depending on time for the boiler that is connected at the input of precipitator with three sections

From this reason, the spectrum currents are changing permanently.

The mass flow rate of steam generated by the boiler, is not constant (for various reasons, i.e.: the type of fuel, the type of burning of fuel, the generator loading), but is stored in certain limits. Figure 5 shows the evolution of the typical mass flow rate of steam over the course of several hours.

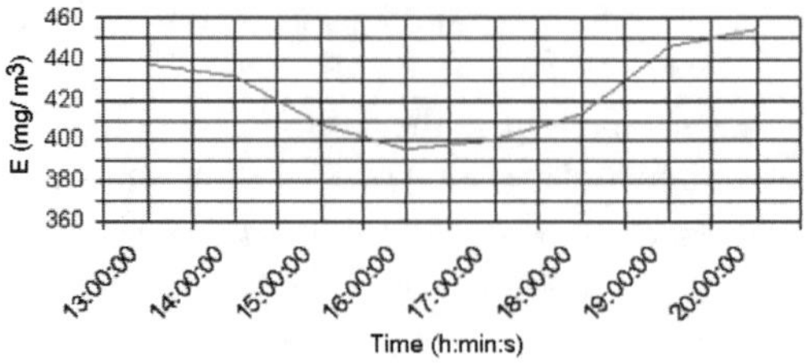

Figure 6. Dust emission E depending on time at the output of precipitator with three sections

The dust emission at the output of the precipitator with three sections, it is not constant (for various reasons, i.e.: type of power sources, type of current control, burning type, the generator loading). A typical evolution over the course of a few hours of operation is shown in Figure 6. Regardless of the factors that influence the performance of the precipitator collection, it must remain within the limits of design in terms of dust emissions (smaller than the designed value).

3. Mathematical Modeling through Least Squares Method

Let consider the mathematical function that will model the measure data:

$$z_t = f\left(x_t, y_t, a_0, a_1, a_2, \ldots a_n\right), \tag{7}$$

where z is the variable dependent, and x, y are the independent variables, and index t represents theoretical values. x, y, and z are electrical and/or technological parameters, and a_0, a_1, a_2, ..., a_n are the coefficients of the function that will be determined. The number of these coefficients depend on the type of function used (i.e., polynomial function of some degree; basically, it can use a function of 2nd or 3rd degree) [6], [7].

A criterion of choice of optimal mathematical model is that of the minimum sum of squares of deviations from the theoretical function compare with real data (made by measurements) which is called the least squares method:

$$F = \sum_{k=1}^{n} \varepsilon_k^2 = \sum_{k=1}^{n} \left(z_{mk} - z_{tk}\right)^2 = \sum_{k=1}^{n} \left[z_{mk} - f_t\left(x_{tk}, y_{tk}, a_0, a_1, a_2, \ldots a_n\right)\right]^2, \tag{8}$$

where, F is the sum of squares of deviations ε_k^2, index k is order number of measuring data, index n is number of measuring data, and index m represents measured values. Mathematical modelling is considered by this method when the function F from Eq. (8) is minimal:

$$F = F_{\min}. \tag{9}$$

For the determination of the coefficients of function (a_0, a_1, a_2, ..., a_n), the system of equations must be solved:

$$\frac{\partial F}{\partial a_0} = 0; \frac{\partial F}{\partial a_1} = 0; \frac{\partial F}{\partial a_2} = 0; \ldots \frac{\partial F}{\partial a_n} = 0, \tag{10}$$

and, after solving, the coefficients a_0, a_1, a_2, ..., a_n are replace in Eq. (8).

It defined two parameters, used in statistics, which are used in this method.

The root-mean square deviation (R_{MSD}) is:

$$R_{MSD} = \sqrt{\frac{1}{n} \sum_{k=1}^{n} \left[z_{mk} - f_t\left(x_{tk}, y_{tk}, a_0, a_1, a_2, \ldots a_n\right)\right]^2}. \tag{11}$$

If R_{MSD} is smaller, with the theoretical function f_t modelled better the real measured data.

The correlation coefficient (ρ) can be compute with:

$$\rho = \sqrt{1 - \frac{\sum_{k=1}^{n}\left[z_{mk} - f_t\left(x_{tk}, y_{tk}, a_0, a_1, a_2, \dots a_n\right)\right]^2}{\sum_{k=1}^{n}\left(z_{mk} - \overline{z_m}\right)^2}}, \tag{12}$$

where $\overline{z_m}$ represents the average values of data:

$$\overline{z_m} = \frac{1}{n}\sum_{k=1}^{n} z_{mk} . \tag{13}$$

Correlation coefficient represents the intensity of dependence among the parameters (dependent and independent), and can has values between 0 and 1. A value of ρ nearby 1 value indicates a direct connection between the parameters, meaning a growth of independent parameters determines increasing of dependent parameter.

Mathematical modelling has been used measured data (Table 1) from an industry precipitator (the main data in the appendix) from S.C. Electrocentrale Deva S.A. (thermoelectric power plant), Romania. In Table 1, P – total active power, Q – total reactive power, S – total apparent power (P, Q, S are sums for all three power sources), M - flow rate of steam from boiler, and E – dust emission at the output of precipitator.

Table 1. Measurement parameters used at mathematical modelling

No.	P (kW)	Q (kVAR)	S (kVA)	M (t/h)	E (mg/m^3)
1	103.11	181.91	209.86	286.47	397.58
2	112.45	184.46	215.97	279.85	404.29
3	124.02	190.65	227.57	280.95	431.81
4	104.93	178.05	207.29	285.73	406.98
5	117.89	179.43	214.95	302.64	445.9
6	108.61	171.16	203.29	287.2	398.92
7	137.91	192.85	237.11	284.63	409.66
8	111.29	170.51	204.1	283.89	412.34
9	115.9	179.85	214.22	284.26	434.15
10	120.68	177.15	214.76	302.64	448.59
11	120.33	179.28	216.55	280.22	425.1
12	112.21	171.55	205.97	285.73	396.24
13	110.45	168.21	201.81	278.01	412.34
14	109.27	164.67	198.42	282.42	433.15
15	117.55	164.08	202.53	287.94	399.59
16	129.17	175.11	217.6	279.11	419.73
17	118.01	169.34	206.95	280.95	402.28
18	123.4	163.99	205.51	283.16	397.58
19	127.73	165.96	209.83	277.64	407.65
20	119.14	162.45	201.54	303.75	449.93
21	119.3	165.72	204.89	286.47	403.62
22	130.5	175.69	218.93	289.41	421.74
23	123.07	170.24	210.21	285	396.24
24	125.44	170	211.3	293.08	429.12
25	127.82	160.83	205.62	298.23	439.19

It was determined the mathematical modelling through least squares method for $E=f(M,P)$, with three types of equations: 1st degree equation Eq. (14) and Figure 7, 2nd degree equation Eq. (15) and Figure 8, and 3rd degree equation Eq. (16) and Figure 9.

The regression's equation for 1st degree equation is:

$$E = 0.365 \cdot P + 1.413 \cdot M - 31.704 . \quad (14)$$

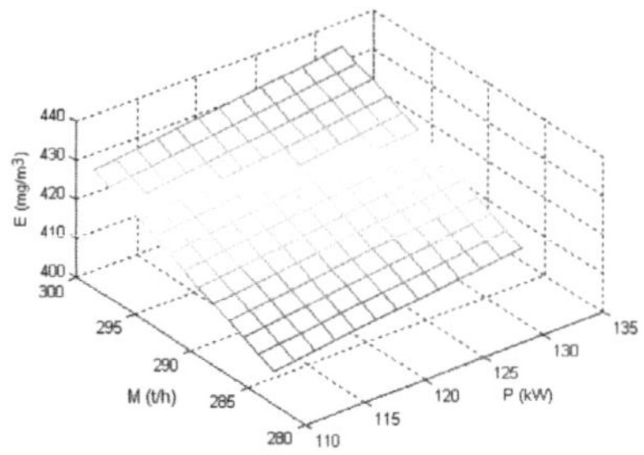

Figure 7. Dust emission *E* depending on mass flow rate of steam *M* and active power *P*, mathematical modelling 1st degree equation

The regression's equation for 2nd degree equation is:

$$E = -0.005 \cdot P^2 + 0.09 \cdot P \cdot M + 0.135 \cdot M^2 - 24.35 \cdot P - 87.788 \cdot M + 14397.975. \quad (15)$$

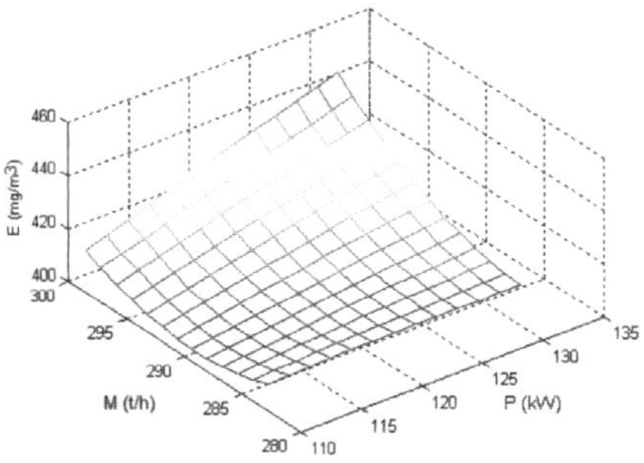

Figure 8. Dust emission *E* depending on mass flow rate of steam *M* and active power *P*, mathematical modelling 2nd degree equation

The regression's equation for 3rd degree equation is:

$$\begin{aligned} E = &-0.001 \cdot P^3 - 0.006 \cdot P^2 \cdot M + 0.006 \cdot P \cdot M^2 + \\ &+ 0.01 \cdot M^3 + 2.318 \cdot P^2 - 2.1 \cdot P \cdot M - 9.2 \cdot M^2 + \\ &+ 3.939 \cdot P + 2743.301 \cdot M - 259231.362. \end{aligned} \quad (16)$$

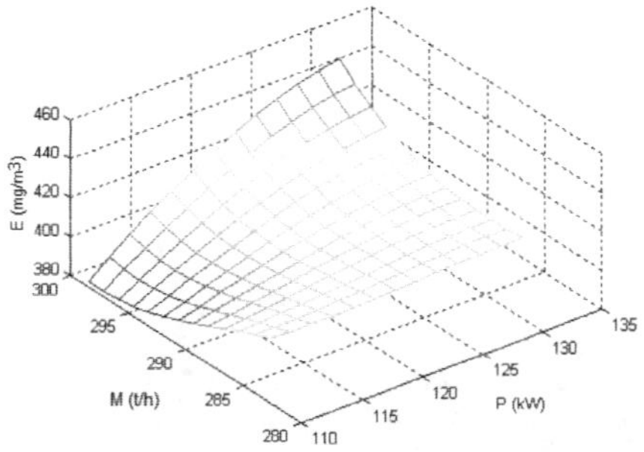

Figure 9. Dust emission *E* depending on mass flow rate of steam *M* and active power *P*, mathematical modelling 3rd degree equation

Table 2. The root-mean squares deviations and correlations coefficients for E = f(M,P)

No.	1st degree equation	2nd degree equation	3rd degree equation
R_{MSD} (mg/m^3)	13.157	10.672	10.43
ρ (-)	0.651	0.788	0.799

The root-mean square deviations and correlations coefficients are presented for the three cases (Eqs. (14), (15), (16)) in Table 2. The best mathematical modelling is made with 3rd degree equation. The lower deviation and the greatest correlation coefficient are for 3rd degree equation.

Root mean square deviation R_{MSD} measure the differences between values predicted by a model or an estimator and the values actually measured. R_{MSD} is a good measure of accuracy. These individual differences are called residuals, and R_{MSD} serves to unit them into a single measure of predictive power. Correlation coefficient ρ is a quantity that gives the quality of a least squares fitting to the measure data [6]. The burning of a large coal amounts, due a greater mass flow rate of steam (*M*) and higher dust emissions (*E*). With increasing amounts of dust is required a greater amount of active power of supplies (Figures 7-9).

In Table 3, are given the minimum, average and maximum *P*, *Q*, *S*, *M* and *E* sizes measured.

Table 3. Measure minimum, average, and maximum values P, Q, S, M and E

No.	P (kW)	Q (kVAR)	S (kVA)	M (t/h)	E (mg/m^3)
Minimum	103.11	160.83	198.42	277.6	396.24
Average	118.81	173.32	210.67	286.8	416.95
Maximum	137.91	192.85	237.11	303.8	449.93

Regression hypersurface equation *E=f(M,P,Q)* is:

$$E = -0.029 \cdot P^2 - 0.024 \cdot Q^2 + 0.131 \cdot M^2 + 0.03 \cdot P \cdot Q - \\ - 0.036 \cdot Q \cdot M + 0.104 \cdot M \cdot P - 27.939 \cdot P + 15.817 \cdot Q - \\ - 80.949 \cdot M + 12217.273. \quad (17)$$

For Eq. (17) R_{MSD}=9.736 and ρ= 0.827 (that indicates a better correlation between parameters *E,M,P,Q*). The stationary point is a stable area of operation, where the parameters change little (their derivatives tend to 0). Coordinates of stationary point for Eq. (17) are: *M*=284.146 *t/h*, *P*=130.585 *kW*, *Q*=193.01 *kVAR*, *E*= 418.885 *mg/m*^{3}.

From Eq. (17) results Eqs. (18), (19) and (20). To three-dimensional representation, the third independent parameter *M*, *P*, or *Q*, the average value is used. In Figures 10-12 show the dependencies

E=f(M,P,Q). In the Eq. (18) and Figure 10 was presented *E=f(M,Q)* for P_{avg}, in the Eq. (19) and Figure 11 was presented *E=f(P,M)* for Q_{avg}, and in the Eq. (20) and Figure 12 was presented *E=f(P,Q)* for M_{avg}.

$$E = -0.024 \cdot Q^2 + 0.131 \cdot M^2 - 0.036 \cdot Q \cdot M + \\ + 19.45 \cdot Q - 68.5679 \cdot M + 8493.276 \tag{18}$$

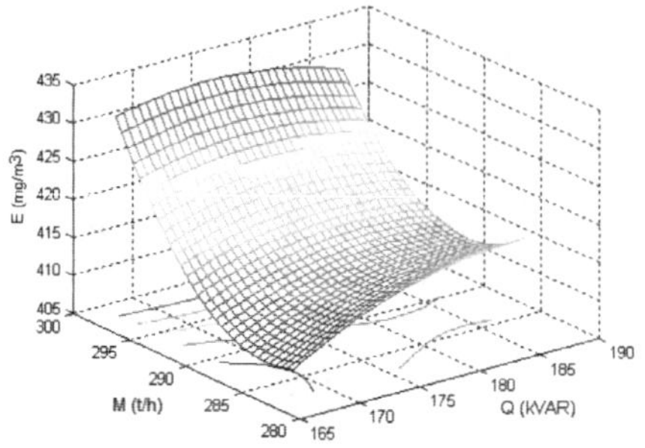

Figure 10. Dust emission *E* depending on mass flow rate of steam *M*, reactive power *Q*, and average active power P_{avg}, mathematical modelling 2nd degree equation

$$E = 0.131 \cdot M^2 - 0.029 \cdot P^2 + 0.104 \cdot M \cdot P - \\ - 87.301 \cdot M - 22.718 \cdot P + 14232.135 \tag{19}$$

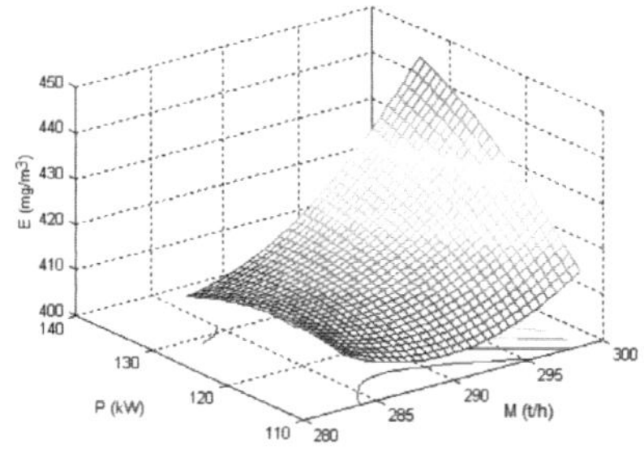

Figure 11. Dust emission *E* depending on active power *P*, mass flow rate of steam *M*, and average reactive power Q_{avg}, mathematical modelling 2nd degree equation

High dust emissions are influenced the most by increasing the mass flow rate of steam, and implicit, a larger quantity of burning coal. With increasing of dust emissions, increases, also, reactive power sources.

$$E = -0.029 \cdot P^2 - 0.024 \cdot Q^2 + 0.03 \cdot P \cdot Q + \\ + 1.948 \cdot P + 5.306 \cdot Q - 228.09 \tag{20}$$

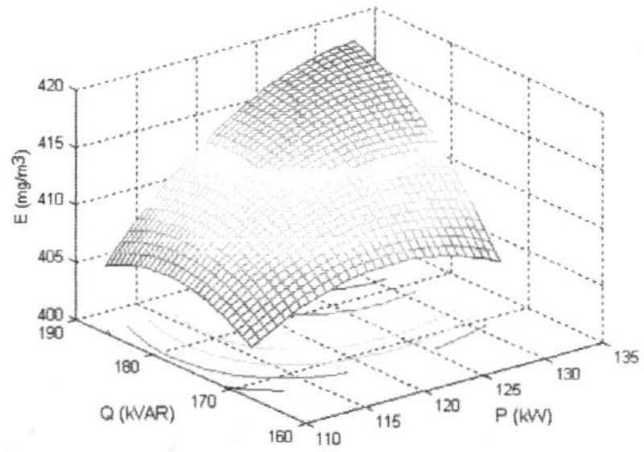

Figure 12. Dust emission E depending on reactive power Q, active power P, and mass flow rate of steam M_{avg}, mathematical modelling 2nd degree equation

Higher dust emissions due growth of active and reactive powers, up to a certain value (Figure 12). The evolutions of active and reactive power are due by type of power sources and the control current inside sections. With increasing of dust emissions should increase powers.

Regression hypersurface equation $E=f(M,P,S)$ is:

$$E = -0.06 \cdot P^2 - 0.04 \cdot S^2 + 0.131 \cdot M^2 + 0.081 \cdot P \cdot S - \\ -0.048 \cdot S \cdot M + 0.131 \cdot M \cdot P - 40.017 \cdot P + 21.428 \cdot S - \\ -80.219 \cdot M + 11941.467. \quad (21)$$

For Eq. (21) R_{MSD}=9.727 and ρ= 0.828 (that indicates a better correlation between parameters *E,M,P,S*). Coordinates of stationary point for Eq.(21) are: M=284.203 *t/h*, P=130.386 *kW*, S=231.954 *kVA*, E=418.677 mg/m^3. The stationary point coordinates are in graphical limits.

From Eq. (21) results Eqs. (22), (23), (24). For three-dimensional representation, the third size M, P, or S, is used as an average value. In Figures 13-15 are presented the dependency $E=f(M,P,S)$. In the Eq. (22) and Figure 13 was presented $E=f(M,S)$ for P_{avg}, in the Eq. (23) and Figure 14 was presented $E=f(P,M)$ for Q_{avg}, and in the Eq. (24) and Fig. 15 was presented $E=f(P,S)$ for M_{avg}.

$$E = -0.04 \cdot S^2 + 0.131 \cdot M^2 - 0.048 \cdot S \cdot M + \\ +31.045 \cdot S - 64.752 \cdot M + 6334.347 \quad (22)$$

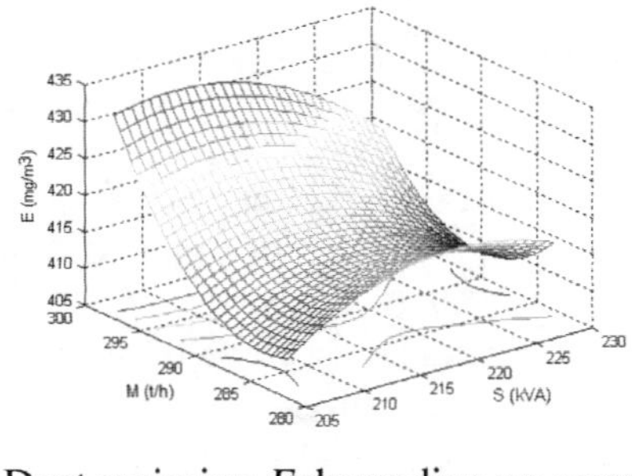

Figure 13. Dust emission E depending on mass flow rate of steam M, apparent power S, and average active power P_{avg}, mathematical modelling 2nd degree equation

$$E = 0.131 \cdot M^2 - 0.06 \cdot P^2 + 0.13 \cdot M \cdot P - \\ -90.308 \cdot M + -22.964 \cdot P + 14698.23 \quad (23)$$

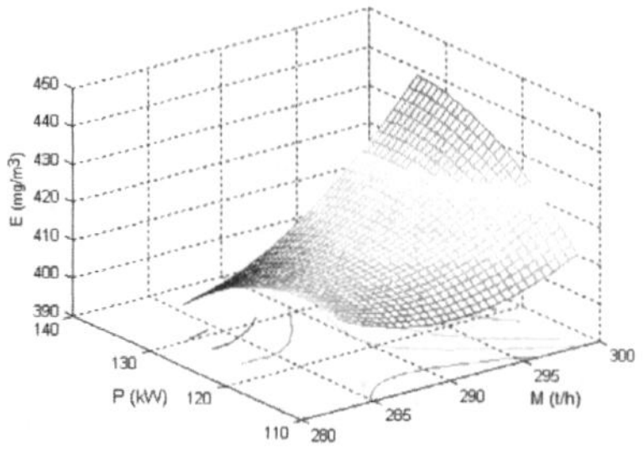

Figure 14. Dust emission E depending on active power P, mass flow rate of steam M, and average apparent power S_{avg}, mathematical modelling 2nd degree equation

$$E = -0.06 \cdot P^2 - 0.04 \cdot S^2 + 0.08 \cdot P \cdot S - \\ - 2.685 \cdot P + 7.693 \cdot S - 305.181 \qquad (24)$$

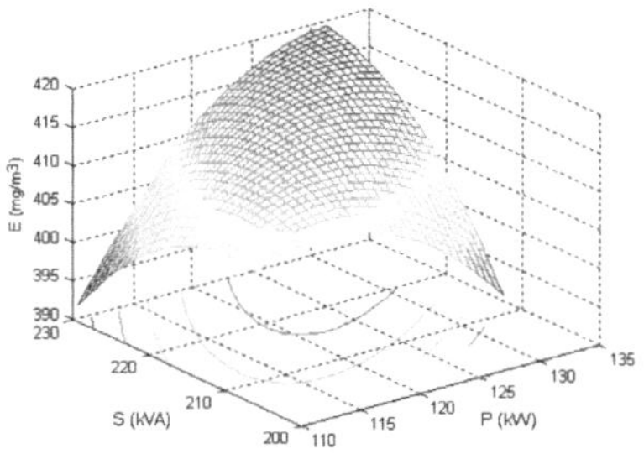

Figure 15. Dust emission E depending on apparent power S, active power P, and average mass flow rate of steam M_{avg}, mathematical modelling 2nd degree equation

High level of dust emissions determines the increasing of apparent power and active power (up to a certain value) of power sources (Figure 15).

4. Conclusions

Using PQA complete data acquisition of main data can be made. It can be acquire the data through a period of time in normal and transient mode operation of industrial precipitators from thermoelectric power plants.

To establish a mathematical model of a dependent parameter on other parameters, it can use the least squares method, which establishes minimum sum of square deviations for a certain mathematical function.

For a precipitator with three sections establishes the evolution of dust emissions depending on the amount (for the all three power sources) of active, reactive, and apparent powers, and depending on the mass flow rate of steam from boiler.

It was found that the mass flow rate of steam from boiler influences the most dust emissions. The levels of active, reactive, and apparent powers depend on the type of power sources and the type of control current inside the section.

Appendix

The main data of precipitator:

- Precipitator with three sections;

- The rise-up transformer: S=166 kVA, 0.4kV/65kV;
- Network voltage: V_{source}=400 V, 50 Hz;
- The maximum high rectification voltage: V_{ESPmax}=92 kV;
- The rated high rectification voltage: V_{ESP}=60 kV;
- Gases flow: F=650,000 m^3/h;
- Maximum design output dust emission q_{out}= 750 mg/m^3_N;
- The gases nominal temperature: t=150 ^0C.

References

[1] Osaci M 2010 Dynamic Modeling Of Ferrite Nanoparticle Systems in Radiofrequency Magnetic Fields, *Proceedings of the Romanian Academy, Series A,* **11**(2) 137-141

[2] Pajak J R 2001 *Improving ESP Performance by Optimizing Power Supply – Industrial Applications,* the VIIIth ICESP, Birmingham, Alabama, USA, May 14-17, no. A2-4, 15 pp

[3] Popa G N, Vaida V, Deaconu S I and Şora I 2010 *An Analysis on the Optimal Fields Number of the Plate-Type Electrostatic Precipitators Used in a Thermoelectric Power Plant*, 12th International Conference on Optimization of Electrical and Electronic Equipment OPTIM 2010, Braşov, Romania, May 20-22, pp 232-239

[4] Prasad N V P R et al. 1999 Automatic Control and Management of Electrostatic Precipitator, *IEEE Transactions on Industry Applications* **35** (3) 561-567

[5] Parker K R 1997 *Applied Electrostatic Precipitation*, Chapman and Hall, London, England

[6] Popa I and Popa G 2000 *Electrical Installations*, Mirton Publishing House, Timişoara, Romania (in Romanian)

[7] Popa G N, Diniş C M, Deaconu S I and Iagăr A 2012 *A Prospective on Power Quality Analyse of Three Sections Plate-Type Electrostatic Precipitator Supplies*, Proceedings of the 11th WSEAS International Conference on Instrumentation, Measurement, Circuits and Systems, IMCAS 2012, Rovaniemi, Finland, April 18-20, pp 49-54

[8] Ionescu T G and Pop O 1998 *Distribution Systems Engineering of Electrical Energy*, Technical Publishing House, Bucharest, Romania (in Romanian)

[9] Maksay Şt 2002 *Numerical Computing*, Politechnica Publishing House, Timişoara, Romania (in Romanian)

[10] Sayan H H, Kahraman M and Koşalay İ 2010 Simulation of Frequency Spectrum of Electric Power Signal, *Electronics and Electrical Engineering* **4** (100) 21-24

[11] Marciulionis P and Zebrauskas S 2011 Numerical Analysis of Electric Wind in Corona Field, *Electronics and Electrical Engineering* **10** (116) 15-18

[12] Nikolic A and Stevanovic I 2009 *Power Quality Measurement Analysis of the Electrostatic Precipitator,* XIX IMEKO World Congress, Fundamental and Applied Metrology, Lisbon, Portugal, September 6-11, pp 509-513

[13] Popa G N, Vaida V, Abrudean C, Deaconu S I and Popa I 2009 *A Case Study of ESP Electrical Characteristics from a Thermal Power Station*, IEEE 44th IAS Annual Meeting, Houston, Texas, U.S.A., Oct. 4-8, 6 pp

[14] ***2007 Three Phase Power Quality Analyzer CA 8334B, *User's Guide*, France

[15] Schröder K 1962 *Grosse Dampfkraftwerke. Die Lehre vom Kraftwerksbau*, Vol. II, Springer-Verlag, Berlin (in German)

[16] Osaci M 2007 The MCAMC algorithm implementation (Monte Carlo with absorbing Markov chains) in the study of magnetic relaxation processes in nanoparticle system, *Proceedings of the Romanian Academy, Series A* **8**(3) 200-203

[17] Răduleţ R 1975 *Electrotechnics Basic. Problems*, Vol. II, The Didactic and Pedagogic Publishing House, Bucharest, Romania, (in Romanian)

IOP Conf. Series: Materials Science and Engineering **106** (2016) 012012 doi:10.1088/1757-899X/106/1/012012

Astable multivibrator circuits made with low capacity PLC

I Popa[1], G N Popa[1], C M Diniş[1] and S I Deaconu[1]

[1]Politehnica University of Timişoara, Department of Electrical Engineering and Industrial Informatics, 5 Revolution Street, 331128 Hunedoara, Romania

E-mail: gabriel.popa@fih.upt.ro

Abstract. The paper presents three versions of astable multivibrator structures that can be made with PS-3 (Klöckner-Moeller) PLCs. They have, in their composition, two time-on delay timers (TR type) that can be found in PS-3 PLC. For each astable multivibrator circuits present the principle accompanied by graphic representations of the main signals of the astable multivibrator circuits and the mode of operation of these circuits using time diagrams. These astable multivibrator strucures can be used, special, for teaching PLCs.

1. Introduction

Astable multivibrator circuits (*AMCs*) is used as clocks, in optical signalling installations, electronic switches when are using with binary counting, binary-decimal decoders, etc. [1], [2], [3].
The paper presents three types of AMCs, their implementation programs on the *PS-3* (Klöckner-Moeller) *PLC* [4], [5] and a simple application of such circuits is to achieve command of a two-inputs multiplexer.

On the two inputs apply signals of different frequencies generated by two *AMCs* that simulates two points of measurement of electrical parameters or non-electrical signals converted to their corresponding frequency values [6], [7]. The third *AMC*, from application, has a much lower frequency than the frequencies of the signals collected from the measuring points. In this way, the output half a period of multiplexer (if *AMC* command of multiplexer is symmetrical) the signal has an input signal, and in the next half cycle, at the output signal is transmitted from the second input. All circuits have a teaching character for signals from the input of these circuits must have low frequency (less than 5 Hz).

2. Astable Multivibrator Circuit, Logically Controlled, Made with RS Flip-Flop Circuit and Two Time-On Delay Timers

This electronic circuit schematic diagram is given in Figure 1.a. Signals marking from schematic diagram is recognized signals from *PS-3 PLC*.

Before starting the *AMC*, which is made with *I0.1* signal on the second input circuit *OR-1* for a short period of time t_{a0} (Fig. 1b), apply signal *I0.0*=1. In this state, the input and output signals of *RS* flip-flop circuit (*RSFFC*) have values: *M0.4*=0 (*S*=0), because *I0.1*=0, *M0.3*=1 (*R*=1), because *I0.0*=1, so *Q0.1*=0 (*Q*=0). The output signal *Q0.1* remains 0 logic, after changing the signal value *I0.0* (*I0.0*=0). When *Q0.1*=0, after *INVERTER-2* gate *NQ0.1*=1, following after a set time period t_{a0} of *TR0* time circuit, the output signal *M0.0*=1. It applies to the first input of the *AND-4* gate.

Since *Q0.1*=0 and the output of the on-delayed circuit *TR1*, has logical 0 value (*M0.1*=0), the signal that is applied to the third input of *AND-6* gate, so *M0.2*=0. After *I0.0*=0, at the output circuit *OR-1* circuit *M0.3*=0 (*R*=0), and after the *INVERTER-3* gate, *NM0.3*=1. This signal is applied to the second input of the *AND-4* gate.

If after the time $t_1(t_0< t_1 < t_{a0}$, Figures 1.b,c,d) measured from when the signal logical *I0.0* increases from 0 to 1 (positive edge signal at *I0.0*=f$_1$(t)) is applied to the input *AND-4* and *AND-6* gates, the signal *I0.1*=1, *AND-6* gate remains closed (locked) because *M0.1*=0 and open *AND-4* gate.

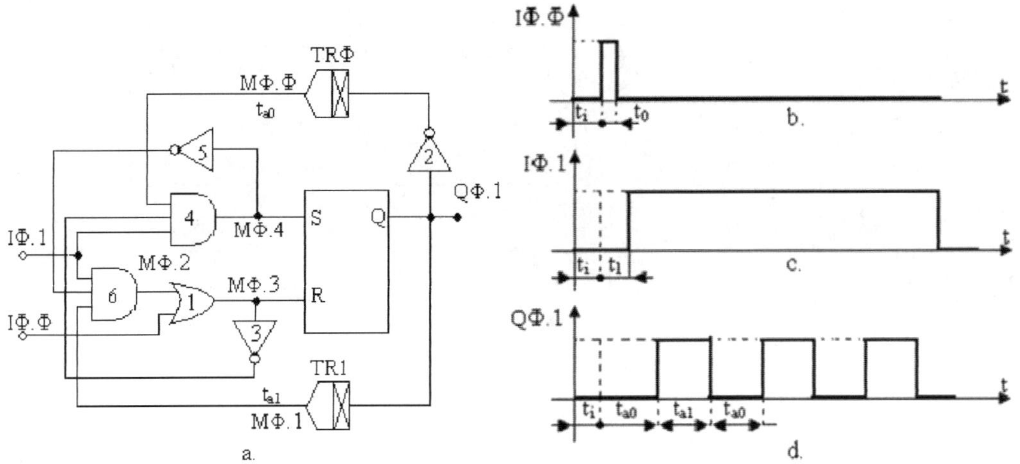

Figure 1. a. The fundamental circuit of *AMC*, logically controlled, made with *RSFFC* and two time-on delay timers; b, c and d: timing diagrams *I0.0*=f$_1$(t), *I0.1*= f$_2$(t) and *Q0.1*= f$_3$(t)

It follows that *M0.4*=1 (*S*=1); as *M0.3*=0 (*R*=0), *Q0.1*=1 (*Q*=1). This condition occurs after the time t_{a0} elapsed since positive front of *I0.0* signal. Now changing signal values *NQ0.1* and *M0.0* (*NQ0.1*=0, *M0.0*=0), so it has a new state of the *RSFFC* namely *M0.4*=0 (*S*=0), *M0.3*=0 (*R*=0), but the output *Q0.1* retain their value; *Q0.1*=1 (*Q*=1). Enter the timer function *TR1* and after the set time t_{a1}, *M0.1*=1, the signal from the output circuit *M0.2* from *AND-6* gate is logical 1 and *M0.3*=1 (*R*=1); because *M0.4*=0, *Q0.1*=0. Immediately after the timer *TR1*, *M0.1*=0 which determines *M0.2*=0 and *M0.3*=0. Now the *RSFFC* state is: *M0.4*=0, *M0.3*=0 and *Q0.1*=0 (*S*=0, *R*=0, *Q*=0).

TR0 timer is activated because of the input signal is a logical 1 (*NQ0.1*=1). After the set time t_{a1} to the output of *TR0* circuit obtains *M0.0*=1, the *AND-4* gate opens and results *M0.4*=1 (*S*=1); because the *AND-6* gate is blocked, *M0.2*=0 and *M0.3*=0 (*R*=0) changes the logical value of the output signal (*Q0.1*=1) and the operation is repeated. If $t_{a0}=t_{a1}$ the *AMC* is symmetrical and *Q0.1* output signal changes as in Figure 1.d. The *AMC* operation stops when *I0.1*=0. In the following is the *PLC* program of *AMC*.

The first program of *AMC* on *PS-3 PLC*, using *RSFFC* with time-on delay:

```
000 TR0
    TR0 * S : NQ0.1
    TR00 * STP :
    TR0 * IW : KW20
    TR0 * EQ : M0.0
001 TR1
    TR1 * S : Q0.1
    TR1 * STP :
    TR1 * IW : KW20
    TR1 * EQ : M0.1
002 L NM0.4
003 A I0.1
004 A M0.1
005 = M0.2
006 L M0.2
007 O I0.0
008 =  M0.3
009 L M0.0
010 A NM0.3
011 A I0.1
012 = M0.4
013 L M0.4
014 S Q0.1
015 L M0.3
016 R Q0.1
```

3. Astable Multivibrator Circuit, Logically Controlled, Made with Two Time-On Delay Timers (1st version)

Schematic diagram of the *AMC* is shown in Figure 2.a, and graphic representations of control *I0.2* and output *Q0.2* signals are given in Figures 2.b and c. In the initial state, control signal *I0.2*=0, so the output *Q0.2* of *AND-1* gate has a logical 0 value (*Q0.2*=0) to determine, after *TR2* timer, *M0.5*=0, and after *INVERTER-2* gate, *NM0.5*=1, signal that is applied to the first input of the *AND-1* gate. After the *INVERTER-3* gate *NM0.2=1*, the timer *TR3* is activated. If the initial condition is maintained for a time longer than the time working t_{a3} of timer *TR3*, after this circuit signal *M0.6*=1 and *M0.7*=1 (from the output *OR-4* gate).

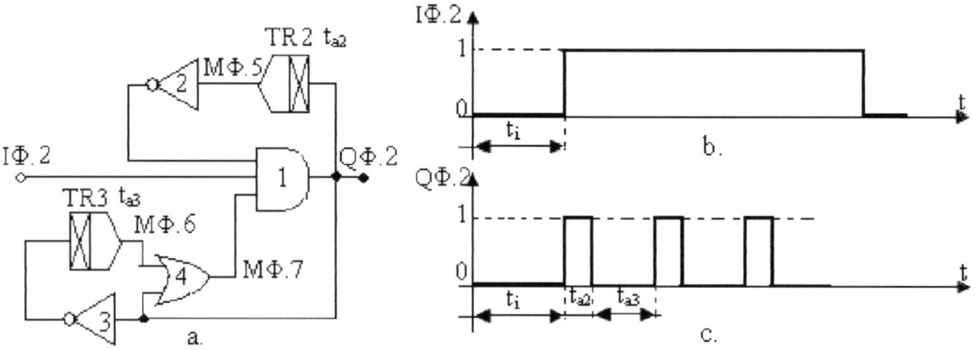

Figure 2. a. The fundamental circuit of *AMC*, logically controlled, made with two time-on delay timers; b and c: timing diagrams *I0.2*=f_4(t) and *Q0.2*= f_5(t)

For commissioning of *AMC*, unlocking control signal *I0.2*, *AND-1* gate (*I0.2*=1). Now, the output signal *Q0.2*=1, signal that applies simultaneously on timer *TR2*, on *INVERTER-3* gate and the second input of the *OR-4* gate. It follows that to change the signal values at the outputs of the circuits 3 and *TR3* (*NQ0.2*=0 and *M0.6*=0), but *M0.7* is maintained at the logical value 1. The signal *M0.5* from output *TR2* circuit is maintained at logical 0 another time t_{a2} from *Q0.2* changing value (from 0 to 1) the period after *INVERTER-2 NM0.5*=1. After t_{a2}, *M0.5*=1, so *NM0.5*=0. It follows *Q0.2*=0, and after *OR-4* gate *M0.7*=0, because on the second the input of this *Q0.2*=0, and on the first input *M0.6* is kept at a logical 0 another time t_{a3} that is set the timer *TR3*. Also, after *Q0.2* changed the value from 1 to 0, after *TR2* circuit and *INVERTER-2* gate the signals have *M0.5*=0 and *NM0.5*=1. After the set time t_{a3}, *M0.6*=1, resulting *M0.7*=1, *Q0.2*=1 and following repeated operation. From the foregoing, it is concluded that during the *Q0.2*=1 is given by the timer *TR2*, and one in which *Q0.2*=0 is required for circuit *TR3*. The output signal *Q0.2* of *AMC* changes over time as in Figure 2.c. The *AMC* stops when *I0.2*=0.

The second program of *AMC* on *PS-3 PLC*, made with time-on delay timers (1st variant):

```
017 TR2
    TR2 * S : Q0.2
    TR2 * STP :
    TR2 * IW : KW15
    TR2 * EQ : M0.5
018 TR3
    TR3 * S : NQ0.2
    TR3 * STP :
    TR3 * IW : KW30
    TR3 * EQ : M0.6
019 L M0.6
020 O Q0.2
021 = M0.7
022 L NM0.5
023 A I0.2
024 A M0.7
025 =  Q0.2
```

It is noted that this *AMC* circuit $t_{a2} \neq t_{a3}$ ($t_{a3} = 2 \cdot t_{a2}$), so this oscillator is no longer symmetrical.

4. Astable Multivibrator Circuit, Logically Controlled, Made with Two Time-On Delay Timers (2nd version)

This AMC has schematic diagram in Figure 3.a and is easiest done with electronic timers.
In the initial state control, the control signal *I0.3*=0, due to the output circuit *AND-1* gate, *M0.9*=0, and after the timer *TR5*, *Q0.3*=0. After *TR4* time circuit, the signal circuit has a logical 0 value (*M0.8*=0), so the output *INVERTER-2* gate, *NM0.8*=1. This state lasts for a long t_i (Figures 3.b and c).

To start oscillator is required *I0.3*=1, value which is maintained throughout the operation (Figure 3.b). Now at the output of the *AND-1* gate *M0.9*=1, but after *TR5*, *Q0.3*=0 since a time t_{a5} that is set time circuit (*TR5*); after this time *Q0.3*=1, *TR4* timer is activated, after a set time t_{a4}, logical signal changes of *M0.8* signal. It follows *M0.8*=1, after *INVERTER-2* gate *NM0.8*=0 which determines very quickly *M0.9*=0 and *Q0.3*=0, the operation is repeated. In this way t_{a5} during work time circuit *TR5*, *Q0.3*=0, and t_{a4} periods imposed by timer *TR4*, *Q0.3*=1. The variation in time of the output signal *Q0.3* of the oscillator is shown in Figure 3.c.

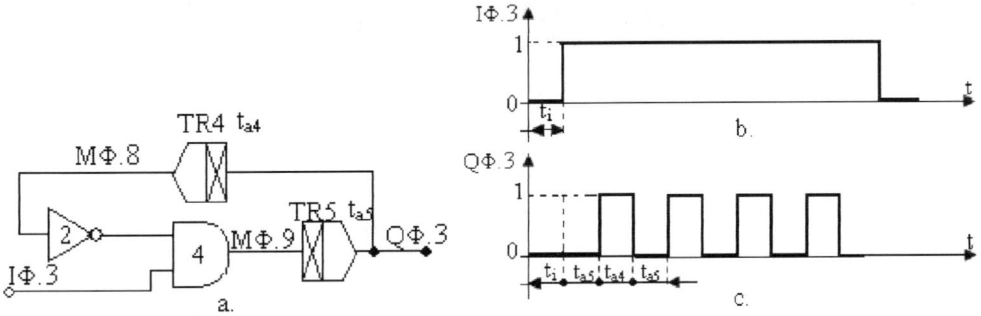

Figure 3. a. The fundamental circuit of *AMC*, logically controlled, made with two time-on delay timers; b and c: timing diagrams *I0.3*=f_6(t) and *Q0.3*= f_7(t)

Stopping oscillator occurs when *I0.3*=0, which value determines lock of *AND-4* gate as a result *M0.9*=0 and *Q0.3*=0. With the *AMC* principle scheme in Figure 3.a, the program could make on *PS-3 PLC*.

The third program achievement *PS-3 PLC* circuit multivibrator, ordered logically made electronic circuit-delayed timers (2nd version):

```
026 TR4
    TR4 * S : Q0.3
    TR4 * STP :
    TR4 * IW : KW15
    TR4 * EQ : M0.8
027 L NM0.8
028 A I0.3
029 = M0.9
030 TR5
    TR5 * S : M0.9
    TR5 * STP :
    TR5 * IW : KW15
    TR5 * EQ : Q0.3
```

5. Application. Two Inputs Multiplexer Controlled by AMC Made with Six Time-On Delay Timers

With three *AMCs*, logically ordered, made with electronic circuit-delayed (1st version) and a two-input multiplexer, made a remote transmission system through a line feed from output of *Q0.4* (Figure 4), data collected from two points of measurement of technological parameters (electric or non-electric), converted into electrical signals modulated in frequency.

The measurement points are simulated by *FFC* (flip-flop circuit) *1* and *2*. The multiplexer with two inputs (*Q0.1*, *Q0.2*) is done with logic circuits *1-4* signal being ordered by output signal *Q0.3* of the *FFC 3*. The output frequency signals *Q0.1* and *Q0.2*, output from these blocks are much higher than the frequency control signal *Q0.3*. It also requires that $t_i > t_{a4}$, $t_i > t_{a5}$, (Figure 5.d). The three circuits

FFC1, *FFC2* and *FFC3* are symmetrical. The control of *FFC1*, *FFC2* and *FFC3* is with the signal *I0.0*.

In its original state, during time t_i as *I0.0* control signal of three *FFC* has logical 0 operation is blocked, so *Q0.1*=0, *Q0.2*=0, *Q0.3*=0, due and output signal *Q0.4* of the multiplexer is set to logical 0, as shown in Figure 5.

After time t_i, *I0.0*=1 and all three *FFC* are un-lock. During the time t_{a4}, *Q0.3*=1, *AND-3* gate opens the gate due to output signal *Q0.4* is sent from the measurement point *2* (*Q0.4*=*Q0.2*), and in the time t_{a5}, *Q0.4*=*Q0.1* (Figure 5.e). The data is transmitted remotely by a line which is not shown in Figure 4, on a frequency-meter calibrated in units of measurement of process parameters or a two-outputs de-multiplexer which are joined together in such devices.

In the first case, in addition to the meter is necessary to have an electronic device showing the number of measurement point, and in the latter case, they must control devices to operate the multiplexer and de-multiplexer synchronously. To stop the operation of the remote measuring technological parameters required *I0.0*=0. With the scheme in principle of two-input multiplexer ordered by *FFC* (1st variant) the 4th program was drawn on *PS-3 PLC*.

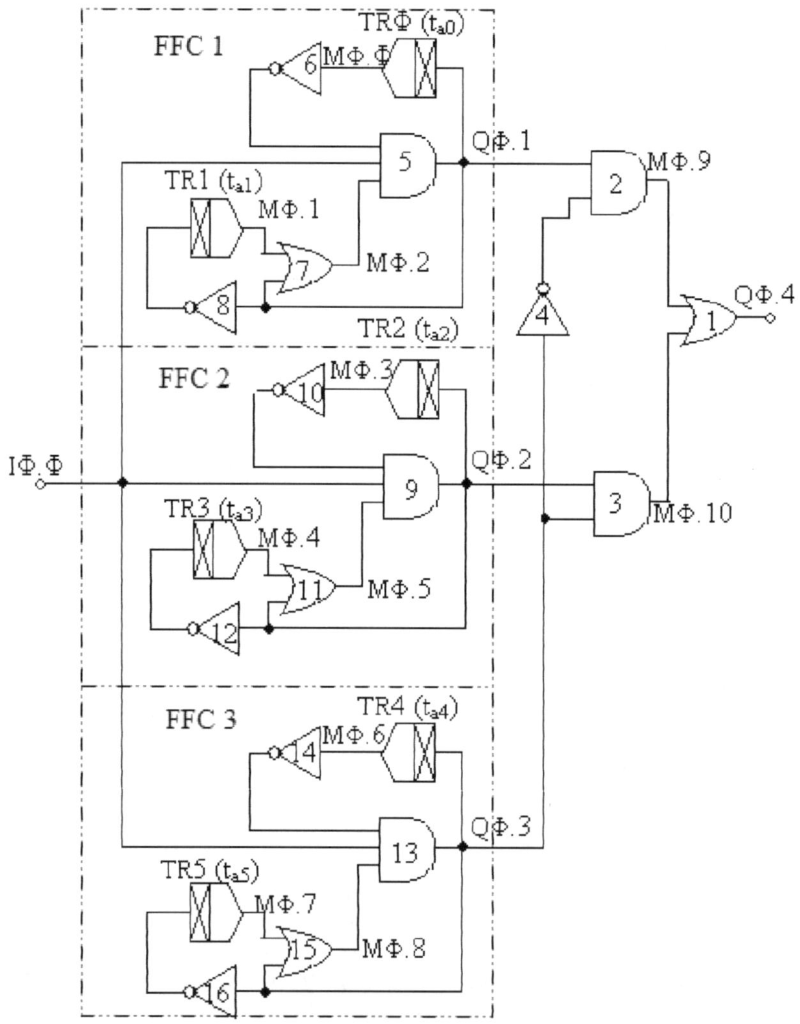

Figure 4. Two inputs multiplexer controlled by *AMC* made with time-on delay timers (1st version)

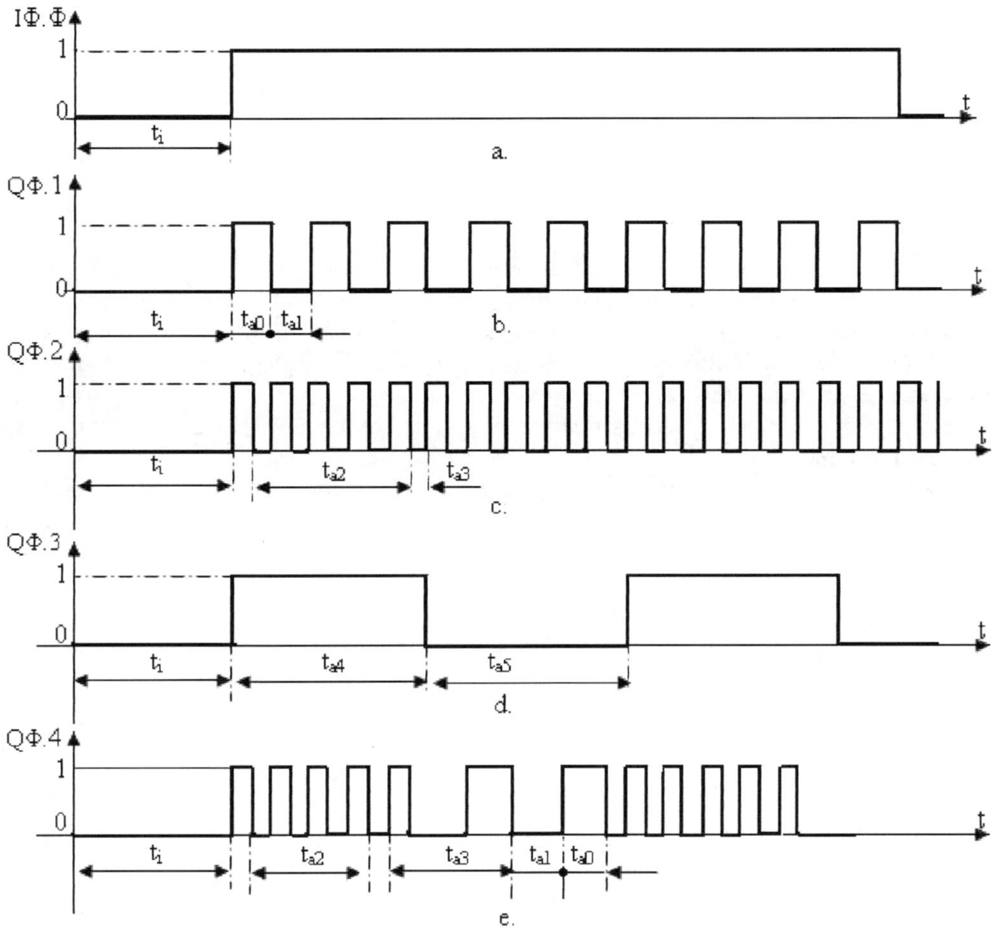

Figure 5. Timing diagrams of the signals: a. *I0.0* = f_1(t), b. *Q0.1*= f_2(t), c. *Q0.2*= f_3(t), d. *Q0.3*= f_4(t), e. *Q0.4*= f_5(t)

The fourth program achievement on *PS-3 PLC* with two inputs multiplexer, ordered by *FFC* made with time-on delayed (1st version):

```
000 TR0
    TR0 * S : Q0.1
    TR0 * STP :
    TR0 * IW : KW10
    TR0 * EQ : M0.0
001 TR1
    TR1 * S : NQ0.1
    TR1 * STP :
    TR1 * IW : KW10
    TR1 * EQ : M0.1
002 L M0.1
003 O Q0.1
004 = M0.2
005 L NM0.0
006 A I0.0
007 A M0.2
010 TR3
    TR3 * S : NQ0.2
    TR3 * STP :
    TR3 * IW : KW5
    TR3 * EQ : M0.4
011 L M0.4
012 O Q0.2
013 = M0.5
014 L NM0.3
015 A I0.0
016 A M0.5
017 = Q0.2
018 TR4
    TR4 * S : Q0.3
    TR4 * STP :
    TR4 * IW : KW150
020 L M0.7
021 O Q0.3
022 = M0.8
023 L NM0.6
024 A I0.0
025 A M0.8
026 = Q0.3
027 L Q0.1
028 A NQ0.3
029 = M0.9
030 L Q0.2
031 A Q0.3
032 = M0.10
033 L M0.9
034 O M0.10
035 = Q0.4
```

```
008 = Q0.1                  TR4 * EQ : M0.6
009 TR2                  019 TR5
    TR2 * S : Q0.2          TR5 * S : NQ0.3
    TR2 * STP :             TR5 * STP :
    TR2 * IW : KW5          TR5 * IW : KW150
    TR2 * EQ : M0.3         TR5 * EQ : M0.7
```

6. Conclusions

The paper was presented three *AMC* with possibility of materializing on the *PLC PS-3* Klöckner-Moeller [4], [5]. They are composed of two electronic circuits on-delayed timers [1], [2], [6], [7], [8]. For each multivibrator circuit schematic diagram presented accompanied by graphic representations of the main signs of *AMC*, the operation of these oscillators and finally realized on the *PLC* program *PS-3* for the analysed signal generators.

From the analysis in the paper, results that the latter oscillator with the output signal rectangular configuration is the simplest.

Schematic diagram of the *AMC* (Figure 3.a.) is taken from the literature [3]. The first two variants of the oscillators, namely: *AMC* made with *RSFFC* and with time-on delay circuits and with *FFC* performed with with time-on delay circuits and *AMC* made with time-on delay circuits, 1st variant are original.

Using three symmetrical *AMC* made with time-on delay circuits, 1st variant and a multiplexer with two inputs has been designed with an electronic switch schematic diagram in Figure 4.a, the two measuring points of technological parameters (electrical or non-electrical), with values converted into two frequency-modulated signals.

These measuring points are simulated with *FFC1* and *FFC2*, the control being provided by *FFC3*, which has a signal output less frequently than the frequency of the signals emitted by the first two square wave generators.

At the output of electric switch obtain a complex signal (*Q0.4*) which in the half cycle of *FFC3* has the frequency signal at the output of a measuring point (*Q0.2*), and the other half cycle, frequency of the other measuring point (*Q0.1*). The principle states that the schemes symbols for time circuit was done after [1], [2], [8], which analyses and its operation. Virtually, all circuits were checked on *PS-3 PLC* programs presented in the paper.

The paper first importance in terms of teaching, but electronic switch for two measuring points can also be used with some improvement in industry (specified in sub-chapter 5).

The principle schemes of *AMCs* and electronic switch from the paper (Figures 1.a, 2.a, 3.a, and 4) could make and the other *PLCs* [9], [10].

References

[1] Popa I and Popa G N 2000 *Electronic Devices with Wired and Programmed Structure, for Protection of Low Voltage Three-Phase Motors*, Mirton Publishing House, Timişoara, Romania (in Romanian)

[2] Popa I, Popa G N and Deaconu S 2008 Electronic Time Relay with Eight Functions, Made with Integrated Logic Circuits, *Science and Engineering* **13** 249 – 254

[3] ***1988 *AP 201. User's Manual*, Bucharest, Romania (in Romanian)

[4] ***1987 *Handbuch Teil 2. Software*, Programmbeispiele, Klöckner-Moeller, Germany (in German)

[5] ***1987 *Handbuch Teil 3. Software*, Programmbeispiele, Klöckner-Moeller, Germany (in German)

[6] Rajput D S, Gwalior I and Singh R 2015 *Timer Circuit Using OTRA and its Application as Astable and Monostable Multivibrator*, 2nd International Conference on Electronics and Communication Systems (ICECS), Coimbatore, India, Feb. 26-27, pp1047-1050

[7] Telles A C C, Renato A, Finco S and Pomilio J A 2014 Modelling of a MOS Ultralow Voltage

Astable Multivibrator for Energy Harvesting, *IEEE Transactions on Circuits and Systems II: Express Briefs* **61**(3) 168-172

[8] Popa G N, Popa I and Deaconu S 2006 *PLC in Applications*, Mirton Publishing House, Timişoara, Romania (in Romanian)

[9] Basile F, Chiacchio P and Gerbasio D 2011 *Progress in PLC Programming for Distributed Automation Systems Control*, 9th IEEE International Conference on Industrial Informatics (INDIN), Caparica, Lisbon, Portugal, July 26-29, pp 621-627

[10] Bolton W 2009 *Programmable Logic Controllers*, 5th Edition, Newnes, Elsevier, UK

Self-oscillating inverter with bipolar transistors

I Baciu[1], C D Cunţan[1] and M Floruţa[2]

[1]Politehnica University of Timisoara, Department of Electrical Engineering and Industrial Informatics, 5 Revolution Street, 331128 Hunedoara, Romania
[2]Key Safety Systems, Ribita, no. 237, Hunedoara, Romania

E-mail: corina.cuntan@upt.ro

Abstract. The paper presents a self-oscillating inverter manufactured with bipolar transistors that supplies a high-amplitude alternating voltage to a fluorescent tube with a burned filament. The inverter is supplied from a low voltage accumulator that can be charged from a photovoltaic panel through a voltage regulator.

1. Introduction

In electric systems, converters are static equipment (electric plants), which through semiconductor devices, enable the conversion of electric energy between two or more systems by modifying the current, the frequency, the voltage, the number of phases, etc.

Hence, there are two types of converters, ones that convert the electric energy between AC and DC systems and others that convert between systems of the same kind only changing the electric parameters (amplitude, voltage, frequency, etc.).

Static converters use only two distinct states of conduction in the power-electronics elements that compose them: open (saturated) or cut-off. This mode of operation results in lower power dissipation on the devices in comparison to linear electronic circuits, which use intermediate states too. Converter efficiency is between 0.85 - 0.95%.

A fluorescent lamp operating at a frequency of 50 Hz goes out twice for every period of the supply voltage, this being when the supply voltage passes through zero, causing the flicker phenomenon (stroboscopic phenomenon) with a frequency of 100 Hz. This phenomenon has unpleasant effects, sometimes leading to work accidents.

If the lamp operates at high frequencies, a continuous light is produced since the response time of the discharge is too low to allow the lamp to extinguish [1].

Electronic ballast comprises of a full-wave rectifier, a self-oscillating inverter, with field effect transistors or bipolar transistors, and inductive ballast, for the high frequency determined by the inverter. In addition, the self-oscillating inverter may be provided with different correction and protection circuits [2].

The supply voltage is directly applied or, for higher power, through a power factor correction circuit.

Modern lamps include a filter to minimize the harmonic currents, in order to comply with EMC regulations [1].

2. Problem Formulation

2.1. Powering fluorescent tubes with DC powered electronic ballasts

The discharge lamp is characterized by a static characteristic $U_L=f(I_L)$ with a negative (falling) slope, having a negative dynamic resistance:

$$r_{d_L} = \frac{dU_L}{dI_L} < 0 \tag{1}$$

Stable operating discharge lamps must be supplied from a source of alternating current with high impedance output.

The sum of the output impedance of the power source, which is positive, and the negative dynamic resistance must be positive in order to satisfy the condition of stability in operating when a constant current flows through the lamp.

The electronic inverters that work as class D (resonating) voltage switches have the resistive load of the lamp r_{dL}. They are a class of high-frequency circuits with high efficiency and are regarded as appropriate sources for these types of applications [2].

To ensure ignition and maintain the discharge of the lamp, DC energy needs to be transferred to the secondary winding of the transformer.

The power supply must have a high output reactance so that the sum of the positive source impedance and the negative reactance of the lamp (as the lamp voltage increases, the current flowing through the lamp drops) is positive, in order to ensure stable operation of the lamp [2].

A power supply installation for a fluorescent lamp through a "flyback" type self-oscillating inverter with isolation is shown in Figure 1.

The Q_1 transistor becomes conductive through a circuit consisting of a resistance R_1 and the 8-5 winding. The changes in the current amplitude through the 7-6 winding of the T_1 transformer induces voltage in the 2-1 secondary winding, which is connected to the lamp through the capacitor C and the 8-5 feedback winding, causing a negative base voltage to the Q_1 transistor. The Q_1 transistor is blocked and the current through the 7-6 winding becomes void. This cycle resumes when there is a positive voltage in the base of the Q_1 transistor. The frequency of the oscillations is determined by the capacitor C_1 and the inductance of the 8-5 winding.

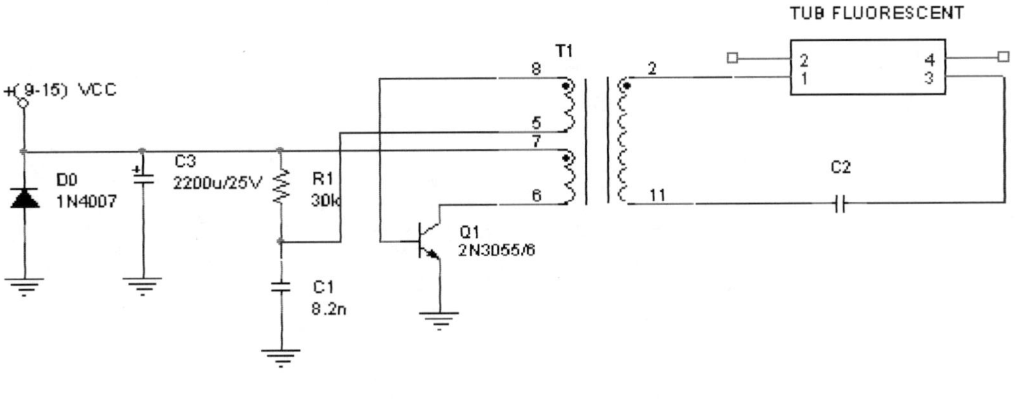

Figure 1. A "flyback" type self-oscillating inverter with isolation

The electric pulses, which control the Q_1 power transistor, are made with the integrated circuit βE 555 connected as an astable circuit. The frequency of the pulses is modifiable within the 5 to 15 kHz range with the help of the variable resistor R_v. By using transistor Q_1 in switch mode, variations will appear in the amplitude of the current flowing through the primary winding 1-2 of the T_1 step-up transformer. The voltage induced in the 3-5 winding is applied to the fluorescent lamp.

IOP Conf. Series: Materials Science and Engineering **106** (2016) 012013 doi:10.1088/1757-899X/106/1/012013

One of the connections commonly used as electronic ballast for discharge lamps is the push-pull connection (Figure 2) with a resonant circuit, which includes the resistance of the lamp, as a dynamic load [3]. The AC power is supplied to the dynamic resistor [4].

The Cc capacitor (of high capacity) is used to cut off the DC component of the current. The DC component of the current would cause the saturation of the magnetic core, destroying the T_1 and T_2 power transistors.

The voltage on the Cc capacitor is approximately equal to $U_a/2$. The two static switches (T_1, T_2) ensure that current flows through the parallel resonant circuit L-C_p-r_d [5].

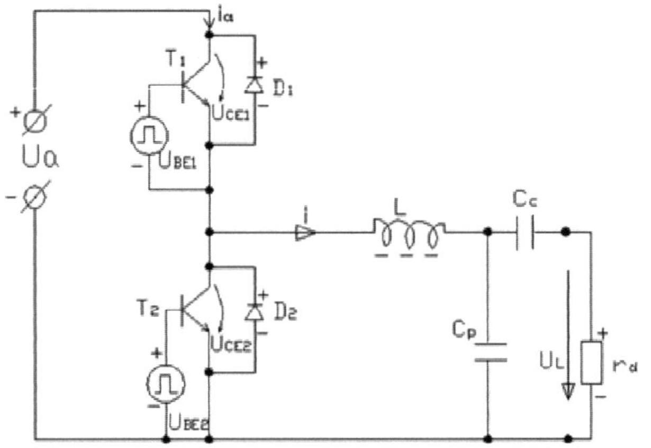

Figure 2. Electronic balast in a „push-pull" connection

Antiparallel diodes enable the circulation of negative currents, but the collector-emitter voltage, in this case, will remain at -0.7÷-1.5V. The transistors are driven with base-emitter voltages (U_{BE1}, U_{BE2}), which are square waves (rectangular with a 50% duty cycle). The attack voltage of the resonant circuit is rectangular (U_{CE1}, U_{CE2}) with values between 0 ÷ U_a (Figure 3) [6].

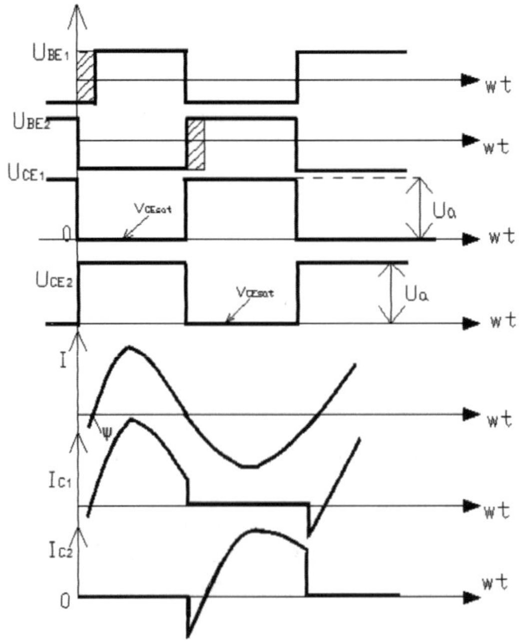

Figure 3. The operation mode diagram of the push-pull circuit

The analysis of the circuit in operation mode is simplified assuming a sinusoidal current flow through L-C_p-r_d. This approximation is indeed valid if the quality factor of the load circuit satisfies the following condition:

$$Q_L = \omega_0 \cdot C_p \cdot r_d = \frac{r_d}{\omega_0 \cdot L} = \frac{r_d}{Z_0} > 2.5 \tag{2}$$

where:

$\omega_0 = \frac{1}{\sqrt{L \cdot C_p}}$ - the angular frequency for natural resonance for the un-dampened circuit.

$Z_0 = \omega_0 \cdot L = \frac{1}{\omega_0 \cdot C_p} = \sqrt{\frac{L}{C_p}}$ - the impedance of the resonant circuit.

In this case, a non-linear load can be precisely modeled by linear impedance. Results show that, for a frequency higher than the resonance frequency, the output voltage and the output power of the inverter increase with load resistance (at $f \geq f_{rezonanta}$, the parallel inverter can function very well during the open-circuit test output and cannot function at all during the short-circuit test). Therefore, the inverter can generate a high enough voltage to ignite the lamp and limit the current after the ignition of the discharge lamp.

The output power, voltage and current through the load can be controlled by changing the frequency of operation, which can be achieved by the continuous modification of the light output from 100% to 10%.

If the $Q_L < 2{,}5$ waveform of the current through the resonant circuit is not sinusoidal, it is more difficult to obtain an exact analytical solution.

Due to the delay in the entry into conduction and especially due to the delay of coming out of conduction, the duty cycle of the control signal (U_{BE1}, U_{BE2}) must be less than 50% (Δt - a few μs) to avoid short-circuiting the source.

Each transistor is turned on when the switching current is negative and it passes through the diode.

The transistor can be turned off through its Base-Emitter voltage while it can only be turned on by turning off the opposing transistor. Hence only the cut-off is directly controllable through base biasing.

If T_1 is turned off by the U_{BE1} voltage, the U_{CE1} voltage increases, causing the U_{CE2} voltage to decrease. When the U_{CE2} voltage reaches -0.7V, diode D_2 is turned on and the current is diverted through diode D_2. Then T_2 is turned on, so the energy stored in the L coil causes the negative alternation of current i, etc. When $f = f_0$, the amplitude of the current and voltage on L-Cp become dangerously high, so high that it can destroy transistors and resonant components. Operating at this frequency is, therefore, to be avoided.

For $f < f_0$ the phase shift is negative $\psi < 0$, the resonant circuit represents a capacitive load (the current through the coil leads the fundamental of the voltage). This is not an advantageous operating mode (diodes are cut-off for large di/dt and cause current spikes through T_1 and T_2).

For $f > f_0$, the phase shift is positive $\psi > 0$, the resonant circuit represents an inductive load (the current lags the fundamental voltage and thus the switching current is negative after entering into conduction and positive before cut-off).

The new generation of electronic ballasts is based on MOS-FET static power switches because they have the following advantages:

- the ability to switch high currents at frequencies of a few MHz;
- reduced switching losses;
- very low power consumption for control signals;
- the ability to withstand high reverse drain-source voltages.

The new requirements of the standards in this domain have encouraged the design of ballasts having a high power factor, diminished harmonic content, a high efficiency (electronic ballasts more efficient by 25% compared to the classic ones), increased reliability and lifetime.

An electronic ballast, with two MOS-FET static power switches in push-pull configuration (half bridge connection), and a self-oscillating type inverter having a control transformer are shown in Figure 4. Current flows through the primary winding of the Tr transformer from the lamp circuit and operates at the resonant frequency given by L-$C_2(\omega_0)$ [3].

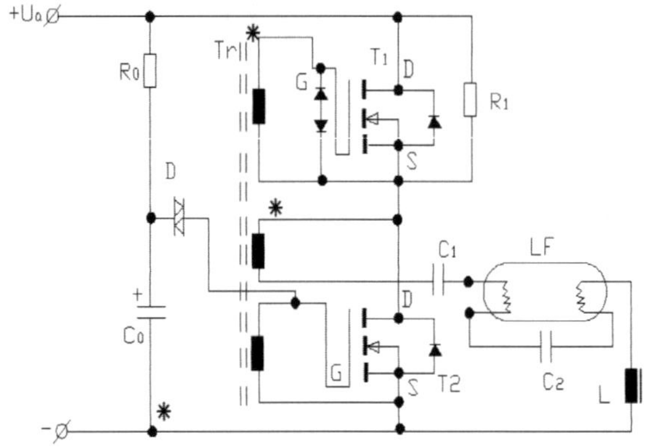

Figure 4. Electronic ballast with 2 MOSFET switches

The disadvantages of this circuit are:

- the oscillations are not damped;
- transistor (T_2) must be pulsed by the DIAC D tied to the gate circuit.

The oscillations are maintained. A high frequency square wave (30 to 80 kHz) excites the resonant circuit L-C2. The sinusoidal voltage across C2 is amplified by QL or during resonance, it being able to produce a magnitude sufficient to ignite the lamp [3]. Then the fluorescent lamp functions at a higher frequency than the resonance frequency and provides illumination without pulsation (flicker) [7].

This circuit represented a "standard configuration" for electronic ballasts for several years, yet it has many disadvantages: no self-ignition, bad switching (control using signals with slowly increasing edges causing relatively high switching losses, the negative portion of the gate-drain voltage waveform is not usable), laborious construction, too expensive for large-scale manufacture, inappropriate for varying brightness output, etc. These disadvantages have been fully resolved using drivers - monolithic integrated circuits for controlling two MOS-FET/IGBT power transistors in half bridge configuration [8].

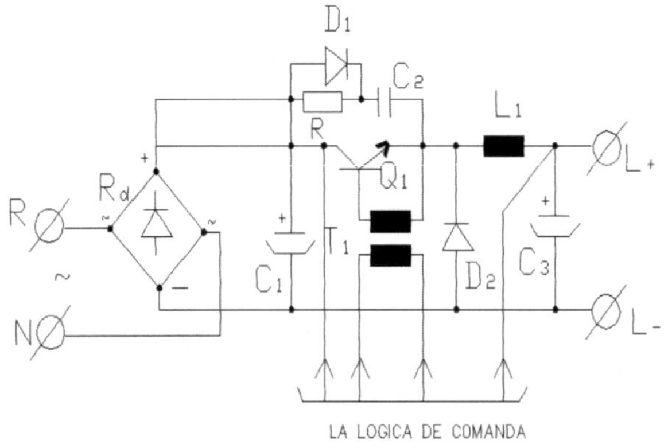

Figure 5. Conventional switching regulator

Widely used as a replacement for incandescent lamps, discharge lamps have better efficacy and reduce electricity consumption. In spite of these advantages, discharge lamps have a low power factor, high total harmonic distortion, both with electronic and electromagnetic ballast. So the impact of these systems on the quality of power supply is of primary interest.

The diagram of a conventional switching regulator (Figure 5) contains a very small capacitor C_1, the network frequency being filtered by a capacitor with high capacity C_3.

The output voltage is controlled by the duty cycle of the switching frequency Q_1 and presents a decent power factor correction and harmonic distortion correction. The circuit is of major complexity, small size and high cost.

2.2 Self-oscillating inverter with bipolar transistors used for powering fluorescent lamps with burned filaments

This inverter is also known as a Royer inverter.

2.2.1 The DC-AC function. The main component of this inverter model is the Royer oscillator. This is a very simple self-oscillating circuit, being made, in accordance with the diagram in Figure 6, out of NPN bipolar transistors, Q_1 and Q_2, capacitor C_1 connected to the collectors of the transistors and step-up transformer T_1 [9].

The step-up transformer is composed of two identical sections (3-4, 4-5); a feedback coil (8-7) and a secondary coil (2-6), on which the fluorescent lamps with burned filaments are connected serially with capacitor C2.

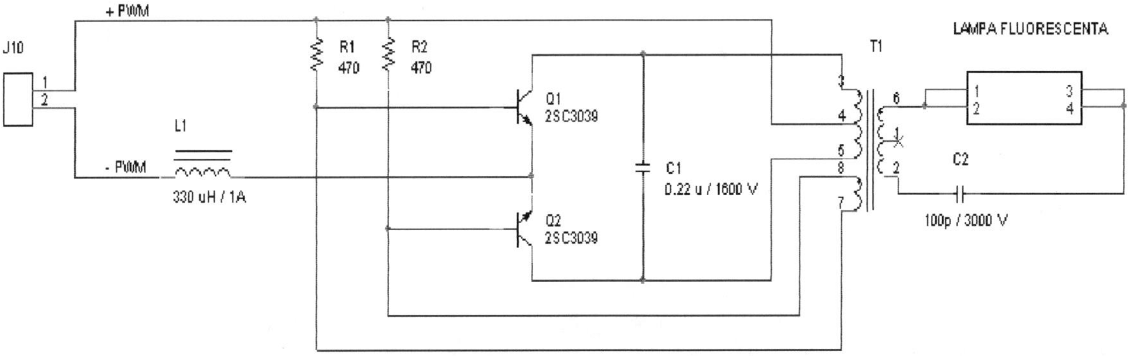

Figure 6. Self-oscillating inverter with bipolar transistors used to power fluorescent lamps with burned filaments

Supply voltage $+V_{cc}$ is applied to its center tap from the voltage generator circuit PWM (PWM = Pulse Width Modulation, it is a pulse width modulated signal, modulated in duration). The transformer's role is to ensure oscillation through the feedback coil and to increase the size of the impulses in the primary coil, ensuring a high voltage in the secondary winding required for the lamp to ignite [10].

When the supply voltage $+V_{cc}$ is connected through resistors R_1 and R_2, the bases of Q_1 and Q_2 are polarized. They open and establish a current through the two primary windings of the transformer, current whose intensity varies rapidly from zero to maximum value. Due to these rapid variations in current, an induction voltage, with the positive potential in the base of transistor Q_1 and the negative potential in the base of transistor Q_2 will appear in the feedback coil. This voltage will cut off transistor Q_2, will maintain transistor Q_1 on and will allow charging the capacitor C_1 through the following path: $+V_{cc}$, winding center tap (4) - Q_2 collector, C_2, Q_1 collector, Q_1 emitter, inductance, ground. Now, Q_2 is blocked.

The current pulse resulting from the rapid charging of the capacitor maintains the inductive voltage through the feedback coil, of the polarities specified above, in such a way as to keep Q_1 on and Q_2 off.

When capacitor C_1 is fully charged, the current will have the reverse change through the primary winding, decreasing rapidly from the maximum value to zero. This current pulse in the winding will induce a voltage of reverse polarity sided as described above: the plus in Q_1's base and the minus in Q_2's base.

As a result, Q_1 is cut-off and Q_2 turns on, allowing the loading and unloading of C_1 according to the following route: + V_{cc}, winding center tap (4)-collector Q_1, C1, collector Q_2, emitter Q_2, inductance, ground. Now, Q_1 is blocked.

After unloading C_1 and loading C_1 with voltage of opposite polarity, Q_2 turns off and Q_1 turns on, unloading and loading C_1 again and the cycle repeats.

Thus, by turning off and on the two transistors, Q_1 and Q_2, capacitor C_1 is charged and discharged continuously through the two primary windings of the step-up transformer, generating a high voltage in the secondary winding, required to ignite the lights.

The voltage obtained in the secondary winding is a sinusoidal AC voltage, as is the supply voltage, thus chosen for the correct operation of fluorescent lamps [11].

This increases the life of the lamp. It is mandatory to use a sinusoidal voltage to control them; otherwise the resulting ions will accumulate at one end of the tube causing irreparable damage to it.

2.2.2 Adjusting the lamp lighting. The amount of light emitted by a fluorescent lamp depends on the voltage that is supplied to it and, if this voltage is low, the lamp can reach the extinction-ignition threshold, resulting in a flickering; a tremor of its brightness. This is the main disadvantage of adjusting lamp brightness by applying a varying voltage.

A variation of brightness of the lamp is adjustable through the application of a voltage of variable value; therefore, a low voltage would lead to a low light and a high one to maximum illumination [12].

Due to the characteristics of fluorescent lamps, they do not suddenly go out when power failure occurs. They have certain inertia, so they continue to produce light for short intervals of time, even without power.

This feature is used to control the brightness of the lamps through the intermittent supply method: multiple voltage pulses with constant amplitude. So, the second option of controlling the brightness of the lamps is the intermittent supply method.

A constant voltage is supplied to the lamp, and then the power is disconnected for a short time.

The supply and the break periods depend on the value of the brightness adjustment by the user, so that their average value would lead to a desired brightness [13]. Large break periods translate into lower brightness and large power periods result in maximum brightness.

2.3 PWM voltage generating circuit

PWM voltage generating circuits can be used to adjust the brightness of fluorescent lamps [14]. Such a signal can be seen in figure 7.

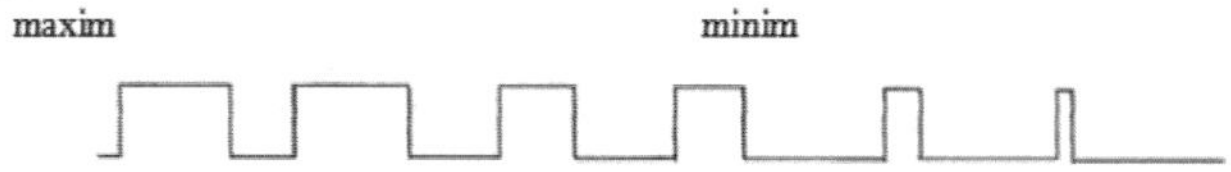

a) The PWM waveform for adjusting the brightness from maximum to minimum

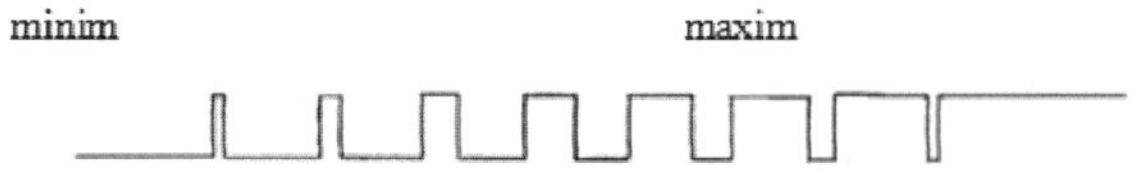

b) The PWM waveform for adjusting the brightness from minimum to maximum

Figure 7. The PWM waveform for adjusting the brightness

The controller presented in Figure 8 supplies a signal as shown above (Figure 7), a signal whose duration (width) depends on the voltage control of the brightness [5].

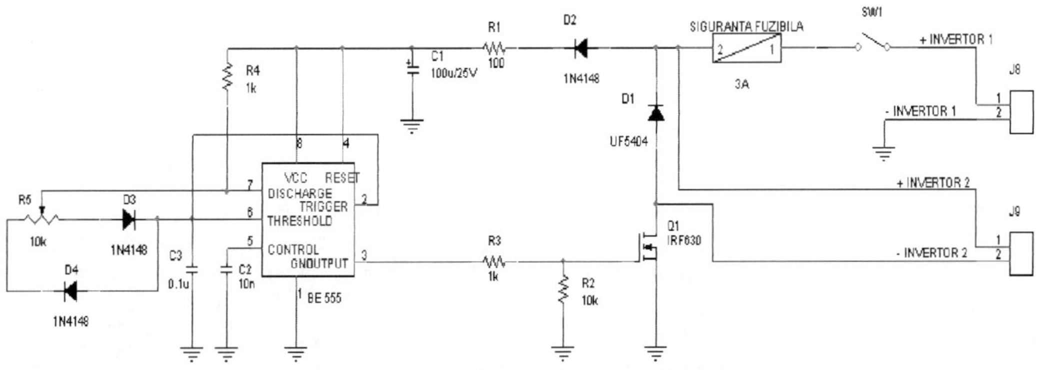

Figure 8. The flow diagram of the PWM circuit

The PWM signal supplied is positive and it is applied to the gate of the Q_1 transistor (IRF630 - n channel), which turns on putting the emitters of transistors Q_2 and Q_3 (2SC3039) to ground through an inductance:

$$L_1 = \frac{330\mu H}{1A} \tag{3}$$

The Royer oscillator is supplied, hence it starts to generate oscillations, the lamp being powered. The oscillator will be powered only during the positive pulse[15]. When the PWM signal becomes zero, the transistor Q_1 is cut-off, the oscillator and the lamp are no longer supplied through transistor Q_1 [10].

A self-induced voltage appears in the inductance L_1, of opposite sign to the original voltage, and rectified by diode D_1. The assembly is further supplied by a lower voltage until the next positive pulse, when transistor Q_1 is opened again; the oscillator will again be powered etc. What follows is an intermittent supply of the lamp; the average voltage pulse sets the brightness level.

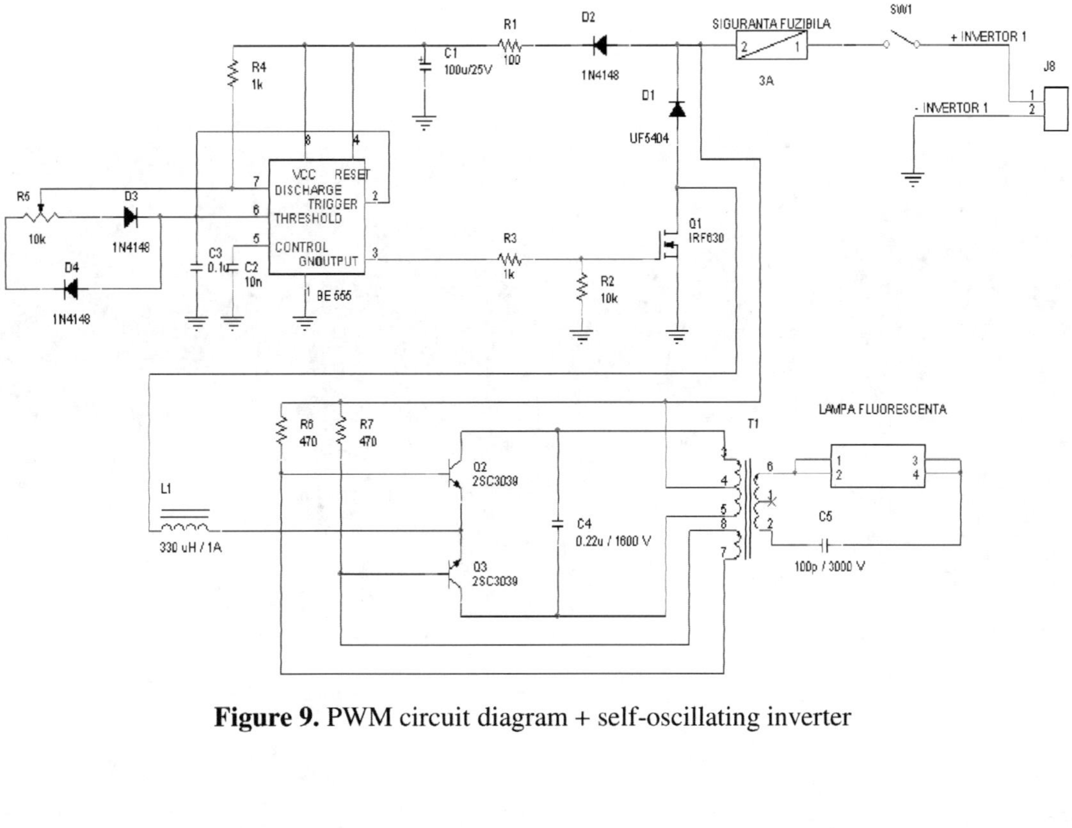

Figure 9. PWM circuit diagram + self-oscillating inverter

The role of the potentiometer R_5 in Figures 8 and 9 is to adjust the brightness of the fluorescent lamp. When the cursor of the potentiometer R_5 moves to the right (towards diode D_3), the brightness of the fluorescent lamp is minimal, and when the cursor of the potentiometer is moved to the left (towards diode D_4), the brightness of the fluorescent lamp is at a maximum.

3. Problem Solutions

For autonomous sources of electric energy gas discharge lamps and LEDs are used.

In the case of the self-oscillating inverter with bipolar transistors used to power the fluorescent lamps with burned filaments, the current consumed by these varies depending on the duty cycle, i.e. the brightness.

Hence, for a duty cycle of:

- 10% of current consumption is approximately 100mA;
- 50% power consumption is about 820mA;
- 100% power consumption is about 1.76 A. This is illustrated in the following figure (Figure 10).

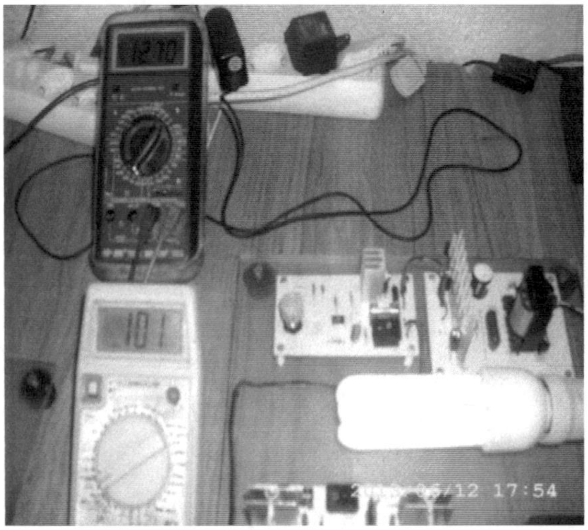

a) $\alpha = 0.1, i = 100mA$

b) $\alpha = 0.25, i = 500mA$

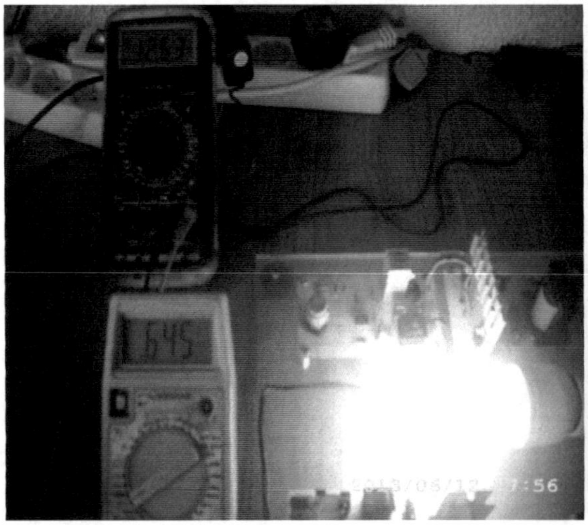

c) $\alpha = 0.4, i = 645mA$

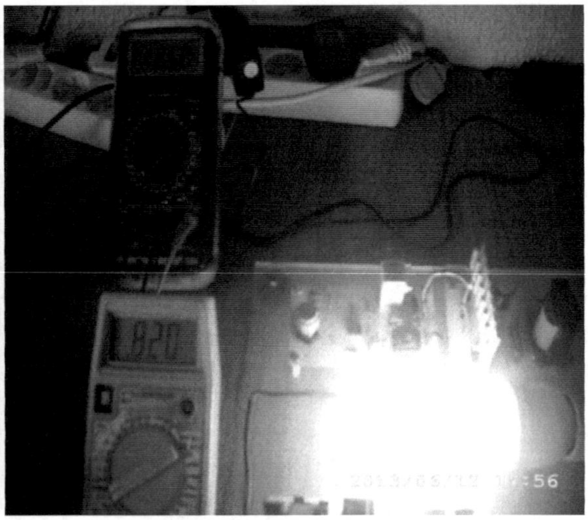

d) $\alpha = 0.5, i = 820mA$

e) $\alpha = 1, i = 1760mA$

Figure 10. The relationship between the current and the duty cycle (α)

4. Conclusions

Using burned filament fluorescent lamps helps increase their life span as well as economizes on energy. This is shown through the brightness value and then relationship of the current and the duty cycle.

This application also presents the possibility that the duty cycle be controlled by a light sensor and so, automatically adjusts the level of brightness according to exterior requests.

References

[1] Tamer T N Khatib, Mohamed A, Amin N and Sopian K 2010 An efficient maximum power point tracking controller for photovoltaic systems using new boost converter design and improved control algorithm, *WSEAS Transactions on power systems* **5**(2) 53- 63

[2] Xingwu Y and Guoqiang W 2013 Research on Direct Power Control of Modular Multilevel Converter Based VSC-HVDC for Offshore Wind Farm, *WSEAS Transactions On Systems* **12**(9) 419-432

[3] Ulrich R, Werner B and Both L R 1994 Transformerless Fluorescent Lamp Operating Circuit, Particularly for a Compact Fluorescent Lamp, with Phase-Shifted Inverter Control, *Patent-Treuhand-Gesellschaft F. Elektrische Gluehlampen mbH*, **935**(004) 1-5

[4] Chang L Y, Chao K H and Chang T C 2013 Application of High Voltage Ratio and Low Ripple Interleaved DC-DC Converter for a Fuel Cell, *WSEAS Transaction on Circuits and Systems* **12**(1) 26-35

[5] Yueh-Ru Y and Chern-Lin C 1999 Steady-State Analysis and Simulation of a BJT Self-Oscillating ZVS-CV Ballast Driven by a Saturable Transformer, *IEEE Transactions On Industrial Electronics* **46**(2) 249-260

[6] Mehdi O, Meridjet M S, Merwan S S and Talbi N 2013 Increase Efficiency of Photovoltaic Pumping System Based BLDC Motor Using Fuzzy Logic MPPT Control, *WSEAS Transaction on Power Systems* **8**(3) 104-113

[7] Xingwu Y and Guoqiang W 2013 Research on Direct Power Control of Modular Multilevel Converter Based VSC-HVDC for Offshore Wind Farm, *WSEAS Transactions On Systems* **12**(9) 419-432

[8] Robert H K and Robert M M 1996 Electronic Ballast With Lamp Current Correction Circuit, Unites *States Patent Assignee: JRS Technology ASSOeiateS* **7**(335, 528) 1-9

[9] Juan Y L and Yang H Y 2013 Hybrid Power System with A Two-Input Power Converter, *WSEAS Transactions on Circuits and systems* **12**(5) 161-170

[10] Chill C and Yorktown H 2000 Regulated Self-Oscillating other Publications Resonant Converter with Current Feedback, *Philips Electronics North America C t. N Y k N.Y.* **09**(221,953) 1-8

[11] Gallo D, Landi C, Luiso M and Fiorucci E 2013 Analysis of a Photovoltaic System: AC and DC Power Quality, *WSEAS Transaction on Power Systems* **8**(4) 45-55

[12] Tsekouras G J and Kanellos F D 2013 Optimal Operation of Ship Electrical Power System with Energy Storage System and Photovoltaics: Analysis and Application, *WSEAS Transaction on Power Systems* **8**(4) 145-155

[13] Merwan S S, Hadj A A, Saleh K and Mahdi O 2013 Improved Incremental Conductance Method for Maximum Power Point Tracking Using Cuk Converter, *WSEAS Transaction on Power Systems* **8**(3) 124-133

[14] Waheed A and Syed M U A 2013 Comparative study of SVPWM (space vector pulse width modulation) & SPWM (sinusoidal pulse width modulation) based three phase voltage source inverters for variable speed drive, *IOP Conf. Series: Materials Science and Engineering* **51**(2013) 1-8

[15] Joel A. D, P.E. and Milan M. J 1994 The LCC Inverter as a Cold Cathode Fluorescent Lamp Driver, *Applied Power Electronics Conference and Exposition, APEC '94 Conference Proceedings* **9**(1) 427 - 433

Research on temperature control with numerical regulators in electric resistance furnaces with indirect heating

C M Diniş[1], G N Popa[1] and A Iagăr[1]

[1]Politehnica University of Timisoara, Department of Electrical Engineering and Industrial Informatics, 5 Revolution Street, 331128 Hunedoara, Romania

E-mail: corina.dinis@fih.upt.ro

Abstract. The paper is an analysis of two-positions (hysteresis) regulators, self-tuned PID controller and PID controller for temperature control used for indirect heat resistance furnaces. For PID controller was used three methods of tuning: Ziegler-Nichols step response model, Cohen-Coon tuning rules and Ziegler-Nichols tuning rules. In experiments it used an electric furnace with indirect heating with active power of resistance of 1 kW/230V AC and a numerical temperature regulator AT-503 type (ANLY). It got a much better temperature control when using the Cohen-Coon tuning rules method than those of Ziegler-Nichols step response method and Ziegler-Nichols tuning rules method.

1. Introduction

Industrial heating processes aimed at raising the temperature of a body for various purposes, such as changing the shape parts (by casting, rolling, forging), heat treatment (normalizing, annealing, stress relief annealing, tempering, reheating, steel, treatment surfaces), melting, brazing, welding; these applications are carried out at different temperatures: 100-1200 ^{0}C.

Temperature control is an important requirement common to all industrial processes [1-4].

Making heating objects or parts can be made by various methods. Electric furnaces convert electrical energy into heat and are divided into: resistance furnaces, induction, electric arc, plasma, or microwave; may be mentioned and capacitive heating, electron beam, or laser.

If indirect heat resistance, heat is obtained from the heating elements (electric resistance power) electric furnace and then transmits the processed material by convection and radiation (Figure 1) [4-6], which will be subjected during heating or after, the physical or chemical transformation.

An electric resistance furnace contains in addition to the heating (heat source) and a walled enclosure of thermally insulating.

Electric furnaces are classified according strength obtained indoor temperature as follows: low temperature furnaces (up to 400 ^{0}C), medium temperature (400-800 ^{0}C) and high temperature (800-1400 ^{0}C) [7].

Electric furnaces are safe resistant, flexible operation (can change the temperature and the heating medium), are effective and do not pollute the environment (no dangerous gas, no sound). Their main applications are in the treatment of metals, ceramics, electronics and glass industry [8].

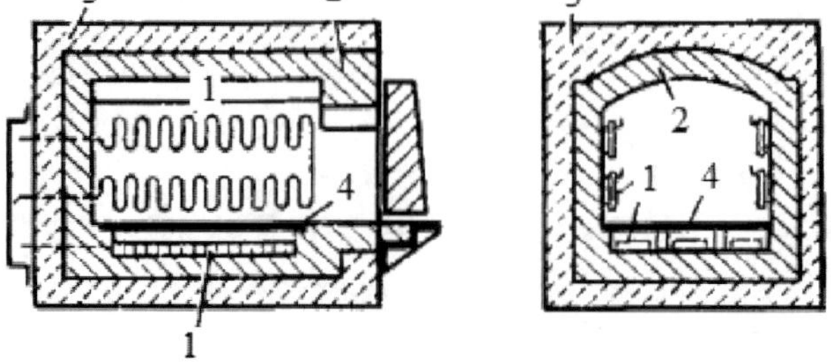

Figure 1. Electric furnace with indirect heating: 1 - resistance heating; 2 - fire brick; 3 - wall thermal insulation; 4 - support pieces

Measuring, monitoring and temperature control are topical in industry and domestic use. Sensors and thermocouples are a wide range of sizes, and typically stainless steel sheath to protect the transducer. For *RTD* temperature range that can be measured is -200 ÷ +600 ^{0}C and thermocouple -40^0 ÷ +1200 ^{0}C. At these sensors output size dependence on the temperature measured is approximately linear [7].

In the case of indirect heating electric furnaces, the processed material is heated at a temperature lower than the melting temperature. The temperature-time characteristic of furnaces for heating of indirect heating is exponential, characterized by three distinct areas (Figure 2): in the low temperature characteristic is approximately linearly increasing with steep growth; in the average temperatures in the temperature is strongly nonlinear evolution; in the high temperature characteristic is saturated with small slope of growth. Theoretically, the electric resistance furnace is a second-order delay element. Its nonlinearity is due, in large part, by convection and radiation heat transmission [9-12].

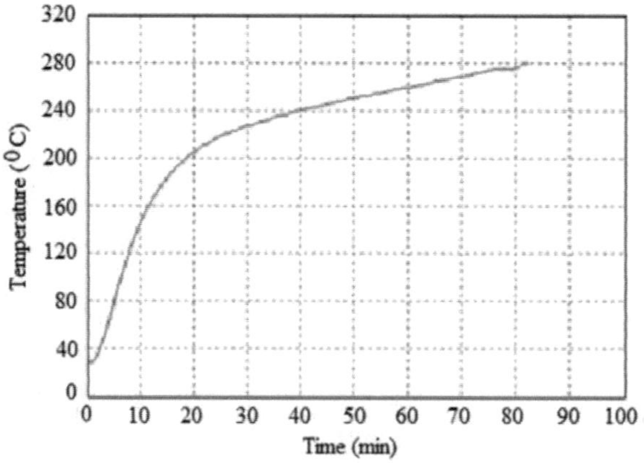

Figure 2. The typical curve of an electric furnace with heating resistors

Some industrial processes require precise temperature control. Getting a temperature as constant as possible in a short time is an important objective of temperature control with indirect heating furnace [6]. Basically, temperature control in closed systems can be implemented with analogue or numerical (digital) controllers [13].

It was found experimentally that determine the parameters electric furnaces with indirect heating (slow process) takes a long time and is done with large errors. Sometimes these parameters must be determined on-line (by using adaptive filters) [1], [12].

The vast majority of applications using temperature control have feedback control systems. The role of the reaction is to work on the system output is adjusted according to the temperature difference between the desired value and the measured value. The most used are type two- positions controllers (on-off controller, with hysteresis, dead band) or *PID* controllers [1]. In general, *PID* controllers are used to control processes that have good performance: high reliability, stable performance, simple algorithm implementation. Setting the *PID* parameters is quite difficult (being more delivery methods) and time.

2. The control algorithms recommended and experimental methods for tuning regulators

Difficulties in identifying precisely the slow processes have led to a limited application of analytical methods and the imposition of practical methods for tuning based on experience gained in exploiting the automatic control systems. Practical methods applied on systems in operation, with reference parameters and disturbance variables held constant.

By changing the tuning parameters of the regulator is reached stability, the regime is determined oscillation and frequency maintained amplitude of the output values. Using these characteristic values, determine the optimal values of tuning parameters controller [14-16].

Method Ziegler-Nichols applies to slow processes that load disturbances are determined and have great times. For an automatic control system with conventional structure provided with a *PID* controller having transfer function of the form:

$$H_R(s) = K_R\left(1 + \frac{1}{T_i \cdot s} + T_d \cdot s\right) \tag{1}$$

with inter-influence factor q=0, the tuning is as follows:

- T_i is fixed parameter at the maximum value $(T_i \to \infty)$, T_d parameter to the minimum $(T_d = 0)$, resulting a *P*-type controller. It changes the amplification factor K_R up to K_{R0} value at which the system reach the stability limit, the output value of the system enters into unamortized oscillation regime with T_0 period. Offreins method is recommended for *PI* regulators to ensure an optimal response to disturbance, namely:

$$K_{Ropt} = 0,5 \cdot K_{R0}\,;\; T_{iopt} = 0,3 \cdot T_{i0} \tag{2}$$

where T_{i0} is the limit value of T_i where the stability limit is exceeded.
Other practical methods for determining the optimal parameters were agreed based on the τ / T_f, characteristic of a process dead time τ and time constant T_f:

$$H_f(s) = \frac{K_f}{T_f s + 1} \cdot e^{-\tau s} \tag{3}$$

Kopelovici relations establish the optimum tuning parameters which provides a minimum duration aperiodic transient response, respectively a response with 20% overshoot maximum.

Chien-Hrones-Reswick relation determining optimal tuning parameters for an optimal behavior when at the input is step signal, ensuring a minimum duration aperiodic response and 20% overshoot.

Adoption of one tuning relation (Ziegler-Nichols, Oppelt, Kopelovici, Chien-Hrones-Reswick, etc.) depends on the specific conditions under automatic control system works. Diversity of practical criteria emphasizes that the tuning of controller for slow processes is very complex and depends on the knowledge of the process and the experience gained in choosing and giving regulators.

To highlight the type of regulator influence the behaviour automatic control systems, responses were plotted in Figure 3 while the output size of an automation control system *y(t)* to a step change in the input, $y_{ref}(t) = 1$, while the used controllers *P*, *PI*, *PD* and *PID* [15], [16].

Comparing the response curves can be made the following remarks:

- *P* controller reduce appreciably overshoot, leading to a short transitional time t_r, but introduces a large stationary error ε_{st};

• by introducing of component *I*, *PI* regulator canceled the stationary error for input step signal, but leads to an overshoot higher than the *P* controller, and a large amount of response time t_r;

• by introducing of of component *D*, the *PD* controller improves the dynamic behavior (the overshoot σ and the transitional time t_r are small), but maintains a large stationary error;

• *PID* controller, combining the effects of *P*, *I* and *D*, provides superior performance, both steady and transient state.

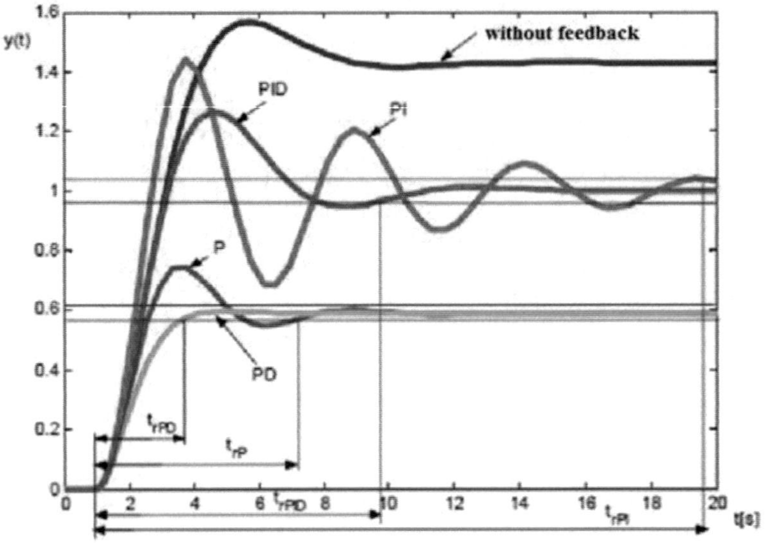

Figure 3. Step responses of an automatic control system for various continuous and linear regulators

Table 1. Recommended control algorithms, depending on the nature of the setting parameter

Regulator type/ Adjust parameter	P	PI	PID
Temperature	YES	YES	YES
Pressure	YES, if there is no too much dead times	YES	In special cases
Flow rate	NO	YES	NO
Level	YES, if there is no too much dead times	YES	YES

Until now, not been determined so far an accurate method to tuning *PID* controllers. There are known temperature control with controllers *P*, *PI* and *PID*. The *P* controller set size is proportional to the error. With this controller obtains a continuous steady error, which can be removed using *PI* controllers by introducing integration effect. A further improvement can be made by the controller derivative action which aims to forecast the error. The *PID* controller can be used to control the temperature, if is right tuned [1].

The tuning of controller represents parameter adjustment for process requirements. If this adjustment takes into account the behavior of the process according to certain criteria (eg. minimum duration transient process, minimal influence of disturbances, etc.), the tuning is called optimal.

Table 2. Recommended control algorithms for different transfer functions of control process

Control type/ Transfer function	P	PI	PD	PID
$\frac{k_f}{T_f\, s+1}$	YES	YES, if requirements are imposed on ε_{st}	YES, if T_f is precisely determined	NO
$\frac{k_f}{(T_{f1}\, s+1)\cdot(T_{f2}\, s+1)}$	YES, with low performance	YES, with restrictions on gain	Rarely used	YES, with restrictions on amplification
$\frac{k_f}{\prod_{1}^{n} T_f\, s+1}$	Rarely used, with low performance	YES	Rarely used	YES
$\frac{k_f}{T_f\, s+1}\cdot e^{-\tau s}$	YES, when $\tau/T_f < 0.1$, ε_{st} is in acceptable limits	YES	Very rarely used	Inconvenient when τ is produced during transport and there is noise
$k_f \cdot e^{-\tau s}$	NO	NO	NO	NO
$\frac{k_f \cdot e^{-\tau s}}{(T_{f1}\, s+1)\cdot(T_{f2}\, s+1)}$	NO	YES	NO	Rarely used, depending on the τ and the effect of component *D*

3. Temperature control experiments at indirect heat resistance furnaces, using numerical regulator AT-503 type (ANLY)

Over time they developed several methods for tuning PID controllers (good gain method, Ziegler-Nichols, Cohen-Coon, Tyreus-Luyben, Oppelt, Kopelovici, etc.) involving different times of tuning [1], [9], [12], [17]. Each of these methods of tuning has not granting maximum performance (overshoot, response time, error stationary). PID controllers can be used for linear systems with known mathematical models. Temperature control in electric furnaces is nonlinear because the delay time is not constant (changes during heating).

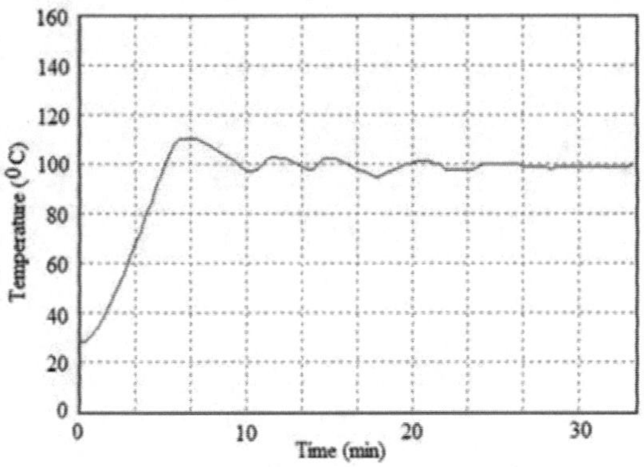

Figure 4. Controlling the temperature in an electric furnace with indirect heating with autotuning controller

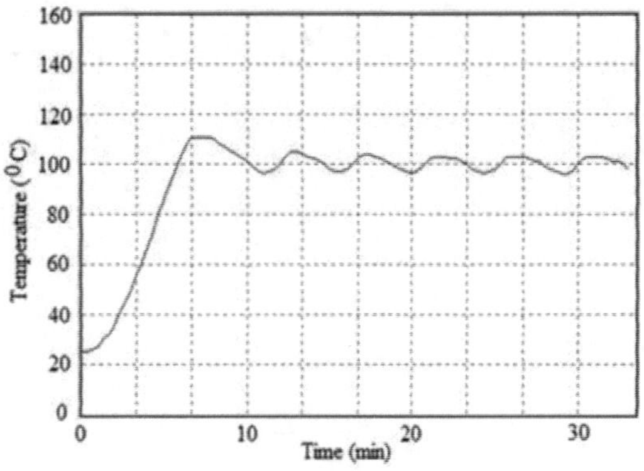

Figure 5. Controlling the temperature in an electric furnace with indirect heating with a two-position hysteresis controller, ±3^{0}C

Usually numerical control temperature controllers are able to use the auto-tuned PID controller that not sometimes offers high performance (Figure 4). The vast majority of digital controllers with auto-tuning do not allow the introduction of the main parameters of furnaces: time delay, time constant, only the set point temperature.

Although two - positions controllers are easier to use (no known constants of electric furnace), have low performance in terms of over-shoot (high values), the response time and error stationary. By using these controllers, controlled size (temperature) oscillates around the setting value (Figure 5). The oscillations occur due to a small change of the system over-reactions of the error. These oscillations can be reduced by using *PID* controllers.

This paper contains an analysis of *PID* controllers used to control temperature resistance furnaces with indirect heating using three methods of tuning regulators: Ziegler-Nichols step response method (*ZNSRM*, Figures 6, 7 and 8), Cohen-Coon tuning rules method (*CCTRM*, Figures 9, 10 and 11) and Ziegler-Nichols tuning rules method (*ZNTRM*, Figures 12, 13 and 14).

In experiments it used an electric furnace with indirect heating with active power of resistance of $1kW$ / $230V$ AC. The furnace enclosure is rectangular, with exterior dimensions: 320 (height)x435 (width)x420 (depth) *mm* and the furnace is coated inside with refractory brick masonry with 45 *mm* thickness. The casing is made of sheet metal with a 5 *mm* thickness. The door dimensions are: 140 (height)x130 (width) *mm*. The furnace temperature was measured using a thermocouple cromel-alumel. In experiments it used a dedicated numerical temperature regulator *AT-503*, ANLY [18].

Ziegler-Nichols step response method

- Measure A and L from step response;
- T_p expected time constant for closed-loop system.

Table 3. Parameters for optimal tuning using Ziegler-Nichols step response model

Controller	K_P ; BP	T_i	T_d	T_p
P	$K_P = 1/A = 36.9$; $BP = 2.7$	-	-	$4 \cdot L$
PI	$K_P = 0.9/A = 33.3$; $BP = 3$	$3 \cdot L = 112$	-	$5.7 \cdot L$
PID	$K_P = 1.2/A = 44.28$; $BP = 2.3$	$2 \cdot L = 74$	$L/2 = 19$	$3.4 \cdot L$

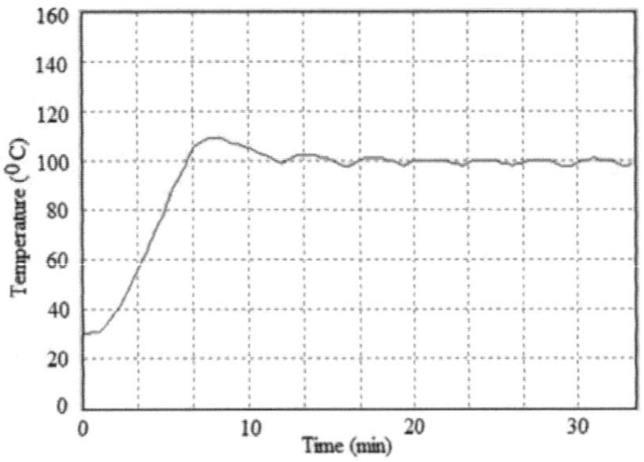

Figure 6. Controlling the temperature in an electric furnace with indirect heating with a *PID* controller using *ZNSRM*

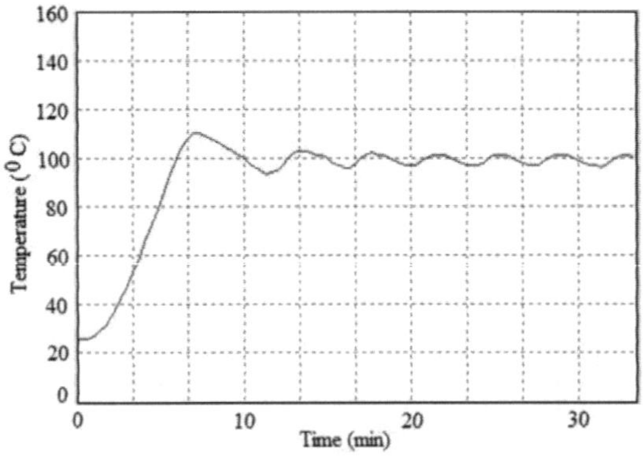

Figure 7. Controlling the temperature in an electric furnace with indirect heating with a *PI* controller using *ZNSRM*

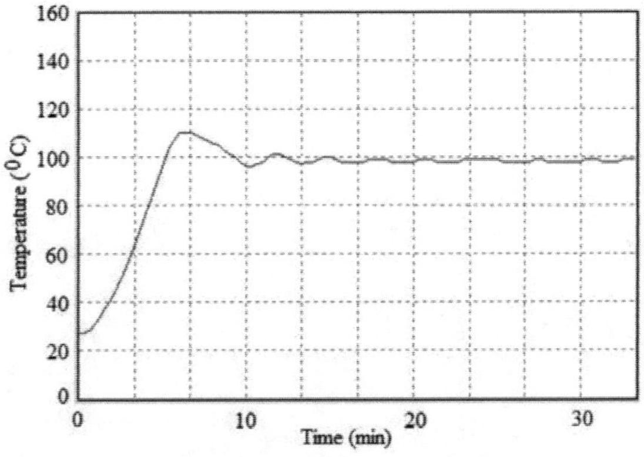

Figure 8. Controlling the temperature in an electric furnace with indirect heating with a *P* controller using *ZNSRM*

Cohen-Coon tuning rules method. Calculate process parameters: t_1, τ, τ_{del}, K, r as follows $t_1 = \dfrac{t_2 - (ln(2))t_3}{1 - ln(2)}$; $\tau = t_3 - t_1$; $\tau_{del} = t_1 - t_0$; $K = \dfrac{B}{A}$; $r = \dfrac{\tau_{del}}{\tau}$.

Table 4. Parameters for optimal tuning Cohen-Coon rules method

Controller	K_c	τ_{int}	τ_{Der}
P	$\frac{1}{rK}\left(1+\frac{r}{3}\right)=6.2$	-	-
PI	$\frac{1}{rK}\left(0.9+\frac{r}{12}\right)=5.4$	$\tau_{del}\frac{30+3r}{9+20r}=390$	-
PID	$\frac{1}{rK}\left(\frac{4}{3}+\frac{r}{4}\right)=8.2$	$\tau_{del}\frac{32+6r}{13+8r}=340$	$\tau_{del}\frac{4}{11+2r}=52$

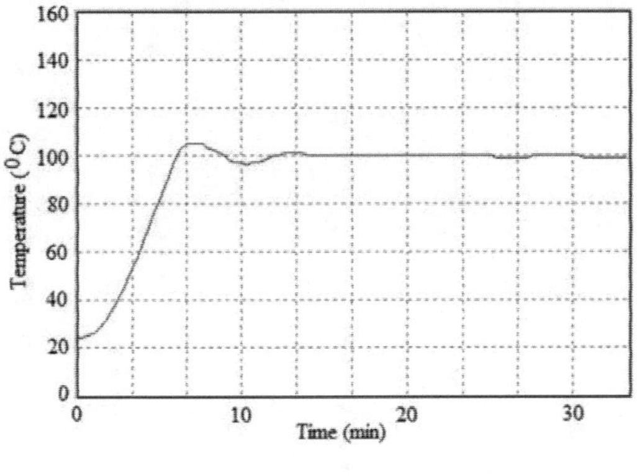

Figure 9. Controlling the temperature in an electric furnace with indirect heating with a *PID* controller using *CCTRM*

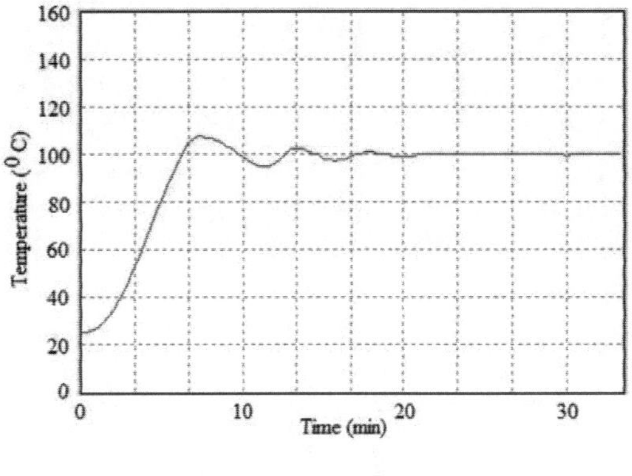

Figure 10. Controlling the temperature in an electric furnace with indirect heating with a *PI* controller using *CCTRM*

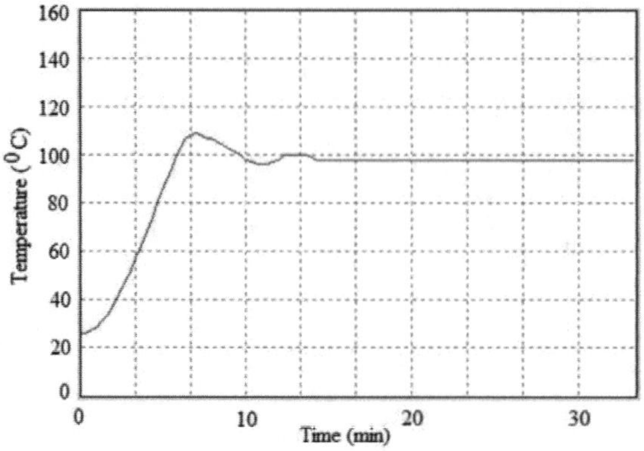

Figure 11. Controlling the temperature in an electric furnace with indirect heating with a *P* controller using *CCTRM*

Ziegler-Nichols tuning rules method

T = 2585.79 s;

L = 113.8 s

Table 5. Parameters for optimal tuning using Ziegler-Nichols rules method

Controller	K_P; BP	K_I	K_D
P	$K_P = \frac{T}{L} = 22.72 \quad BP = \frac{100}{K_P} = 4.4$	0	0
PI	$K_P = 0.9 \cdot \frac{T}{L} = 20.4 \quad BP = \frac{100}{K_P} \cong 4.9$	$0.27 \cdot \left(\frac{T}{L}\right)^2 \cong 139$	0
PID	$K_P = 1.2 \cdot \frac{T}{L} = 27.2 \quad BP = \frac{100}{K_P} \cong 3.6$	$0.6 \cdot \left(\frac{T}{L}\right)^2 \cong 309$	$0.6 \cdot T \cong 1551$

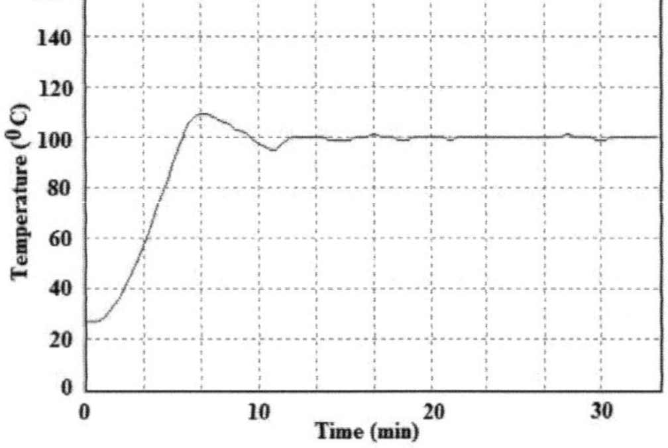

Figure 12. Controlling the temperature in an electric furnace with indirect heating with a *PID* controller using *ZNTRM*

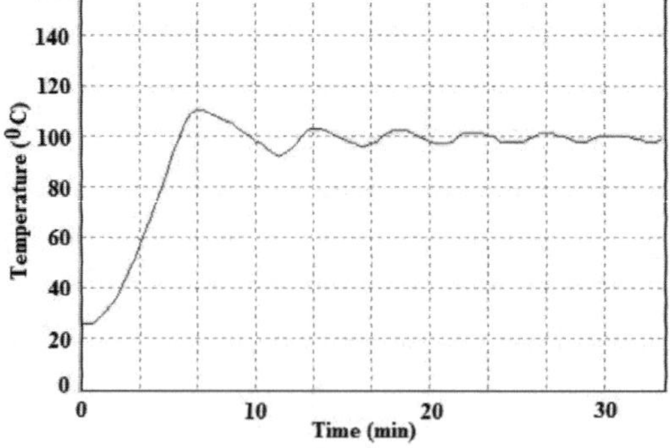

Figure 13. Controlling the temperature in an electric furnace with indirect heating with a *PI* controller using *ZNTRM*

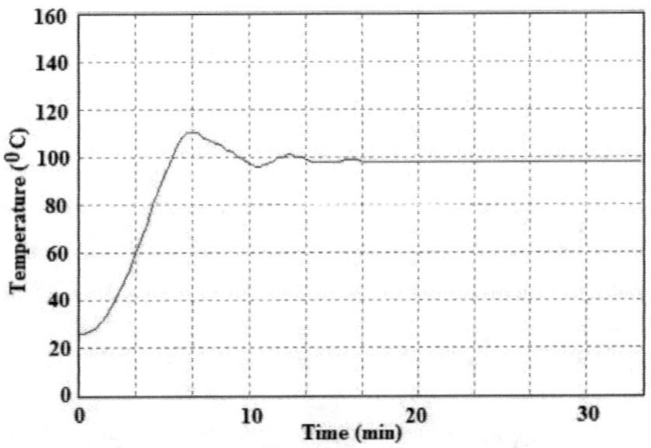

Figure 14. Controlling the temperature in an electric furnace with indirect heating with a *P* controller using *ZNTRM*

PID controllers controlling nonlinear elements and not have outstanding performance [1], [9]. However, the heating curve for electric ovens with indirect heating depends on the operating mode (idle or under load). Determination of the electric furnace parameters is done with imprecision and PID controller parameters can't be set correctly. Sometimes (eg. thermal treatments) temperature overshoot must be eliminated or reduced, in terms of a minimum response time.

4. Conclusions

Comparing the experimental response curves obtained from the use of optimal parameters for tuning Ziegler-Nichols step response method (Table 3), optimal parameters for Cohen-Coon tuning rules method (Table 4) and optimal parameters for Ziegler-Nichols tuning rules method (Table 5), the following observations can be made:

- overshoot response curves obtained with Cohen-Coon tuning rules method is smaller than the overshoot response curves obtained with Ziegler-Nichols step response method and Ziegler-Nichols tuning rules method;
- in curves obtained experimentally Cohen-Coon tuning rules method achieved a response time (minimum 13 *min.* for *PID* controller and up to 20 *min.* for the *PI*) much less than the response time curves obtained from Ziegler-Nichols step response method and Ziegler-Nichols tuning rules method;
- the stationary error obtained in the response curves (Figures 9, 10 and 11) from the Cohen-Coon tuning rules method is much lower than that error obtained by Ziegler-Nichols step response method (Figures 6, 7 and 8) and Ziegler-Nichols tuning rules method (Figures 12, 13 and 14).

It found that get a temperature control with numerical controller *AT-503* with indirect heating furnace much better when using the method parameters for Cohen-Coon tuning rules than those of Ziegler-Nichols step response method and Ziegler-Nichols tuning rules method.

Using fuzzy controllers will be an important alternative to temperature control in electric furnaces with indirect heating. Fuzzy controllers can be implemented to control nonlinear processes, which are difficult to determine mathematical models and the time constants changes in time. Implementing rules in fuzzy controllers will be done correctly to achieve optimum performance [2], [17], [19-21].

References

[1] Astrom K J and Murray R M 2008 *Feedback Systems*, Princeton University Press, USA

[2] Han Y, Jinling J, Guangjian C and Xizhen C 2011 *Temperature Control of Electric Furnace Based on Fuzzy PID*, International Conference on Electronics and Optoelectronics ICEOE 2011, Dalian, China, July 29-31, pp 41-44

[3] Iagăr A, Abrudean C, Diniș C M and Popa G N 2012 Java Program for Computing the Thermal

Losses of Resistance Furnaces, *AWERProcedia Information Technology and Computer Science* **2** 290-294

[4] Iagăr A, Diniş C M and Popa G N 2010 *Study Upon the Thermal Field from a Resistance Oven,* International Symposium on Advanced Engineering & Applied Management, Hunedoara, Romania, November 4-5, pp 231-238

[5] Chook K C and Tan A H 2007 Identification of an Electric Resistance Furnace, *IEEE Transactions on Instrumentation and Measurement* **56**(6) 2262-2270

[6] Kennedy M W, Haaland H and Aune J A, *High Intensity Slag Resistance Furnace Design*, pp.1-10 (http://www.metallurgy.no/1873-Kennedy.pdf)

[7] ***2009 *Resistance Heating Alloys and Systems for Industrial Furnaces*, Kanthal, Sweden

[8] ***1994 Indirect Resistance Heating, *EPRI Center for Materials Fabrication* **3**(7) 1-4

[9] Dreinhoefer L H 1988 Controller Tuning for a Slow Nonlinear Process, *IEEE Control Systems Magazine* **8**(2) 56-60

[10] Grinchuk P S 2010 Mathematical Modelling of Thermal Operating Regimes of Electrical Resistance Furnaces, *Journal of Engineering Physics and Thermophysics* **83**(1) 30-40

[11] Hub U Y and Kim J H 1995 *Adaptive Predictive Control for Electric Furnace*, IEEE IECON 21st International Conference on Industrial Electronics, Control, and Instrumentation, Orlando, Florida, USA, November 6-10, pp 878-883

[12] Pănoiu C, Pănoiu M 2010 *Using Adaptive Filters in Controlling of Electrical Resistance Furnace Temperature Based on a Real Time Identification Method* in *Modelling, Simulation and Optimization*, InTech, Rijeka, Croatia, pp.281-306

[13] Laciak M, Fazekas P and Kacur J 2011 *Monitoring and Control of Temperatures in Electric Furnace in PROMOTIC System*, 12th International Carpathian Control Conference ICCC, Velke Karlovice, Slovenska, May 25-28, pp 239-242

[14] Shehenaz F M, Vaishnavi D, Ramkumar EL, NithyaRani N and Kala H 2014 Comparison of PID Controller Tuning Techniques for Temperature Process, *International Journal of Innovative Research in Electrical, Electronics, Instrumentation and Control Engineering* **2**(8) 1880-1883

[15] Soare C, Făgărăşan I, Arsene P and Iliescu S St 2001 *Theory of Automatic Adjustment. Applications*, Printech Publishing House, Bucharest, Romania

[16] Iliescu S St 2006 *Theory of Automatic Adjustment*, Proxima Publishing House, Bucharest, Romania

[17] Wang S, Li N and Yang F 2012 Resistance Furnace Temperature System on Fuzzy PID Controller, *Journal of Information and Computational Science* **9** 2627-2634

[18] ***2012 *AT03 PID Temperature Controller. User's Manual*, ANLY Electronics Co., LTD

[19] Radakovic Z R, Milosevic V M and Radakovic S B 2002 Application of Temperature Fuzzy Controller in an Indirect Resistance Furnace, *Applied Energy* **73**(2) 167–182

[20] Silaşi C, Cunţan C and Baciu I 2015 Using microcontrollers for orienting a set of mirrors to focus the light beam, *IOP Conf. Ser.: Mater. Sci. Eng.* **85** 012009

[21] Pănoiu C, Rob R and Pănoiu M 2015 Microcontroller interrupts in frequency controlling, *IOP Conf. Ser.: Mater. Sci. Eng.* **85** 012022

IOP Conf. Series: Materials Science and Engineering **106** (2016) 012015 doi:10.1088/1757-899X/106/1/012015

Researches on obtaining composite materials to be used in manufacturing brake pads

A Socalici[1], L Pascu[2], E Ardelean[1] and E Popa[2]

[1]Politehnica University of Timisoara, Engineering and Management Department, 5 Revolution Street, 331128 Hunedoara, Romania
[2] Technical College of Railway Transportation Anghel Saligny, 136 National Street, 335900 Simeria, Romania

E-mail: virginia.socalici@fih.upt.ro

Abstract. The paper presents the laboratory experiments and results meant for obtaining composite materials to be used in manufacturing brake pads for the drive and driven rolling stock. Six samples of composite material have been designed and obtained; each of them has been assessed in terms of tribological behavior using a tribotester and using the friction and wear parameters. Also, determinations have been made on the influence of the material factors and of the main parameters of the work conditions upon the tribological characteristics of the samples under tests and upon the processes that are going on in the superficial strata.

1. Introduction

The noise pollution generated by Rail traffic can be reduced by the application of passive and active measures at the source of disturbance noise. Applying passive methods to reduce rail traffic noise impact on the environment (protection walls and insulation windows) having effect only locally and require major investments [1]. The oriented measures reduce noise source for the entire railway system, if widespread are introduced. This is possible if the cast iron brake blocks will be replaced with composite brake blocks. This issue is examined in the rail industry and currently affects about 370,000 old freight cars and can be used the composite brake blocks K or LL.

K-blocks have higher coefficient of friction than cast iron brake blocks of about 2.5 times their installation on existing wagons requiring changes to the braking system, which is recommended for use in new vehicles. LL-blocks have coefficient of friction similar to cast iron brake blocks, their attachment to the existing wagons do not require major changes to the braking system, which is recommended for use in older vehicles in operation. Currently, there are a variety of composite blocks in operation [2]. The conditions which have been imposed to composites are [3], [4]: high friction coefficient over 0.3; changes in friction coefficient with very low or even zero speed; keeping the friction and wear characteristics at high temperatures.

European Commission recommends obtaining composite materials for brake shoes with superior characteristics to significantly reduce costs and noise [5].

2. Experimental research

The paper presents the laboratory experiments made to obtain composite materials meant for the brake pads to be used by train engines or carriages.

For the preliminary tests of the composite materials the following have been established: the composition of the experimental recipes, the technology of sample manufacturing, the assessment of the tribological behavior of the samples, the analysis of the results [6-8].

For the composite material, 6 experimental test rods were made, using the following components (Figure 1): 19-40% novolac; 3-5% hexamethylenetetramine; 0-7,5% sulphur; 0-10% carbon fiber; 9-15% graphite; 0-20% aluminum; 0-15% brass; 0-30% rubber.

The experimental samples made of composite material have been subjected to tribological tests. The evolution of the friction coefficient has been analyzed by means of a universal UMT-2 tribometer (CETR®, SUA), which allows rod-to-disc tests and also the visualization of the control parameters (the normal force, FN), as well as of the gauged parameters that can be chosen according to the needs (the friction force, the friction coefficient, the acoustic emission, the distance between the two friction elements, marked Z).

Figure 2 shows the device used in the tribological tests, and further on, the shape and dimensions of the steel rod (Figure 3) and of the composite material samples used in the experiment (Figure 4).

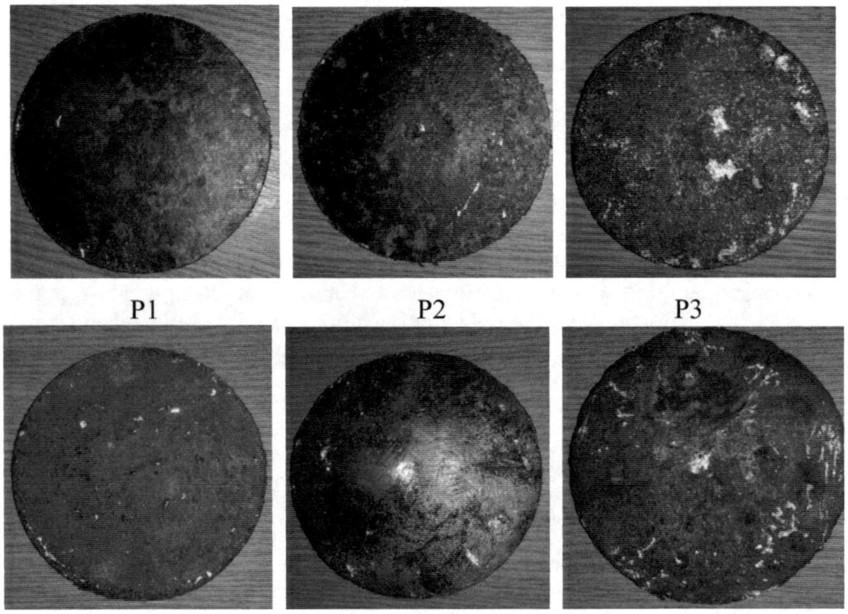

P4 P5 P6

Figure 1. Sample

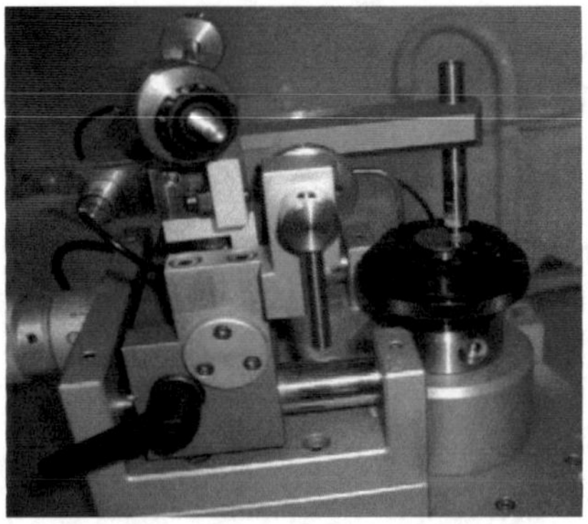

Figure 2. Universal UMT-2 tribometer

Figure 5 shows one example of visualization of the parameters on the tribometer monitor.

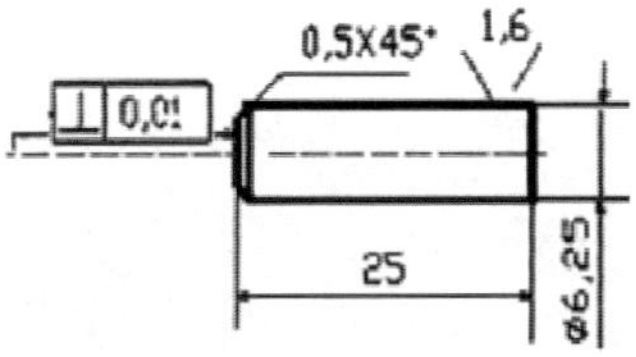

Figure 3. The shape and dimensions of the steel rod

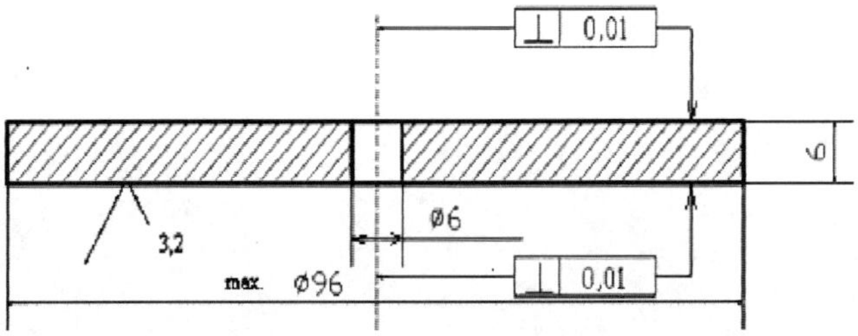

Figure 4. The shape and dimensions of composite samples

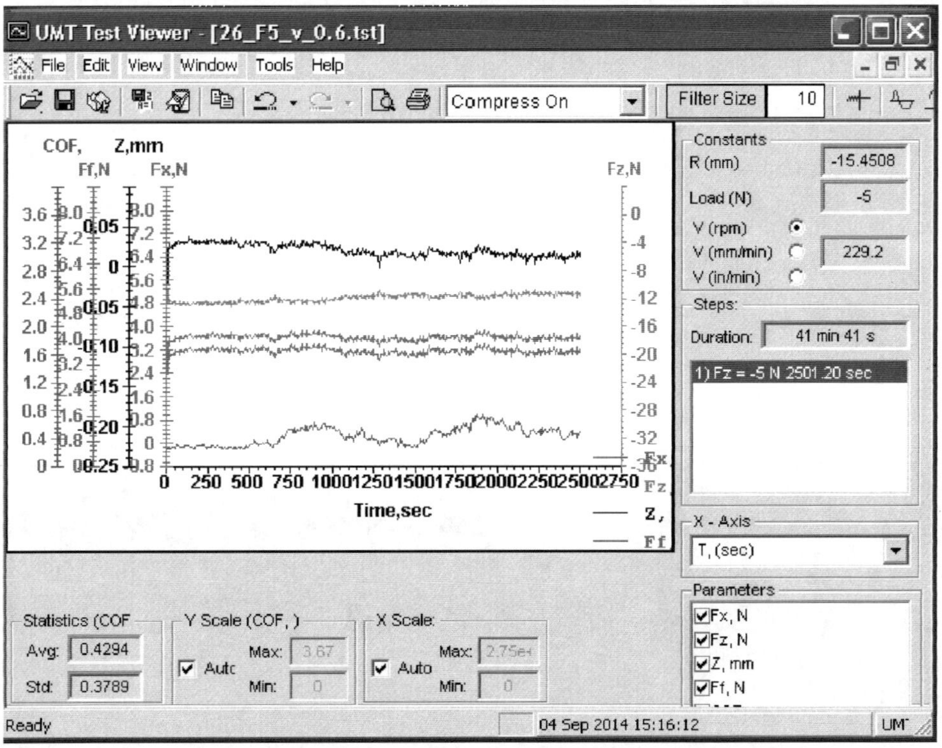

Figure 5. One example of visualization of the parameters on the tribometer monitor

3. Result

The time taken for covering a distance of 1500 m is calculated for each level of sliding rate, under laboratory conditions. Each test is carried out on a radius of 25 mm from the disc center to the rod axis, a single wear trace resulting on each disc. The wear has been measured for each disc separately, as mass loss between the value initially measured and the value measured at the end of the test. Two successive weightings have been done and their average has been used in calculations.

The experimental materials have been tested under dry sliding conditions, for the normal mean pressure of 0.17 MPa and 0.34 MPa and a sliding rate of v = 0.4 m/s, 0.6 m/s and 0.8 m/s using discs made of the pad material (composite material) and rods made of steel for roll bearings.

The results of the tests have been given in text files for the trial data, as well as the graphs of the parameters recorded by the software of the computer the tribometer is endowed with.

Figures 6-11 show the variation of the friction coefficient for a force of 5N and 10N, three different sliding rates.

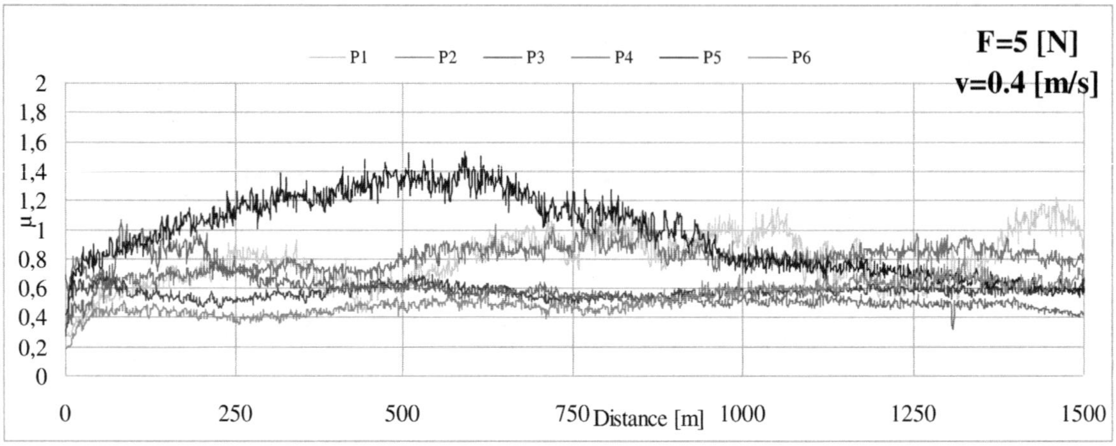

Figure 6. The variation of the friction coefficient for a force of 5 N and a sliding rate of v = 0.4 m/s

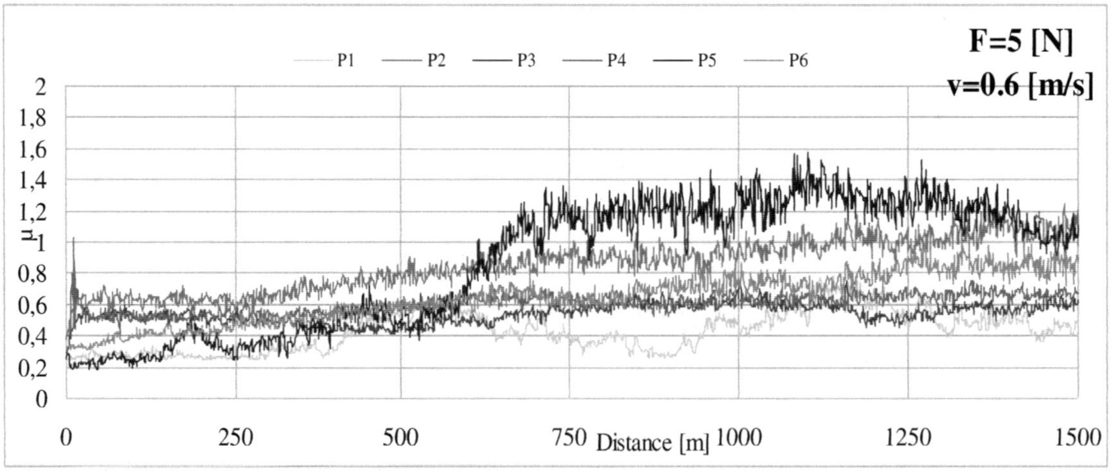

Figure 7. The variation of the friction coefficient for a force of 5 N and a sliding rate of v = 0.6 m/s

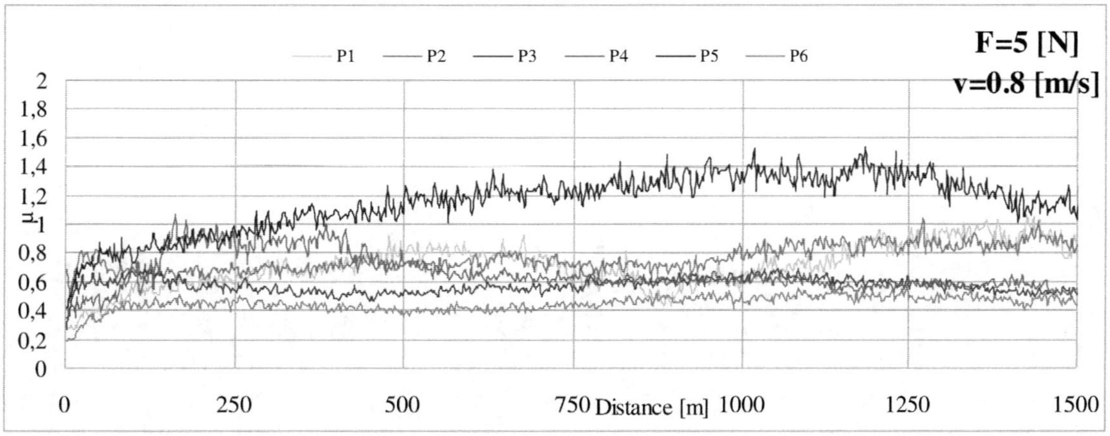

Figure 8. The variation of the friction coefficient for a force of 5 N and a sliding rate of v = 0.8 m/s

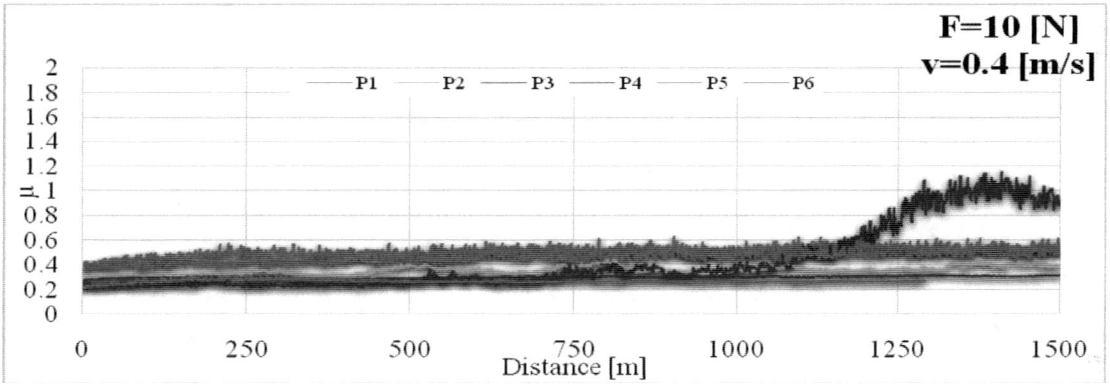

Figure 9. The variation of the friction coefficient for a force of 10 N and a sliding rate of v = 0.4m/s

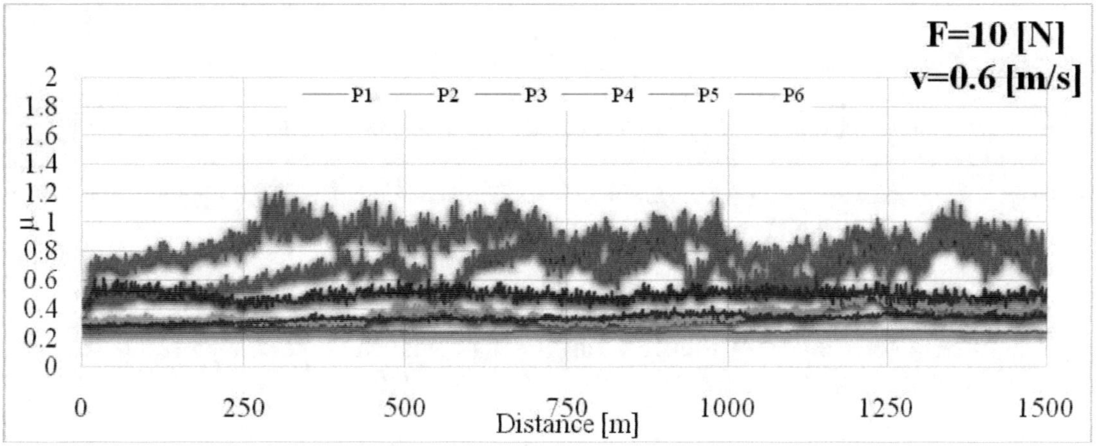

Figure 10. The variation of the friction coefficient for a force of 10 N and a sliding rate of v = 0.6 m/s

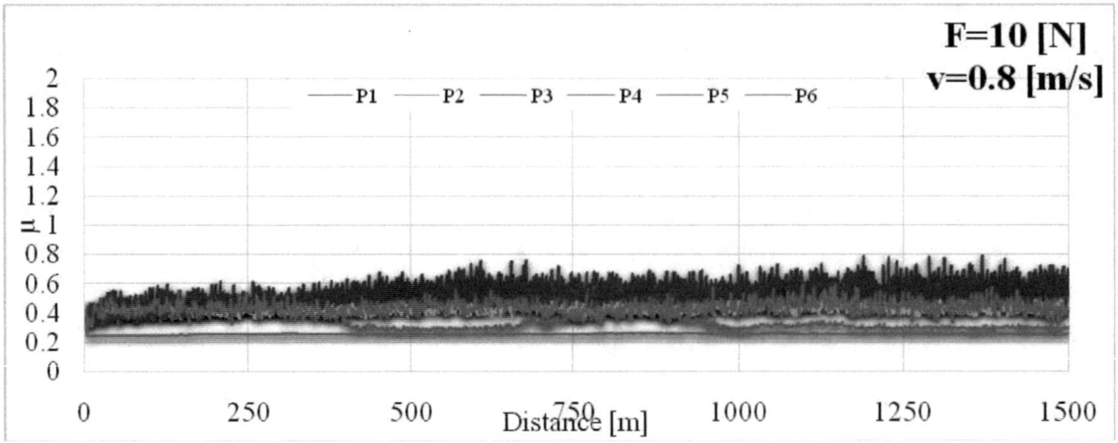

Figure 11. The variation of the friction coefficient for a force of 10 N and a sliding rate of v = 0.8 m/s

4. Conclusions

The analysis of the research results leads to the following conclusions:

- When the sliding rate goes up, for the same load, F = 5 N, one can notice an increase of the friction coefficient as the distance grows, for most samples;
- The evolution of the friction coefficient depending on the work distance is approximately constant;
- When the work load grows, for the same sliding rate, the friction coefficients decrease, but in time, they become stable;
- The friction coefficients obtained by tribological tests on the composite material samples are generally inversely proportional to the increase of the load and of the sliding rate, but they stay above 0,4.

References

[1] ***www.uic.asso.fr (Noise Reduction in Rail Freight - a 2007 report on the state of the art)
[2] ***http://www.agir.ro/buletine/780.pdf
[3] Neharkar Suresh P, Patil R J and Sonawane P R 2014 Study of friction and wear for optimization of disc break material for reduction of brake sound, *International Journal of Research in Aeronautical and Mechanical Engineering* **2**(6) 137-144
[4] ***http://frenoplast.eu/files/SHOES_L_EN.pdf
[5] ***2006/66/EC of 23 December 2005 *Commission Decision concerning the technical specification introperabilitate on the subsystem "rolling stock - noise" of the conventional rail system*, OJ L 37, 8.2.2006
[6] Pascu L 2015 *Research on improving the quality of brake blocks intended for the rolling stock*, University Politehnica of Timişoara, Romania, Doctoral Thesis
[7] Socalici A, Pascu L, Popa E and Heput T 2015 The influence of the cast iron structure upon the hardness of brake shoes meant for the rolling sock, *IOP Conf. Ser.: Mater. Sci. Eng.* **85** 012026
[8] Josan A and Pinca Bretotean C 2015 Using special additions to preparation of the moulding mixture for casting steel parts of drive wheel type, *IOP Conf. Ser.: Mater. Sci. Eng.* **85** 012017

Industrial research on the quality of brake shoes meant for rolling stock

E Popa[1], L Pascu[1], A Socalici[1] and A Lăscuţoni[1]

[1]Politehnica University of Timisoara, Department of Engineering and Management, 5 Revolutiei street, 331128 Hunedoara, Romania

E-mail: erika.popa@fih.upt.ro

Abstract. Brake shoes wear appears as a normal exploitation process and depends both on the braking force and on the material the shoe is made of. Brake shoes are made of molded sulfurous cast iron. The industrial research and experiments aim at determining the specific characteristics of the phosphorous cast iron (chemical and structural homogeneity, hardness) and their optimization in view of improving the quality of the brake shoes meant for the rolling stock.

1. Introduction

The criteria of assessing the rail braking shoes are the following [1]:
- the stability of the braking paths;
- lowest possible specific wear;
- highest possible thermal load capacity;
- lowest possible break sensitivity;
- low sparking tendency.

During braking, the superficial area of the shoe can reach temperatures of 800 – 900^{0}C, and therefore, some structural constituents can melt. On cooling, structural modifications may arise, which increases the hardness.

On manufacturing brake shoes meant for the rolling stock, phosphorous pig irons are largely used. Their use is limited, their friction coefficient diminishing dramatically on braking at high speeds, while their wear, due to the seizing up tendency, is growing when the temperature in the braking coupling goes up [2]. The chemical structure of the braking shoes for railway vehicles running at speeds of up to 120km/h is the following: C=2,9-3,2 %; Si=1-2 %; Mn=0,3-0,7 %; S<0,2 %; P=0,2-1,3 %. Their structure has to be a pearlitic one, without free ferrite.

The high content of phosphor improves the friction – wear behavior of such cast iron. The higher concentration of phosphor grants the formation of phosphorous ternary eutectic, also called Steadite (Fe_3C + Fe_3P + P), hard (650 HB), brittle and with a low melting temperature (950 ^{0}C). The structure of the cast iron made of perlite, graphite and phosphorous eutectic makes for its wear resistance and high fluidity.

The alloying element Ni, Cr, Mo, Cu, etc., influence the basic metallic mass, leading to a finer structure of the graphite.

The brake shoe must have on its lateral surface, as well as on the cross section, a Brinell hardness ranging within 197-225HB [3], [4].

IOP Conf. Series: Materials Science and Engineering **106** (2016) 012016 doi:10.1088/1757-899X/106/1/012016

The resulting experimental data pertaining to the influence of chemical composition on hardness have been processed in Matlab and the results are given both in analytical and graphical form.

2. Experimental researches

The researches that are being done are going to establish the influence of the chemical elements (content of carbon and residuals elements) in the structure upon the physical and mechanical characteristics of the braking shoe material. In order to achieve this, the chemical structure and the physical and mechanical characteristics were analyzed for 25 charges of P10 phosphorous pig iron, elaborated in a trading company.

An important characteristic with a significant influence upon the useful life of brake shoes if hardness. Hardness is tested according to SR EN ISO 6506-1:2006. Hardness is measured at the extreme points of the brake shoe (1 and 2 – Figure 1.a), on its front and at three points located diagonally (s, m and j – Figure 1.b). The brake shoe must have on its lateral surface, as well as on the cross section, a Brinell hardness ranging within 197-225HB [5-8].

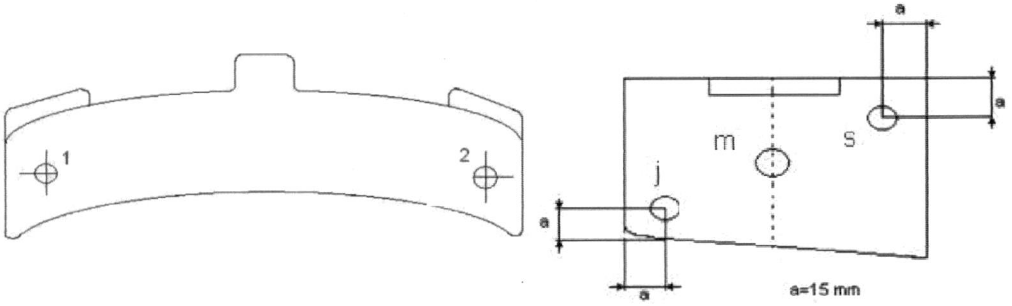

Figure 1. Sampling areas hardness shoe [5]

The experimental data related to the influence of the chemical structure upon the mechanical hardness characteristics have been processed in Matlab and the results are given in graphical and analytical form in Figure 2 - Figure 8.

The processed data resulted in 2nd degree correlation equations between hardness (HB_1, HB_2, HB_s, HB_m, HB_j respectively the means for the shoe surface $(HB_1+HB_2)/2$ and for its cross section $(HB_s+HB_m+HB_j)/3$ – dependent parameters) and the elements present in the chemical composition (C and residual elements – independents parameters). Residual elements present in the chemical composition of pig iron are: *Cr, Mo, Ni, Nb, V, W.* The amount of residual elements is denoted by *R.*

Correlation equations for hardness variation depending on the content of carbon (C) and residuals elements (R) are:

$$HB_1 = 90.5 \cdot C^2 + 90.4 \cdot R^2 + 85 \cdot C \cdot R - 556 \cdot C - 294.2 \cdot R + 1074.6 \quad (1)$$

Correlation coefficient: $R^2 = 0.65$

Point of minimum: C= 2.96%; R=0.23%, HB_1=217.17HB

$$HB_2 = -41 \cdot C^2 - 333.4 \cdot R^2 - 302.2 \cdot C \cdot R + 355.9 \cdot C + 1095.2 \cdot R - 503.8 \quad (2)$$

Correlation coefficient: $R^2 = 0.79$

Inflection point: C= 2.55%; R=0.48%, HB_2=216.15HB

$$HB_s = 134.4 \cdot C^2 - 157.5 \cdot R^2 - 120.1 \cdot C \cdot R - 799.6 \cdot C + 450.1 \cdot R + 1397.9 \quad (3)$$

Correlation coefficient: $R^2 = 0.59$

Inflection point: C= 3.08%; R=0.25%, HB_s=220.28HB

$$HB_j = 45.58 \cdot C^2 + 133.96 \cdot R^2 + 274.64 \cdot C \cdot R - 329.72 \cdot C - 896.8 \cdot R + 809.44 \quad (4)$$

Correlation coefficient: $R^2 = 0.44$

Inflection point: C= 3.09%; R=0.17%, HB_j=221.52HB

$$HB_m = -124.9 \cdot C^2 + 43.2 \cdot R^2 + 56.7 \cdot C \cdot R + 783.4 \cdot C - 199.9 \cdot R - 1016.7 \tag{5}$$

Correlation coefficient: $R^2 = 0.50$

Inflection point: C= 3.18%; R=0.22%, HB_m=209.11HB

$$(HB_1 + HB)/2 = -2.33 \cdot C^2 - 561.13 \cdot R^2 - 4.68 \cdot C \cdot R + 8.0 \cdot C + 269.16 \cdot R + 189.96 \tag{6}$$

Correlation coefficient: $R^2 = 0.43$

Point of maximum: C= 1.49%; R=0.23%, $(HB_1+HB_2)/2$=227.44HB

$$(HB_s + HB_m + HB_j)/3 = 284.2 \cdot C^2 + 328.8 \cdot R^2 + 229 \cdot C \cdot R - 1805.8 \cdot C - 883.3 \cdot R + 3100.5 \tag{7}$$

Correlation coefficient: $R^2 = 0.72$

Point of minimum: C= 3.06%; R=0.27%, $(HB_s+HB_m+HB_j)/3$=210.24HB

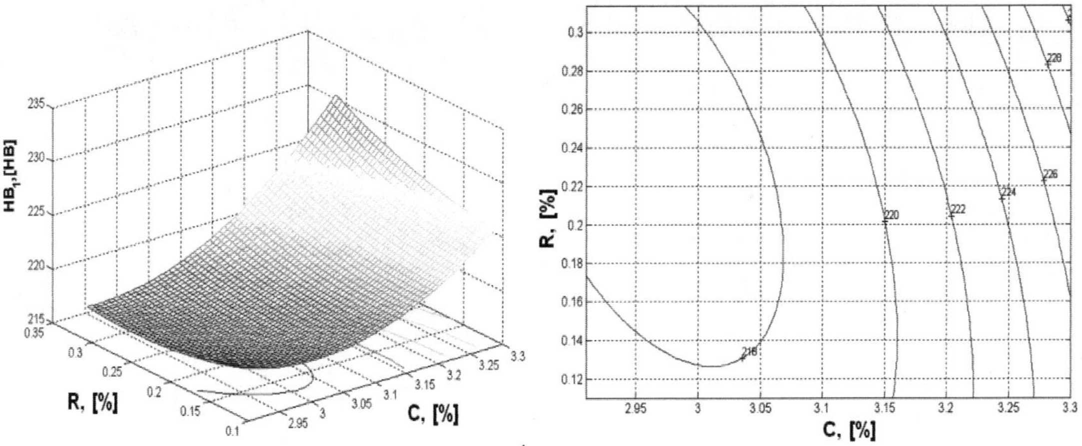

Figure 2. HB_1=f(C,R)-correlation 2nd degree a-surface regression; b-level lines

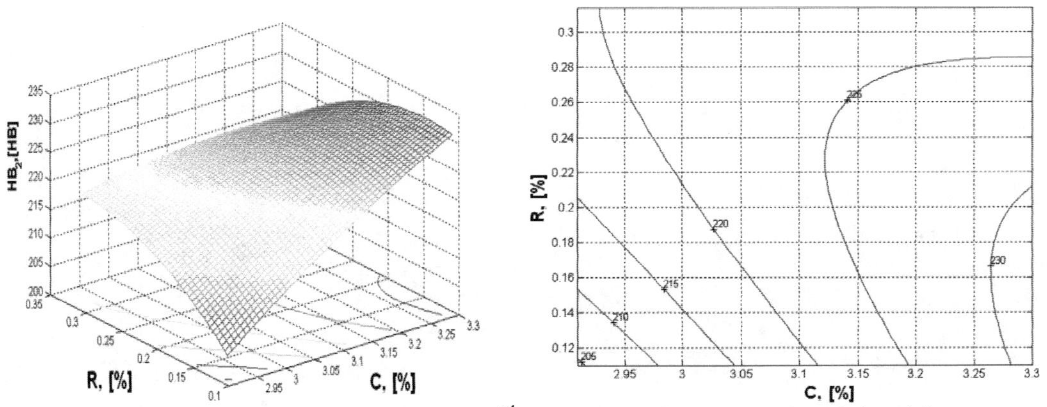

Figure 3. HB_2= f(C,R)-correlation 2nd degree a-surface regression; b-level lines

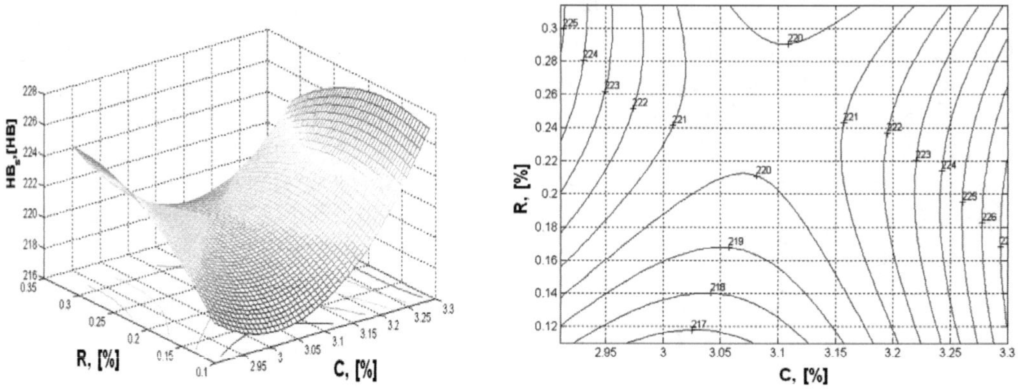

Figure 4. $HB_s = f(C,R)$-correlation 2nd degree a-surface regression; b-level lines

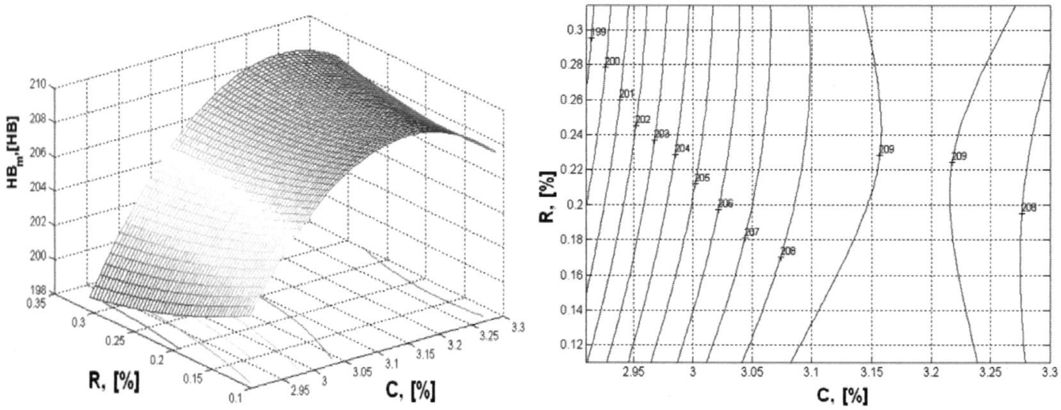

Figure 5. $HB_m = f(C,R)$-correlation 2nd degree a-surface regression; b-level lines

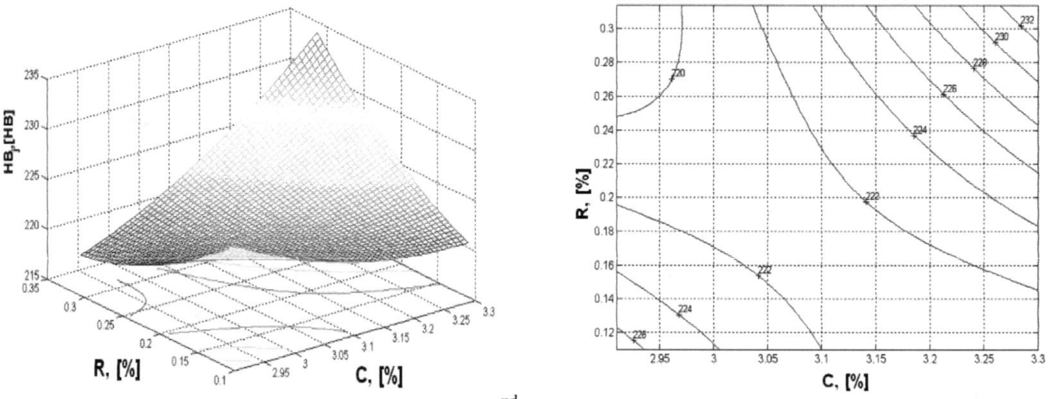

Figure 6. $HB_j = f(C,R)$-correlation 2nd degree a-surface regression; b-level lines

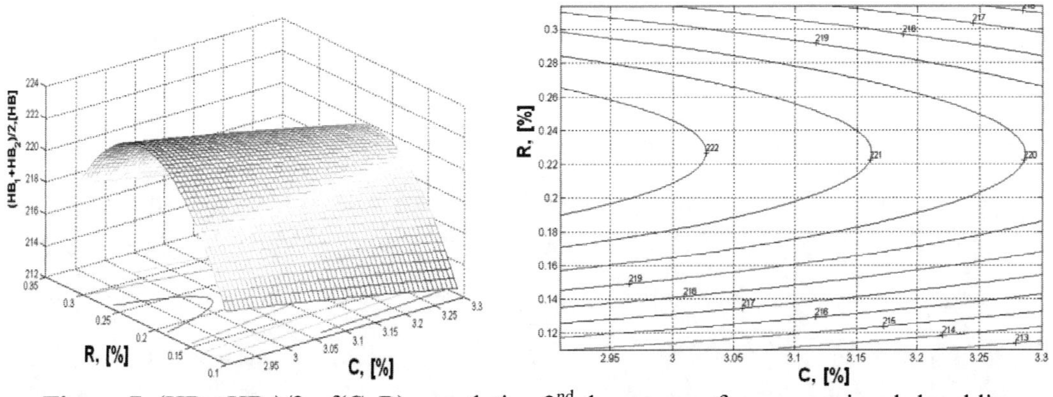

Figure 7. $(HB_1+HB_2)/2= f(C, R)$-correlation 2nd degree a-surface regression; b-level lines

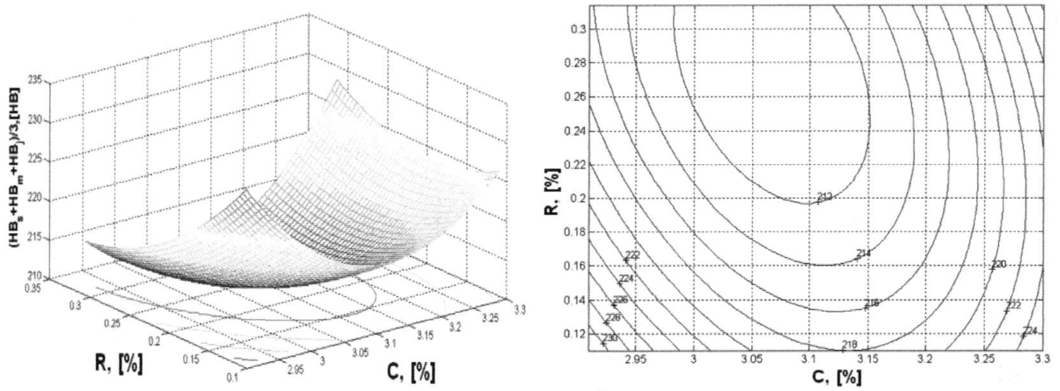

Figure 8. $(HB_s+HB_m+HB_j)/3= f(C,R)$-correlation 2nd degree a-surface regression; b-level lines

3. Conclusions

The analysis of the research results leads to the following conclusions:

- the chemical composition of the cast iron used in making brake shoes grants the ranging of hardness within standard limits;
- there is a hardness difference between the cross section extremities and the centre of its cross section, which is explained by the solidification conditions;
- the level curves obtained in the graphical representation allow the choice of the independent parameters (P, Cr, Mo, Ni, Nb, V, W) so as to obtain the desired value for hardness.

Acknowledgments

This work was partially supported by the strategic grant POSDRU/159/1.5/S/137070 (2014) of the Ministry of National Education, Romania, co-financed by the European Social Fund – Investing in People, within the Sectoral Operational Programme Human Resources Development 2007-2013.

References

[1] ***Specification, nr.1/SFMR/SDT/2000, *Brake shoes for the rolling sock,* approved by Romanian Railway Authority

[2] ***www.uic.asso.fr (Noise Reduction in Rail Freight - a 2007 report on the state of the art)

[3] Cojocaru V and Filipiu C 2006 Preparing load for the development of iron in induction furnaces, Samia Publishing House, Iaşi, Romania

[4] ***Commission Decision 2006/66/EC of 23 December 2005 concerning the technical specification interoperabilitate on the subsystem "rolling stock - noise" of the conventional rail system, OJ L 37, 8.2.2006

[5] ***www.tef.com

[6] Pascu L V, Putan A, Socalici A and Heput T 2013 Researches on the quality of pig iron to be used in making rail braking shoes, *Metalurgia International* **18**(8) 193-199

[7] Socalici A, Pascu L, Popa E and Hepuţ T 2015 The influence of the cast iron structure upon the hardness of brake shoes meant for the rolling stock, *IOP Conf. Ser.: Mater. Sci. Eng.* **85** 012026

[8] ***Staff Working Document European Commission, Brussels, 8.7. 2008, SEC (2008) 2204

Risk analysis for renewable energy projects due to constraints arising

G Prostean[1], C Vasar[2], O Prostean[2] and A Vartosu[3]

[1]Politehnica University of Timisoara, Department of Management, Remus Street, no. 14, 300191 Timisoara, Romania
[2]Politehnica University of Timisoara, Department of Automation and Applied Informatics, Vasile Parvan Street, no. 1-2, 300223 Timisoara, Romania
[3]Politehnica University of Timisoara, Department of of Measurement and Optical Electronics, Vasile Parvan Street, no. 1-2, 300223 Timisoara, Romania

E-mail: gabriela.prostean@upt.ro

Abstract. Starting from the target of the European Union (EU) to use renewable energy in the area that aims a binding target of 20% renewable energy in final energy consumption by 2020, this article illustrates the identification of risks for implementation of wind energy projects in Romania, which could lead to complex technical implications, social and administrative. In specific projects analyzed in this paper were identified critical bottlenecks in the future wind power supply chain and reasonable time periods that may arise. Renewable energy technologies have to face a number of constraints that delayed scaling-up their production process, their transport process, the equipment reliability, etc. so implementing these types of projects requiring complex specialized team, the coordination of which also involve specific risks. The research team applied an analytical risk approach to identify major risks encountered within a wind farm project developed in Romania in isolated regions with different particularities, configured for different geographical areas (hill and mountain locations in Romania). Identification of major risks was based on the conceptual model set up for the entire project implementation process. Throughout this conceptual model there were identified specific constraints of such process. Integration risks were examined by an empirical study based on the method HAZOP (Hazard and Operability). The discussion describes the analysis of our results implementation context of renewable energy projects in Romania and creates a framework for assessing energy supply to any entity from renewable sources.

1. Introduction

The study begins form the fact that Romania, through National Authority for Energy Regulation (ANRE), it was employed in fast mode in the EU Directives regulations, due to its specific characteristics, by being among the first countries in the Member States, which transposed in their legislation EU Directives on renewable energies [1]. The share of renewable energy production is indicated by Directive 2009/28/EC, setting mandatory targets for renewable energy for all EU Member States and which establishes that Romania must increase the amount of renewable in its energy mix from 17.8% at the end of 2005 to 24% by 2020 [2]. According to the European strategy 20/20, through which the European Union (EU) established mandatory targets to be achieved by 2020 for a 20% overall share of renewable energy in the EU [2]. Starting from the 20/20 strategy, aimed at addressing the problem of identifying and exploiting every area of Romania, which has a large potential to achieve this target. Romania, however, has a highly diversified geographic relief and consequently to achieve the 20/20 strategy creates challenges for wind energy development projects, operating in geographically isolated regions, but potentially suitable for wind. Implementation of wind power projects lead to a complex process based on best techniques with the implications of quality raw materials that correspond to specific technical requirements. Bottlenecks in wind power supply chain supply chain can be overcome by applying the right technologies. The risk arising from the development and implementation of such projects is very high, due to the combination of several determinant factors. The study addressed in this paper examines the risks that may occur along the life cycle of such projects. Based on the experience of implementing such projects and given the fact that these risks are diversified, the research team has identified and systematized them in a conceptual model. As well, the research team identified bottlenecks in wind power supply chains based on a wind farm project developed in Romania in isolated regions with different particularities, hill and mountain locations in Romania, having difficult climatic conditions at certain times of the year.

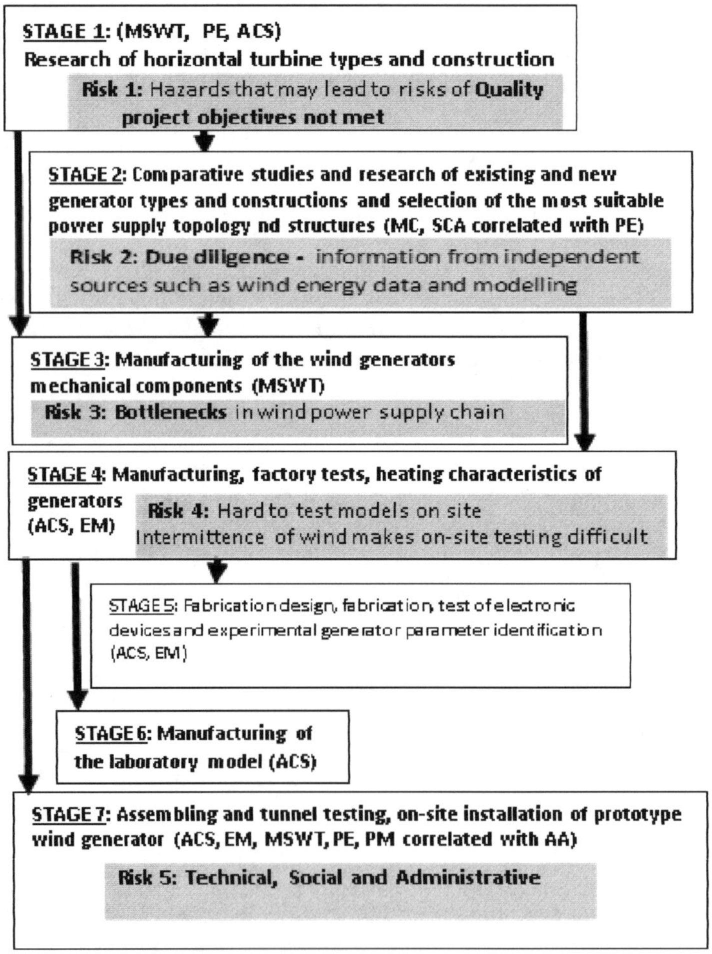

Figure 1. Risk assessment – conceptual model

The development of renewable energy projects require complex knowledge and particularly much more experience in many fields: Mechanical structures and wind turbines (MSWT); Electrical machines (EM); Power electronics (PE); Automatic controlling strategies (ACS); Project management and administrative problems (PM). All these teams are specialized in developing and implementing specific integrated conceptual model for identifying major risks, which may be expressed over the project life cycle (Figure 1).

Each risk factor was identified and analyzed in particular. The study of this paper focuses on the stage 3 of the conceptual model shown in Figure 1, specifically on how to identify constraints in the supply chain of manufacturing of generators, respectively on the right methodology of analyzing this risk factor.

2. Constraints arising

Usually made of fiberglass, the nacelle contains: 1. the gearbox that increase the low rotational speed of the rotor shaft in several stages to the high speed needed to drive the generator; 2. the brake system that consists in disc brakes that bring the turbine to a halt when required; 3. the generator that converts mechanical energy into electrical energy; 4. the yaw system, consisting into a mechanism that rotates the nacelle to face the changing wind direction; 5. the power converter that converts direct current from the generator into alternating current to be exported to the grid network; 6. the transformer that converts the electricity from the turbine to higher voltage required by the grid; and 7. the pitch system that adjusts the angle of the blades to make best use of the prevailing wind. The study is focused on the generator manufacturing, taking into account that the size of turbines has been increasing dramatically from less than 1 MW up to 2 MW and larger [3]. This means, that the number of generator suppliers who can satisfy the demand reliably is reduced considerably. Increasing capacity involves a huge investments and bottlenecks, and in addition to that, these kind of bottlenecks represent a high risk, taking into account that the generator manufacturing represent 3,44% from the total project cost. The generator (G) is a standard equipment whose main components are steel and copper. It's not as easy to increase the generator manufacturing capacity as it is with other components, because there are necessary a lot of equipment [3].

In this approach the research team after strong an action research in wind power projects implemented and based on the methodology Theory of Constraints (TOC) using Drum-Buffer-Rope (DBR) strategy identified bottlenecks for the generator manufacturing supply chain, more exactly for some important materials, taking into account the distribution for raw materials (copper and steel), necessary for the generator manufacturing.

The bottlenecks represent an excess of inventory or insufficient or poor raw materials because the fluctuations are inconsistent.

A solution to avoid bottlenecks in a wind power supply chain is done using an intermediary buffer at every link in the chain. In this way each link will be able to manage individually an intermediary stock, placing orders at the right time. An addition to the proposed solution is given by the fact that the links in the supply chain must focus more on obtaining sustainable cooperation between supply activities regarding the movement of necessary and special raw materials for wind components in the chain. The impact of all bottlenecks can largely be mitigated with appropriate combinations of industry and policy measures. In order to mitigate supply chain bottleneck is applied TOC-DBR philosophy which implies dimensioning some intermediary buffers at each link such this intermediary stock can ensure continuous high precision materials according to design specifications.

2.1. Drum-Buffer-Rope philosophy

The DBR method is designed to optimize a flow process, obtaining the full capacity of the most constrained capacity link (CCL) in the supply chain. The rhythm of CCL represents the drum for the rest of the system. The rope represents the mechanism of releasing the raw material into the flow process, protecting the CCL from being swamped with "work in progress" [4]. The rope regulates the rate of inserting the raw material into the flow process. The inserting rate is no faster than that impose by the drum. The rope is connected with the drum with the help of the safety buffer that protects the CCL from starving because of the work during the process. The TOC method through DBR helps to move important raw materials (cooper and steel) in supply chain in safety stocks that are placed and dimensioned to protect the rhythm of each individual link in the chain. The purpose of using DBR solution for dimensioning intermediary buffer has the role to protect the links in supply chain against bottlenecks variation occurred within the chain, with the condition to keep the movement rhythm of the project at each link in the supply chain. Dimensioning intermediary buffer through DBR philosophy is done according to the following steps [4], [5]:

1. Is identified each link rhythm (D-Drum) in the study of this paper is about test medium consumption / link supply chain;
2. Is dimensioned an intermediary stock (B-Buffer) in front of each link in the chain (Figure 2) which can create link protection, regardless of orders variation that can occur;
3. Buffer dimensioning is made on three levels (green, yellow and red) (Figure 2) in such a way that placing orders decision to refill the buffer must fulfill the following:

- while the consumption rhythm keep the buffer on green is not created another command:
- when the buffer reaches yellow the order is created and validated (it is send to refill the buffer) so that by correlating link consumption rhythm (Drum) with delivery lead time (DLT), the buffer must be refilled during the period when he is at the red level. This is the stage where the replenishment valve rope is pull, also called the rope mechanism (R-Rope). DBR helps provide at every time raw materials for a wind power creating safety stocks that are placed and dimensioned in order to ensure fluency tests included in the phase of project cycle development. The levels divided into colors have the result that technical testing of pieces that need to be integrated in component manufacturing must fulfil correctly the specific requirement.

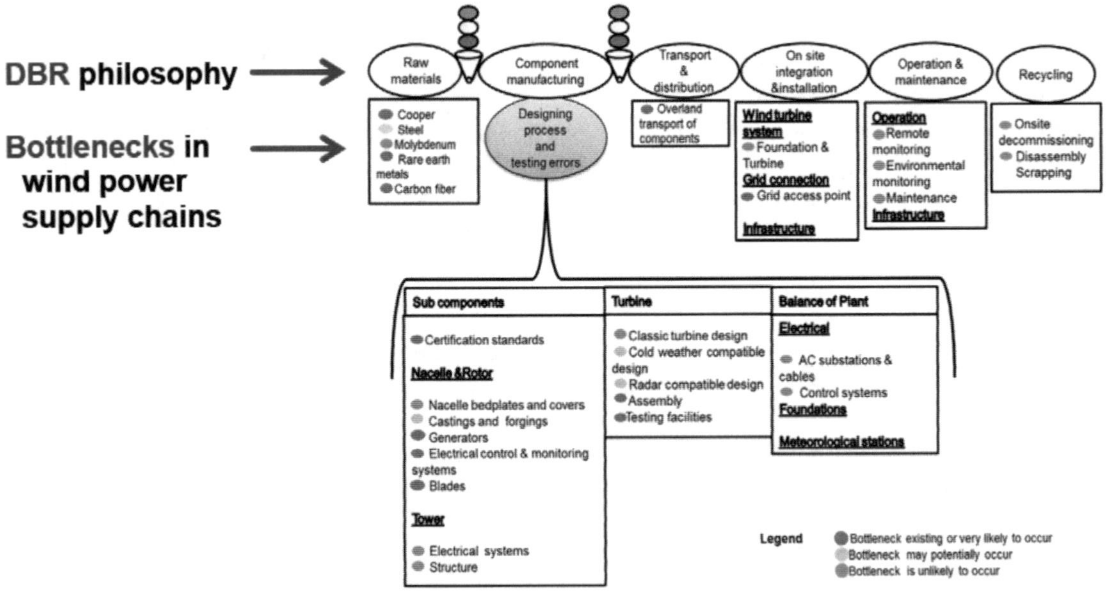

Figure 2. Wind power supply chain and its bottlenecks

To illustrate further the philosophy DBR it is configured a buffer dimensioning for two components (copper and steel). In this regard the DBR method avoid the bottlenecks through adjustment created by each link rhythm (D-Drum), respectively a dimensioning algorithm of the three decision levels (Figure 3) (B-Buffer, R-Rope). Buffer dimensioning is divided into three areas: the top has green status, the middle has yellow status and the bottom has a red status (Figure 3).

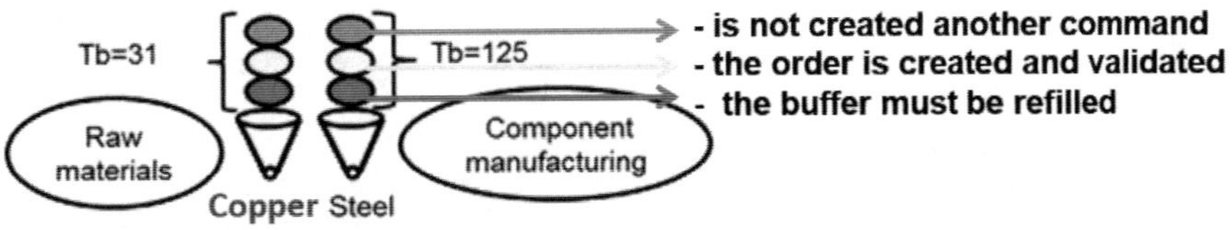

Figure 3. Buffer dimensioning for copper and steel

Target buffer (Tb) represent the maximum level of the buffer, where:

$$Tb = DLT \cdot \frac{MC}{UDLT} \cdot SP \qquad (1)$$

DLT - delivery lead time
MC - medium consumption, on
UDLT - measuring unit delivery lead time
SP - safety percentage

The Status buffer (Sb) refers to the buffer stock level at a time. Further status buffer is calculated for yellow in percentage rate (Sb_{yellow}), which will generate the order release, respectively for red status buffer (Sb_{red}) in percentage rate. Sbred has the significance of buffer level in which pulling is generated „ROPE", more specific, the entrance of the new stock in buffer. The status buffer for red (Sb_{red}) is calculated based on the test medium consumption daily (TMC_{daily}), (Table 1).

Table 1. Buffer Dimensioning for copper and steel

	MC	DLT	SP	Tb	Sb_{yellow}	Drum	Sb_{red} Rope
Copper	25	1week	125%	31	80,64%	3,57/day	11,52%
Steel	100	1week	125%	125	80%	14,28/day	11,42%

3. Risk analysis

The scientific study of this work is further supported by the method HAZOP (hazard and operability) to analyze the degree of risk for each factor separately identified. HAZOP method was introduced to define and delineate the principles required to perform interoperability studies and risk analysis, due to the growing complexity of processes that cannot be examined using conventional methods [6]. For example the first step in Table 3 shows the probability of occurrence for each risk factor separately, for example, the factor no. 6 (Table 2) - "Manufacturer can delay the completion of the works". According to the analysis performed in this study, this is a high risk factor for the projects, and also taking into account the relatively high probability of occurrence, it must be considered.

Regarding the probability of occurrence of this risk factor 6, after the analyses resulted, it has been found the appearance of some delay for the completion of the works, in this phase to one of 5 projects studied, reason which is why it was considered that this factor has probability of occurrence "Moderate/Possible", namely value 4 according to the scale from Table 3.

Table 2. The project risk factors

Nr.	Risk Factor
1	The cost insurance for extreme weather
2	Increase cost components in short term (panels photovoltaic, wind generators)
3	Component availability in short term
4	Repair key components during operation / warranty
5	Malfunction of turbines or solar panels during construction or putting into operation
6	Manufacturer can delay the completion of the works 9for various reason)
7	Changes in legislation which may have an impact on the project
8	Financing cancellation
9	Delays in obtaining construction permits
10	Thefts may lead an increase in insurance prices
11	Regulation of installations cancel/recycling
12	The lack of qualified labor force
13	Contract interruption with the company which provides maintenance, which lead to a renegotiation of its subsequent
14	Modules functioning under the parameters declared by the manufacturer
15	A too high rate of breakdowns can have the effects of increasing the operating costs
16	Major damage to the turbines/solar panels which can occur during operation

In the final step the research team analyzed in an individual manner each risk factor of 16, listed in Table 2 and established through estimations based on brainstorming and experience, severity classification Table 4. Through severity classification it is estimated worst potential, consequences which result from a failure and even affected accomplishment of the project. For exemplification, it is also considered that factor no.6.

Table 3. Probability of occurrence for risk factor no. 6

Scale	Probability of occurrence	Explanation	Probability of occurrence assigned to risk factor
6	Almost certain	>=75% (1 in 1.33)	
5	Likely	>=50% (1 in 2)	
4	Moderate/Possible	>=25% (1 in 4)	**4**
3	Unlikely	>=20% (1 in 5)	
2	Rare	>=10% (1 in 10)	
1	Almost never	>=5% (1 in 20)	

The analysis of these RES projects is identified the position of each risk factor in the risk matrix (Figure 4) in accordance with individual scales established for the probability of occurrence and severity.

Table 4. Severity classification appearance for factor no. 6

Scale	**Severity classification**	**Factor 6 classification**
6	Sever impact	
5	Major impact	
4	Moderate impact	**4**
3	Minor impact	
2	Insignificant impact	
1	No impact	

Starting from the 16 factors described within the paragraph, they have to be analyzed as a system. Analyzing only one factor would lead to misjudgment. Furthermore there are some more factors which have to be included in the logic of decision. These factors are chosen based on the project experience of the authors.

Significance		Severity (Consequence)					
		1 No impact	2 Insignificant impact	3 Minor impact	4 Moderate impact	5 Major impact	6 Severe impact
Probability of occurrence	1 Almost never	Very low	Very low	Very low	Low	Moderate	High
	2 Rare	Very low	Low	Low	Moderate	Moderate	Very high
	3 Unlikely	Very low	Low	Moderate	Moderate	High	Very high
	4 Possible	Low	Low	Moderate	High	Very high	Sever
	5 Likely	Low	Moderate	High	High	Very high	Sever
	6 Almost certain	Low	Moderate	High	Very high	Sever	Sever

Figure 4. Risk Matrix for the factor no. 6

4. Conclusions

Risk analysis method presented in this article can be used to evaluate any supply energy entities from renewable sources, useful in order to achieve feasibility study, or even to identify possible measures necessary to minimize exposure to certain risks.

According with the Risk matrix, the factor 6, detailed above is framed in the category of „High" level, it is observed for probability of occurrence (catalogued 4) and impact or (catalogued 4) towards higher levels, which would lead to the classification as "High" factor.

This factor risk reduction can be accomplished effectively through a variety of measures such as:

- To achieve a very detailed monitoring of intermediary buffers, that requires a control mechanism based on constrained capacity link (CCL);
- Orders prioritization based on consumption;
- Signal through color when order must be created, respectively for the moment in which must enter into buffer.

Acknowledgement

This paper was developed through the Partnerships in Priority Areas - PN II, with the support of ANCS, CNDI - UEFISCDI, Project no. 36/2012.

References

[1] ****ANRE Ordinance 43 /2011, Regulation for GC issue*, www.anre.ro/download.php?id=4035

[2] ***Directive 2009/28/EC of European Parliament and of the Council of 23 April 2009 on the promotion of the use of energy from renewable sources and amending and subsequently repealing Directives 2001/77/EC and 2003/30/EC. *Official Journal L140*, 16–62; 5 June 2009

[3] Keith Hays of EER 2007 *Wind directions*, http://www.altenergystocks.com/assets/Wind%20Directions.pdf

[4] John H Blackstone Jr 2010 *A Review of Literature on Drum-Buffer-Rope, Buffer Management and Distribution*, McGraw-Hill, New York, USA

[5] Goldratt E M 2003 *Production the TOC way (revised edition)*, North River Press, Great Barrington, USA

[6] Lawley H G 1974 Operability studies and hazard analysis, *Chemical Engineering Progress* **70** (4) 45–56

Web server with ATMEGA 2560 microcontroller

E Răduca[1], D Ungureanu-Anghel[2], L Nistor[1], C Hațiegan[1], S Drăghici[2], C Chioncel[1], E Spunei[1] and R Lolea[1]

[1]Eftimie Murgu University of Resita, Department of Electrical Engineering and Industrial Informatics, P-ta Traian Vuia, no. 1-4, 320085 Resita, Romania
[2]Politehnica University of Timisoara, Department of Automatic and Applied Sciences, Vasile Parvan str., no. 1-2, 300223 Timisoara, Romania

E-mail: e.raduca@uem.ro

Abstract. This paper presents the design and building of a Web Server to command, control and monitor at a distance lots of industrial or personal equipments and/or sensors. The server works based on a personal software. The software can be written by users and can work with many types of operating system. The authors were realized the Web server based on two platforms, an UC board and a network board. The source code was written in "open source" language Arduino 1.0.5.

1. Introduction

With the spread of the Internet, computer networks have been continuously developed and in this context and network servers aimed mainly communication network clients. Current servers are able to provide solutions and information services increasingly various users (customer network) and the main reason for choosing to study and build a Web server, is to obtain a device that is capable of offering both services and information to command, control and monitoring other devices that require this.

One of the main objectives of the work, is the data communication network, whether local or the Internet, with equipments a [1], [2] and/or sensors [3] or at a distance [4]. This communication will be indirect, because between the equipment to be controlled and user is interleaved Web server based on microcontroller. The protocol used for this server in this project is the HTTP protocol. Language for building web pages is HTML that allows viewing pages with most browsers and thus permits communication with Windows, Linux, Android etc. The server is designed to enable users not having spot devices like PC, Tablet, Smartphone etc. be able to control and monitor remote the server peripherals connected to the devices it manages.

2. Design Server

2.1. Figures Block diagram of the server

The server is accomplished on base of two platforms, one [5] built around ATMEGA 2560 microcontroller [6] and other [7] based W5100 network controller [8]. Network controller interface is provided of W5100 circuit manufactured by WIZnet and command and control network controller will be in the load ATMEGA 2560 microcontroller, product in AVR architecture of Atmel. Interconnecting the two platforms, having regard to the objectives pursued was performed according to Figure 1.

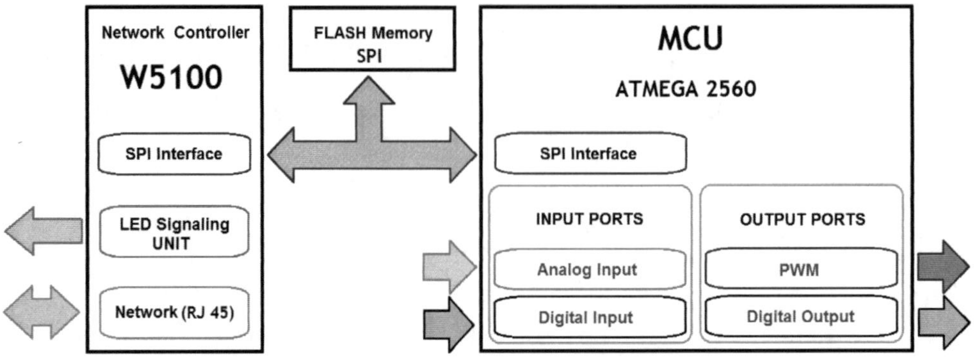

Figure 1. Block diagram of ATMEGA 2560 Web Server

Making server was designed in order to use it as a command entity, control and remote monitoring through networks, of some equipment from industry and personal privat. The user who handles the said equipment has the function of client and, from network platform for own internet, connects from anywhere to the Web server ATMEGA 2560 with cable UTP and RJ45 connector. The Web server, through its communication components with the outside, input and output ports are connected from the UC platform [5], to the interest equipment at existing local networks (Figure 1).

From the beginning it was intended that the server made to provide maximum flexibility, high reliability and a low price.

This led to the choice components used, mentioning specifically that server utilization is independent of the presence of the networks it is connected to a particular operating system, be it Windows, Linux, Mac OS X, Android or another; the server was thought, that it can work with any of these operating systems, which is possible by the design Atmel microcontroller ATMEGA 2560 and UC platform for manufacturers, but also network controller W5100.

2.2. Working principle

The client , connects to the Internet platform: PC, smart device, the Web server in question and hereinafter ATMEGA 2560 Web server, the algorithm is shown in Figure 2.

Based on a request it receives real-time information requested but at the same time can transmit commands terminated by validation or invalidation of input and output ports of the server, thus exerting control functions.

The Web server will connect it to the data network via the ethernet network block that provides a set of SPI command ports, a signalling unit with LED, to display Ethernet communication activity level and physical bidirectional port with RJ45 jack for connecting itself to support copper Internet network.

The network block is managed via command line interface SPI specific, meaning the microcontroller having server function will answer HTTP requests issued by network clients by generating HTML pages, pages that must contain information in real time "collected "from its ports, analog or digital. There are also regularly read logical status of ports or voltage values present at the input port pins corresponding analog or PWM output.

However, the server executes commands generated by the network client, commands that will be included in client request that can be like: state transition logic 1 or 0 digital output ports or the increment or the decrement the voltage of output PWM ports; maximum voltage applied to the microcontroller port is 5V.

3. Design Hardware Server's

3.1. Figures Block diagram of the server

The W5100 circuit [8] is an ethernet controller able to work at speeds of 10/100 Mb / s, designed for embedded applications characterized by ease of integration, stability, performance, low cost. W5100 has been designed to facilitate easy implementation connectivity of Internet, with TCP / IP protocol and Ethernet MAC & PHY without requiring an operating system.

The integrated module,TCP / IP , of the W5100 controller allows a wide range of Ethernet protocols: TCP, UDP), IP4V, ICMP, ARP, IGMP, and PPPOE, which have proved successful in various applications for many years.

The Ethernet controller block diagram [8] is shown in Figure 3.

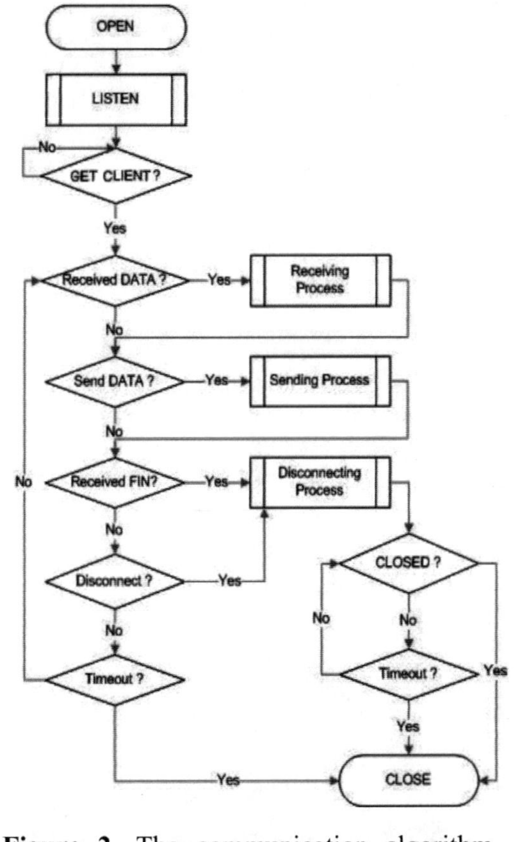

Figure 2. The communication algorithm, Web Server ATMEGA 2560 - user Server

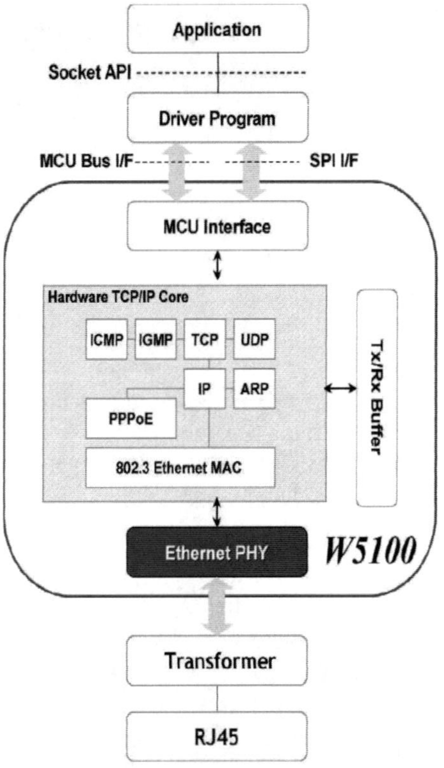

Figure 3. Ethernet Controller Block Diagram

3.2. Setting network parameters

Programming the network parameters [8], are set in the Main through four registers existing in the W5100 network controller. That registers intended network parameters each containing four bytes and the maximum value allowed is 255 byte and such resolution is 32-bit addresses for IPv4 domain.

The GWR register (Gateway IP Address Register), (Figure 4) sets up the default gateway address.
Example in case of "192.168.0.1"

The SUBR register (Subnet Mask Register), (Figure 5). sets up the subnet mask address.
Example in case of "255.255.255.0"

The SHAR register (Source Hardware Address Register), (Figure 6) designed for configuring the Ethernet physical interface MAC type and the default is "00.08.DC.01.02.03".

The SIPR register (Source IP Address Register), (Figure 7) set the most important value, to the ethernet interface, ie IP address, which initially contains a Class C address, public class, with the value: "192.168.0.3".

0x0001	0x0002	0x0003	0x0004
192 (0xC0)	168 (0xA8)	0 (0x00)	1 (0x01)

Figure 4. GWR register

0x0005	0x0006	0x0007	0x0008
255 (0xFF)	255 (0xFF)	255 (0xFF)	0 (0x00)

Figure 5. SUBR register

0x00	0x08	0xDC	0x01	0x02	0x03

Figure 6. SHAR register

0x000F	0x0010	0x0011	0x0012
192 (0xC0)	168 (0xA8)	0 (0x00)	3 (0x03)

Figure 7. SIPR register

For home users who have already created a local network, this class of IP addresses is ideal. A local network can be created by introducing a router to generate the LAN, beside the provider network of Internet, which is usually a network by type WAN.

Network addresses can be set to your choice by adding appropriate values for variables that keeps the field of network parameters.

3.3. Setting network parameters

The microcontroller used [6] (Figure 8) is a next-generation processor suitable for the intended purpose for this project in that:

- contains a UC fast enough, for the range of applications designed;
- has analog and digital ports, bidirectional, totaling 86 programmable line, which can be directly connected equipment that require ADC, PWM, digital communication series or parallel, in the range 0-5 V;
- allows a flexible dialogue, direct, indirect or SPI bus. This last option is used in project, to communicate with the W5100 circuit;
- internal memory is sufficient, 256 kB Flash,4 kB EEPROM, 8kB SRAM which can be supplemented with 64kB external;
- can be programmed in AVR Studio [9] or Arduino IDE 1.0.5 [10], languages "open source", followed widespread range of applications;
- with the W5100 circuit does not require a standardized operating system to communicate with "clients" who use Windows, Linux, Mac OS X, Android, etc.
- the program can be transferred via USB controller on a classic support;

4. Software Server's

The Web server software [11] is based on two programming environments, "HTML" and hybrid "C / C ++ and Arduino".

Environment "open source" Arduino makes it easy writing source code, the programmer by providing a package of standard libraries containing explained and illustrated instruction sets, memory for command and control, communication interfaces, LCD display drivers and so on.

It also provides programming environment compiling C language program and can achieve data transfer program in the microcontroller.

As an integrated development environment, the application Arduino IDE 1.0.5 version can run on platforms like Windows, Mac OS X, Linux and newer Android. He successfully supports ATMEGA 2560 microcontroller.

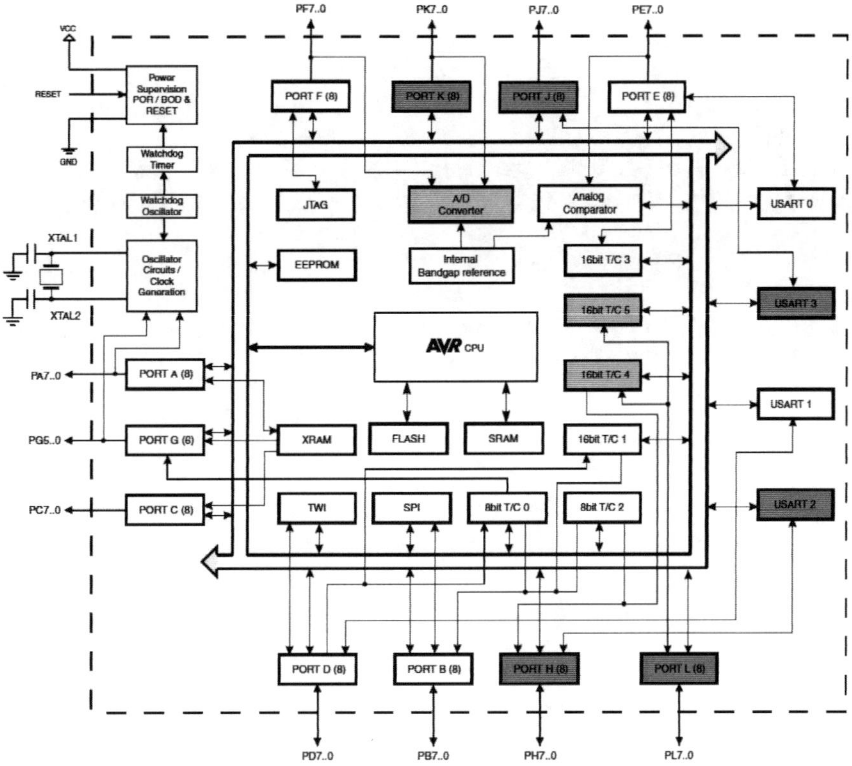

Figure 8. The block diagram of the ATMEGA 2560 microcontroller

4.1. Software interface

Scheme programming interface is built around a ATMEGA 16 microcontroller [12], ATMEGA 2560 microcontroller [6] positioned on the platform [5], which has the task of converting information from the USB port, the data communication interface compatible with the microcontroller (PE0 and PE1 ports) lines TxD RxD and according to the protocol or SPI programming ICSP port.

Program in source code is loaded into the programmer hardware software interface where the compiler is software as machine language and after connecting programming interface. The operation of the microcontroller and load the program, microcontroller accumulates information transmitted from software interface.

4.2. The source code

The program is edited in Arduino programming environment and contains a total of 730 lines of program.

The program can access additional instructions for SPI protocol, Ethernet interface and EPROM memory included in bookstores SPI.h, Ethernet.h and EEPROM.h.

#include <SPI.h> *// SPI command library instruction*
#include <Ethernet.h> *// Ethernet interface library instruction*
#include <EEPROM.h> *// library instruction EPROM*

As soon as the libraries necessary to define a variable of type Buffer for data communications network,

```
// buffer size used to capture HTTP requests
#define REQ_BUF_SZ   15    // maximum capacity of 15 characters
```

Network parameters are defined

```
// NETWORK PARAMETERS
byte ip[] = {192,168,1,110};            // IP address
byte gateway[] = {192,168,1,1};         // Gateway address
byte subnet[] = {255,255,255,0};        // network mask
byte mac[] = {0xDE, 0xAD, 0xBE, 0xEF, 0xFE, 0xED};              //MAC address
EthernetServer server = EthernetServer(80);  // Ethernet port
```

Variables and constants are declared then:

```
// DEFINING VARIABLES
char HTTP_req[REQ_BUF_SZ] = {0};         // reset
char req_index = 0;   // index in the buffer HTTP (req_index) starting from zero
char c = 0;
   ......
```

Follow boot:

```
// BOOT
void setup()
{
  Ethernet.begin(mac, ip, gateway, subnet);  //-boot ehternet interface
  server.begin();        // start "clients detect"

  for (int var=0; var < 8; var++){      //- reset Boolean area
    stareIesire[var] = false;
  }
  // loop declaration 8 output ports
  for (int var = 0; var < 8; var++){    //- setting selected as output ports
    pinMode(outDig[var],OUTPUT);      // output mode
    digitalWrite(outDig[var],LOW);      //-reset ports
  }
  // loop declaration of 8 input ports
  for (int var = 0; var < 8; var++){    //- setting selected as input ports
    pinMode(inDig[var],INPUT);        // input mode
  }
}
```

and loop program:

```
// LOOP PROGRAM
void loop()
{
  EthernetClient client = server.available(); // Check for cient!
if (client) {                                          // get client?
     boolean currentLineIsBlank = true;
     while (client.connected()) {
      if (client.available()) { // data can be read from the client?
      char c = client.read();  // read one byte (character) of the client
   .....
```

In this section are essential *// Interpretation subroutines on client requests to access HTML pages as they are received from the client* interprets commands: *"spaţiu", "main" , "log" , "ret" , "por" , "sec" , // Interpretation and requests from the client as the port command is received from the client interprets commands type "outHk", "outLk" , k= 0...7*

Difficulty of the program is to process the information received from the client combined with information "collected" from the microcontroller peripherals.

Also, gaining control over the Internet server requires combining the two programming languages: Arduino (based on C / C ++) and HTML.

4.3. Communication between server and user

The communication between the server and the user is based on the establishment of codes by which to transmit text and execute commands (Table 1) for both control and ports for accessing pages provided by the server.

These codes will be assigned to the buttons on the main page and in the interaction with the user, issued codes will be attached in HTTP requests required the client to the server.

Table 1. Codes for the function of ATMEGA 2560 WEB Server

COD	FUNCTION	COD	FUNCTION	COD	FUNCTION
outH0	P.38 = HIGH	pwmH0	Inc. PWM.2	main	Main Page
outL0	P.38 = LOW	pwmL0	Dec. PWM.2		
outH1	P.39 = HIGH	pwmH1	Inc. PWM.3	log	Login Page
outL1	P.39 = LOW	pwmL1	Dec. PWM.3		
outH2	P.40 = HIGH	pwmH2	Inc. PWM.4	ret	Network Page
outL2	P.40 = LOW	pwmL2	Dec. PWM.4		
outH3	P.41 = HIGH	pwmH3	Inc. PWM.5	por	Ports Page
outL3	P.41 = LOW	pwmL3	Dec. PWM.5		
outH4	P.42 = HIGH	pwmH4	Inc. PWM.6	sec	Security Page
outL4	P.42 = LOW	pwmL4	Dec. PWM.6		
outH5	P.43 = HIGH	pwmH5	Inc. PWM.7		
outL5	P.43 = LOW	pwmL5	Dec. PWM.7		
outH6	P.44 = HIGH	pwmH6	Inc. PWM.8		
outL6	P.44 = LOW	pwmL6	Dec. PWM.8		
outH7	P.45 = HIGH	pwmH7	Inc. PWM.9		
outL7	P.45 = LOW	pwmL7	Dec. PWM.9		

The server will receive the characters in demand, will decode the codes and execute actions assigned to each code. To avoid any reading error codes in the program were not used repetition loops.

5. Implementation. Experiments

5.1. Logging into the server

The server identify potential client and server responds to requests initiated by the firm.

The first request initiated by the client will be connecting to the server. In response generates a web server, the client receives page to view and whose graphical form, corresponding to a standard login page (Figure 9) provided with boxes for placing the keyboard data Login: Username and password, plus a button to validate the data.

Managing data entered from the keyboard, not driven directly from the microcontroller server, which makes the web page content will be inserted login scrip JavaScript (Figure 10) with the task of verifying the data account, entered with the saved microcontroller.

The script shown in Figure 1, the program starts at line no. 4 and ends at line 24. In addition to the role of "guardian", this script is designed to generate alarms JavaScript in the login page by signalling with text messages (displayed in independent windows superimposed login page) login errors, errors that can be generated: account validation request without keyboard input data, account validation application without entering the password, or account validation request by entering false data (Figure 11).

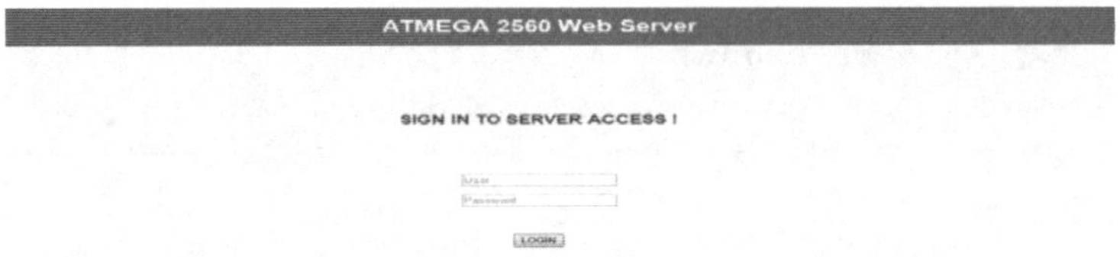

Figure 9. Home Login

http://192.168.1.110/log - Original Source

File Edit Format

```
1 <!DOCTYPE HTML>
2 <html>
3 <head>
4 <script type="text/javascript">
5 function verifica() {
6 var username=document.getElementById("username");
7 var pass=document.getElementById("pass");
8 var isValid=true;
9 var utilizator = "Luci";
10 var parola = "1234";
11 if (username.value=="") {
12 alert("Introduceti un nume de utilizator!");
13 isValid=false;}
14 else if(pass.value=="") {
15 alert("Va rugam sa introduceti parola!");
16 isValid=false;}
17 else if(username.value!=(utilizator)) {
18 alert("Numele de utilizator sau parola sunt incorecte!");
19 isValid=false;}
20 else if(pass.value!=(parola)) {
21 alert("Numele de utilizator sau parola sunt incorecte!");
22 isValid=false;}
23 return isValid;}
24 </script>
```

Figure 10. The Login page script

Figure 11. JavaScript Alert, false data logging

5.2. Main page

Home or "main page", provides the user with an interactive panel with a discount rate which is currently set at 10s. The value of the discount rate can be changed, of course.

It appears that the current application was designed for command, control and monitoring of 32 independent programmable lines, belonging to four 8-bit ports. Two ports are used for inputs or digital outputs; status of each line, 0 or 1 is indicated on the main page via a virtual LED assigned to each line. For output the settings for each line are buttons ON (logical 1), OFF (logical 0). Each of the 8 port analog input lines can be brought voltages in the range 0-5 volts through a successive approximation ADC with internal 10-bit ATMEGA 2560's are converted into digital signals by multiplexing. The eight analog output lines of the port are used to generate PWM signals ordering the idea of a three-phase static converter. Each output can be generated pulse train of amplitude U = 5V with variable duty cycle can be changed from the "+", "-".

The left panel shows a black band (Figure 12), which gives the user full rights auxiliary opening pages dedicated to amend additional names of ports, network configuration, modify data logging and the bottom of the strip is present output path or logout action.

Front panel made physically ATMEGA 2560 Server is shown in Figure 13.

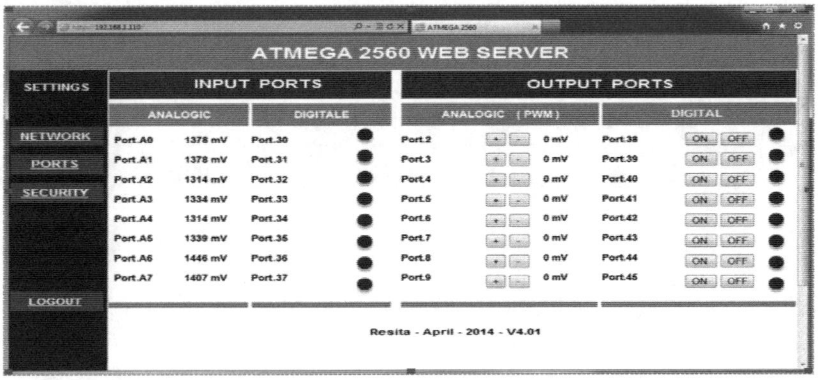

Figure 12. Main page of ATMEGA 2560 Web server

Figure 13. Front panel, ATMEGA2560 Server without LEDs

On the left side you can see the LEDs for communication control. In the central area is observed RJ 45 connector for interconnecting server - client via Internet and left 4 such connectors for interconnecting LAN server with the equipments/sensors.

Conclusions

The Web server with microcontroller ATMEGA 2560 has been designed for distance communication between manufacturing equipment's and/or sensors and users. It is part of the category of modern command, control and monitoring at distance entities and of different equipment and/ or sensors. The communication between equipment/ sensors on the one hand, and the user, on the other hand, is done by means of already classic networks computer: the server Web-user communication is connected via Internet, the server equipment/sensors communication is connected by LAN type networks.

The techniques used for realizing the server have covered a large spectrum of areas such as data transmission by networks communications, implementing interactive Web pages, programming and using items designed with programmable circuits, and generally also microcontrollers, especially in editing their programs, electric and mechanic compatibility of the server elements with the other physical components, using virtual systems.

The server permits communicating with other network components that they are part of, regardless of the operating system: Windows, Android, Mac X, etc. The source code can be written in open source Arduino or Studio.

For carrying out the proposed objectives, the authors have physically created the server using two platforms, an UC board and a network board and they have written the source code in open source language Arduino IDE 1.0.5., which at the moment contains 730 program lines.

The carried out tests, have shown that the program created for the server is correct, efficient and precise assuring the data acquisition, their processing and supplying calculation results on the graphic interface represented by web pages generated in HTML language.

Upon experimenting one has noticed the using the Firefox browser is optimal, lacking the unwanted effect of loading a page with a horizontal sweeping effect.

The carried out experiments have shown that the server is viable and has a stable functioning. The server is part of the "Low Cost" category, its hard components being very cheap, and the software dedicated to the used components is "free".

The signals generated and acquired by the server are signals working at the level of own ports with reduced values of tension and power, this is why it is useful to extend the signal levels, so as to be compatible with equipment's/sensors, fact finalized in the future.

One intends a development of the server [13]. Another direction is the security of our systems of the information transmitted through the network, in parallel with the optimizing of its functioning.

References

[1] Stinean A I, Preitl S, Precup R E, Dragos C A, Petriu E M and Radac M B 2013 *DOF control solutions for an electricdrive system under continuously variable conditions*, 8th IEEE International Symposium on Applied Computational Intelligence and Informatics (SACI), Timisoara, Romania, May 23-25, pp 115-120

[2] Morosabn A D and Sisak F 2013 *An intelligent system designed for controlling the manufacturing process in a flexible manufacturing system*, 8th IEEE International Symposium on Applied Computational Intelligence and Informatics (SACI), Timisoara, Romania, May 23-25, pp 353-357

[3] Vujovic V, Maksimovic M, Perisic B and Milosevic V 2014 *A Grafical Editor for RESTful Sensor Web Networks Modeling*, 9th IEEE International Symposium on Applied Computational Intelligence and Informatics (SACI), Timisoara, Romania, May 15-17, pp 61-66

[4] Chioncel C P, Gillich N, Chioncel P and Gillich G R 2007 *CMS solutions in monitoring an real-time data transfer of photovoltaic plants,* XV-th International Symposium on Electrical Apparatus and Technologies SIELA 2007, Plovdiv, Bulgaria, May 31-June 1, pp 14– 17

[5] ***http://arduino.cc/en/Main/arduinoBoardMega2560

[6] ***http://www.atmel.com/images/doc2549.pdf

[7] ***http://www.dx.com/p/ethernet-w5100-shield-network-expansion-board-w-micro-sd-card-slot-for-arduino-152203#.VNLd7i4jWAQ

[8] ***http://www.wiznet.co.kr/Sub_Modules/en/product/Product_Detail.asp?cate1=5&cate2=7&cate3=26&pid=1011

[9] ***http://www.atmel.com/microsite/atmel_studio6/

[10] ***https://code.google.com/p/arduino/downloads/detail?name=arduino-1.0.5-windows.exe&can=2 , 2014B

[11] Berdie A D, Osaci M, Prostean G and Cristea A D 2011 *Web Programming features on integrated system SAP*, 6th IEEE International Symposium on Applied Computational Intelligence and Informatics (SACI), Timisoara, Romania, May 19-21, pp 227-230

[12] ***http://www.atmel.com/images/ doc2466.pdf

[13] Sebu M L and Ciocarlie H 2014 *Applied process mining in software developement, Case study*, 9th IEEE International Symposium on Applied Computational Intelligence and Informatics, (SACI), Timisoara, Romania, May 15-17, pp 55-60

Metal rolling – Asymmetrical rolling process

V Alexa[1], S Rațiu[1] and I Kiss[1]

[1]Politehnica University of Timisoara, Department of Engineering and Management, 5 Revolution Street, 331128 Hunedoara, Romania

E-mail: vasile.alexa@fih.upt.ro

Abstract. The development of theory and practice related to the asymmetric longitudinal rolling process is based on the general theory of metalworking by pressure and symmetric rolling theory, to which a large number of scientists brought their contribution. The rolling of metal materials was a serious problem throughout history, either economically or technically, because the plating technologies enabled the consumption of raw materials (scarce and expensive) to be reduced, while improving the mechanical properties. Knowing the force parameters related to asymmetric rolling leads to the optimization of energy and raw material consumption. This paper presents data on symmetric rolling process, in order to comparatively highlight the particularities of the asymmetric process.

1. Introduction

The modern rolling theory itself is an attempt to provide a scientific method for calculating the pressure developed by the metal materials on the rolls. It does not introduce any empirical coefficients, and, for performing the calculations for a given passing, in addition to the dimensions of the semi-finished product and rolls, we need two experimentally determined values, which incorporate a deep physical sense [1-3]:

- value of the coefficient of external friction;
- value of the yield point of the metal material, for the passage in question.

Besides these simplifying hypotheses that formed the basis of the classical theory, the following assumptions can be accepted in the modern theory:

- the influence of the areas located in the vicinity of the deformation outbreak is neglected, which makes the theory to be valid only for the cases where $\frac{l}{h_m} > 1$;
- the clamping angles are assumed to be not too large, but, seeing the calculations, it seems that this condition does not introduce great errors even in case of hot rolling, when the clamping angles can reach 22-24%;
- we adopt the assumption that the yield point σ_c is constant along the arc of contact; this hypothesis means that we do not take into account, at cold rolling, the influence of cold hardening during the passing of material and, at hot rolling, the influence of speed and deformation degree, i.e. we adopt the assumption that the yield point does not depend on the compression speed and the relative reduction per pass, which actually are not changing along the arc of contact following a complicated dependence;
- we introduce mathematical simplifications.

Usually, the content issues of the modern theory are given in a complicated mathematical form, which is sometimes hard to understand, although the key points are pretty simple. Therefore, the experimental researches on pressure distribution are important either for the theory or for the practice of rolling processes.

They can serve directly for the experimental confirmation of the correctness of theoretical conclusions regarding the general law of pressure distribution along the arc of contact, or for the influence on pressure of the main rolling factors, i.e. the coefficient of external friction, initial thickness of the metal material, degree of reduction, and temperature.

Depending on the assumptions mentioned above, in the modern rolling theory appeared several solutions for determining the rolling pressure. We are going to analyse some of them, which are more or less consistent with the experimental data or with those recorded in the operational practice.

Next, we present the main theories approached in the symmetric and asymmetric rolling theory, i.e. the mathematical apparatus they are based on:

1.1. Karman's theory

Hereinafter, the following simplifications are allowed [1]:

- the vertical sections of the metal material are analysed as they are separated from each other, without any sliding friction between them, and the displacement strains in these edges are neglected;
- the vertical speed of the metal material particles is constant, so a uniform pressure distribution on the two vertical sides of the elementary section is accepted.

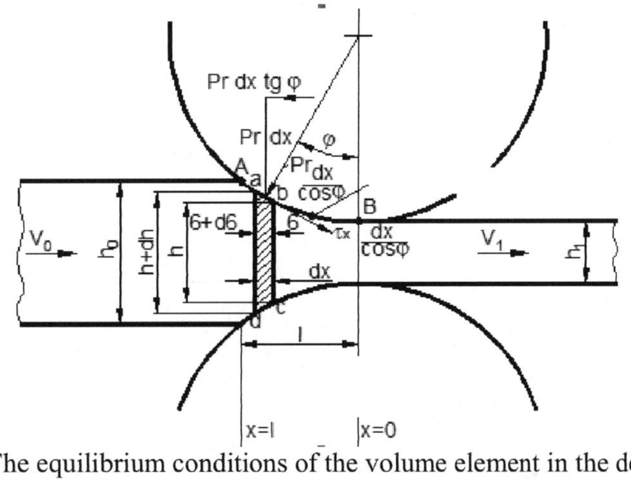

Figure 1. The equilibrium conditions of the volume element in the deformation area

For the delay area, we obtain:

$$p_x = \frac{k}{\delta}\left[(\zeta_0\delta - 1)\left(\frac{h_0}{h}\right)^{\delta} + 1\right] \tag{1}$$

and for the advance area:

$$p_x = \frac{k}{\delta}\left[(\zeta_1\delta + 1)\left(\frac{h}{h_1}\right)^{\delta} - 1\right] \tag{2}$$

where: $\delta = \dfrac{2 \cdot \mu \cdot l}{\Delta h}$;

l - length of the contact arc, equal to $\sqrt{R \cdot \Delta h}$

1.2. Analysis of Siebel's solution

The solution proposed by Siebel to solve more conveniently the Karman's equation accepts two assumptions [1]:

- the frictional force over the length of the contact arc $\tau_x = const = \mu \cdot k$;
- the deformed contact arc has a parabola shape.

In this case, the solutions for solving the equation for the delay area are:

$$p_x = k\left[\zeta_0 - log\frac{z_0^2+1}{z^2+1}\right] + 2\tau_x\sqrt{\frac{R}{h_0}}\,arctg\frac{z_0 - z}{1+z_0 z} \quad (3)$$

and for the advance area:

$$p_x = k\left[\zeta_1 + log\left(z^2+1\right)\right] + 2\tau_x\sqrt{\frac{R}{h_1}}\,arctgz \quad (4)$$

where: $z = \frac{1}{l}\sqrt{\frac{\Delta h}{h_1}} \cdot x;\quad z_0 = \sqrt{\frac{\Delta h}{h_1}}$

1.3. Analysis of Nadai's solution

The solution proposed by Nadai is also based on two assumptions:

- acknowledging the existence of viscous friction, $\tau_x = \eta \cdot \frac{dv}{dh}$, where η represents the dynamic viscosity;
- the deformed contact arc has a parabola shape.

The solution of this equation proposed by Nadai is:

$$p_x = k\left[\zeta_1 + log\left(z^2+1\right)\right] + \frac{A\cdot\tau_x}{2}\left[\frac{z}{z^2+1} - B\cdot arctgz\right] \quad (5)$$

where: $A = \frac{2\cdot l}{\sqrt{h_1 \cdot \Delta h}}$, $B = \frac{1-z_n^2}{1+z_n^2} = \frac{\frac{2\cdot k}{A\cdot\tau_x}\cdot\left[\zeta_1 - \zeta_0 + log\left(1+z_0^2\right)\right] + \frac{z_0}{1+z_0^2}}{arctgz_0}$

and $z = \frac{1}{l}\sqrt{\frac{\Delta h}{h_1}} \cdot x;\quad z_0 = \sqrt{\frac{\Delta h}{h_1}}$

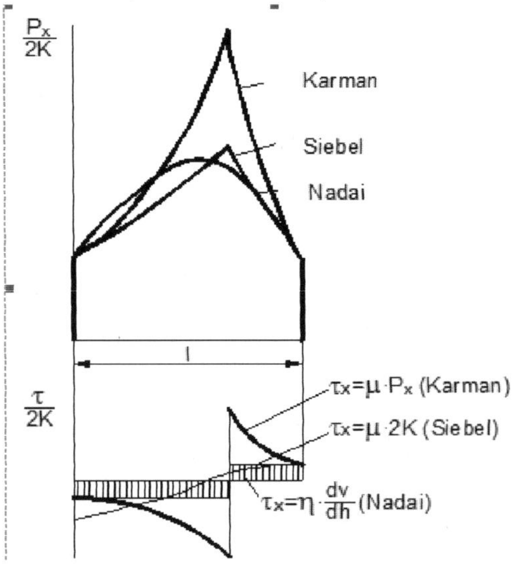

Figure 2. Graphical representation of solutions analyzed

As can be seen in Figure 2, the maximum values of the ratio $\frac{p_x}{2k}$ differ widely from one solution to another.

The attempt to apply one of these solutions to hot rolling leads to considerable errors, mainly because the effect of temperature on the physical properties of the metal material and on the friction coefficient μ is not taken into account.

Secondly, the vertical speed of the metal material particles undergoing deformation in the rolling direction cannot be considered constant, this assumption being not be justifiable. In other words, the plane section assumption cannot be maintained.

From the comparative analysis of the calculation solutions, we can draw the following conclusions:

- the normal pressure distribution over the contact arc length is depending on the hypothesis considered for the distribution of frictional forces between the metal material and rolls;
- all cases result in the presence of portions with sliding friction; therefore, the pressure distribution diagram highlights the existence of delay and advance areas. By using the Nadai's equation, the separation by neutral plan of these areas is not highlighted;
- the analysed solutions are mostly applicable for cold strip rolling with contact angles less than 6…8°.

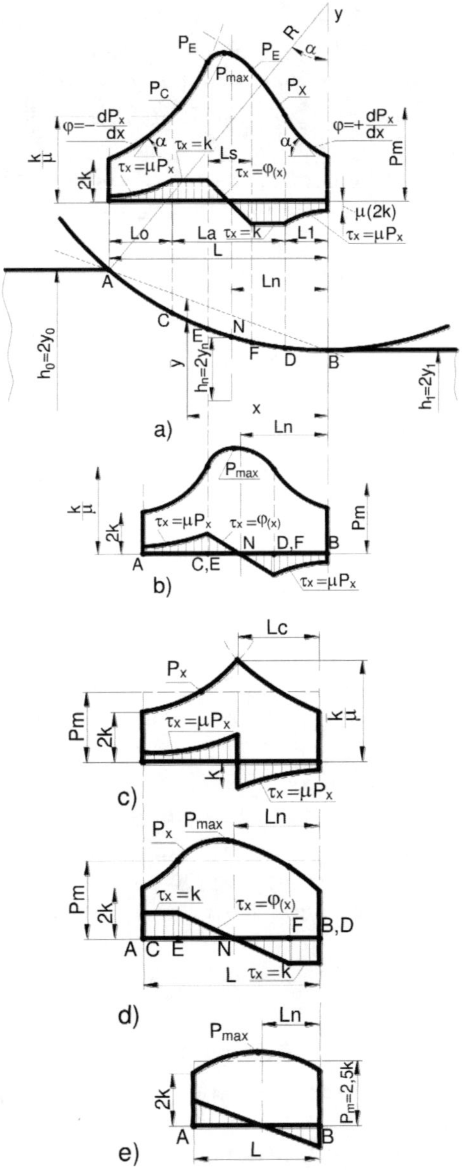

Figure 3. The diagrams of the normal pressure distribution over the contact arc length

The theoretical and experimental research conducted, [4-7] have broadened the knowledge about the phenomena occurring in the rolling deformation area.

In general, when rolling relatively flat strips, $\left(\frac{l}{h_m} > 5\right)$, the geometric deformation area consists of two portions:

- the portion of metal sliding on the rolls, where the contact friction is subjected to Coulomb's law, $\tau_k = \mu \cdot p_x$; there are actually two portions included here, which are located in the immediate proximity of the geometric deformation area, i.e. at the metal material entry / exit between / from the rolls;
- the adherence area (where the material slides on the roll surface) is missing, i.e. the metal surface layer moves with a tangential speed equal to the peripheral speed of the rolls, as though the metal material particles stick to the surface of the rolls.

So, the general relation for calculation will be:

$$p_m = 2k\left\{\frac{1-2\mu}{2\mu^2}\cdot\frac{h_m}{l} + \left[\frac{l}{2\mu} + \frac{l}{4h_l} - \eta\left(1 + \frac{h_m}{l}\left(\frac{1}{\mu} - \eta\right)\right)\right]\right\} \quad (6)$$

In this relation, the first term characterizes the composition of total average pressure determined by the sliding areas, and the second term – the adhesion area component.

2. The experimental plant and testing method

The research for this theme purpose have been made on a 170 mm reversing two-high rolling mill, created and installed in the no conventional technologies and plastic deformation laboratory of the Engineering Faculty from Hunedoara [4].

An experimental installation formed of: special construction rollers, bearings, punctiform captors for lamination pressure, lamination forces captors and lateral pressure captors it was created for research in condition of technological similitude symmetrical and asymmetrical process.

Figure 4. Montages punctifom captors for recording the pressure of contact

In Figure 1 it is presented in overview the mentioned installation, with the way force captors are assembled in order to determine the lateral efforts in the longitudinal asymmetrical lamination but also to show the author's contribution regarding method of experimentation [1].

The bearing holders of the inferior roller were modified for recording the lateral efforts so that the respective captors could be installed incorporated perpendicularity on the bearing' s axis.

On the surface of captors were stuck tensometric stamps bound in deck, stamps that modify their dimension under the action of the effort to be measured. These dimensional modifications of tensiometer stamps are generating variations of their electric resistance, that are proportional to the deformation efforts and the measuring of the forces is limited to the measuring of these resistance variations.

The research conducted on symmetric and asymmetric rolling process aimed the recording, along with other process parameters, of the rolling pressures and actual contact lengths of the arcs with the upper and lower roll.

This enabled us to know the distribution modality of the average pressure compared to the actual length of the contact arc, and to assess the influence of the distorted shape of the deformation area on the key process parameters.

The symmetric and asymmetric rolling process was carried out by equipping the work rolls with segments adjusted with various radii, which have led to the following ratios between the diameters of the upper (D_s) and lower (D_i) rolls:

$$\frac{D_s}{D_i} = \frac{170}{170}; \frac{160}{180}; \frac{150}{190}; \frac{140}{200} [mm]$$

The modality to mount the segments on the rolls (of the rolling mill) for obtaining these combinations is shown in Figure 5 and 6, and schematically, for the symmetric and asymmetric process, the working variant is given in the Figures 5 and 6.

Usually, the tensiometer stamps of a forces captor are bound in deck. The deck has on a diagonal it is measured a electrical signal – proportional of the applied effort – and for recording of the measured values this signal is recorded by an oscillograph.

The oscillograph is a the type N-700, having 14 channels, the impulses on recorded are a scale of 120 mm width, heaving 4 cm/s moving speed of the paper band.

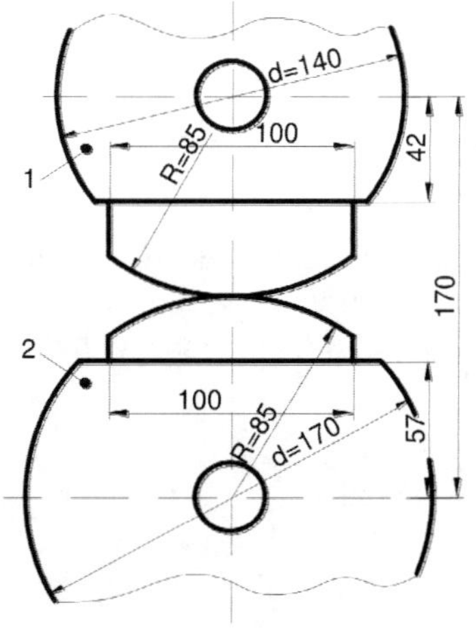

Figure 5. Research variant of the contact pressure on the symmetric longitudinal rolling

$$\frac{D_s}{D_i} = \frac{170}{170} [mm]$$

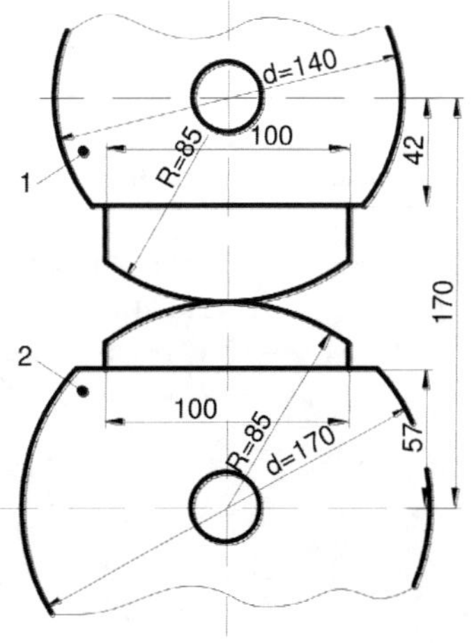

Figure 6. Research variant of the contact pressure on the asymmetric longitudinal rolling

$$\frac{D_s}{D_i} = \frac{140}{170} [mm]$$

1-upper roller
2-lower roller

3. The analysis of experimental data related to the rolling pressure in the symmetric and asymmetric processes

Since the intended purpose was to study the qualitative aspect of phenomena related to the asymmetric longitudinal rolling, for eliminating the inevitable influence of the iron oxides (scale) on the process, the tests were carried out on samples of aluminium and copper, with the following dimensions:

$$h_0 = 12; 6; 2 \; and \; 1 \; mm$$

$$b_0 = 40 \, mm \quad l_0 = 150 \, mm$$

The mechanical properties of the materials used to carry out the experimental tests are given in Figure 7.

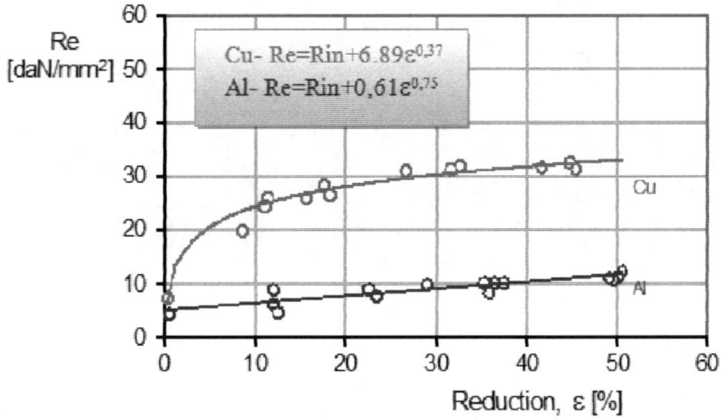

Figure 7. The mechanical properties of the materials used to carry out the experimental tests

The samples of aluminium and copper were cut from the same rolled strip.

To ensure the isotropy of the properties, before rolling, the aluminium samples were subjected to annealing for recrystallization at the temperature of 420^{0}C, and the copper samples at 750^{0}C, for two hours, and then carefully cleaned with a fine emery paper. Before rolling, each sample was washed with chemically clean acetone [1].

The segments of the rolls were also washed with acetone, before each sample was individually rolled.

In this paper, by analyses and calculations, we are going to use the experimentally obtained values for pressure, and the actual contact lengths of the arcs with the upper and lower roll.

In the first part of researches, we carried out the rolling by using rolls with equal diameters

$$\frac{D_s}{D_i} = \frac{170}{170} [mm]$$

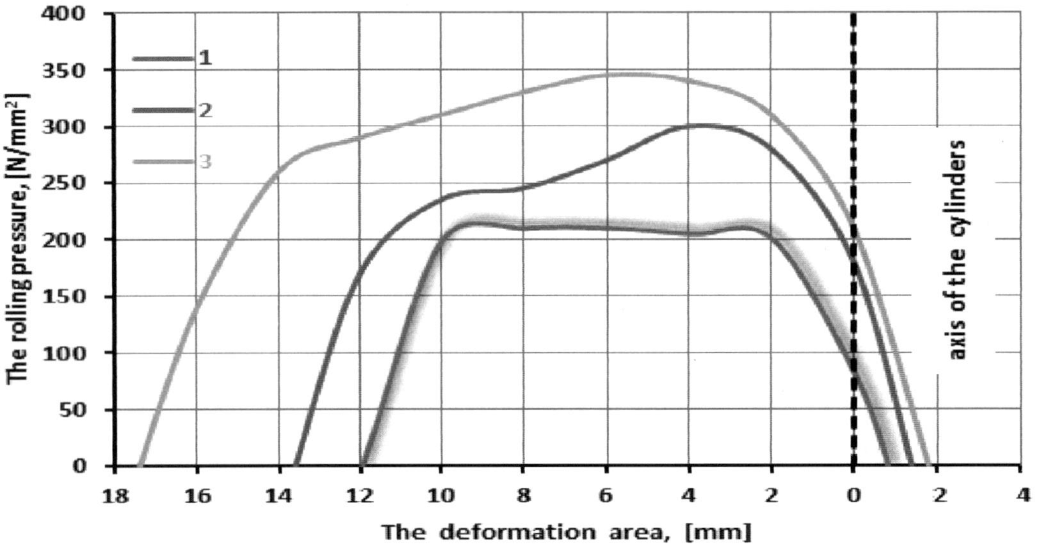

Figure 8. Variation of the pressure in the rolling contact arc length of copper, $h_0 = 12\,mm$ samples between same diameter cylinders $\frac{D_s}{D_i} = \frac{170}{170} [mm]$, applying various reductions

1-oscillogram number 35, ε=9,15[%]
2-oscillogram number 37, ε=13,75[%]
3-oscillogram number 39, ε=22,1[%]

Through a careful execution and adjustment of the transducers, as well as choosing the right power supply for the measuring strain gauge bridges, we succeeded to obtain at calibration absolutely identical features of the upper and lower transducers.

The asymmetric process, related to the roll diameters, was performed by changing the various- radii segments mounted in the lower and upper roll.

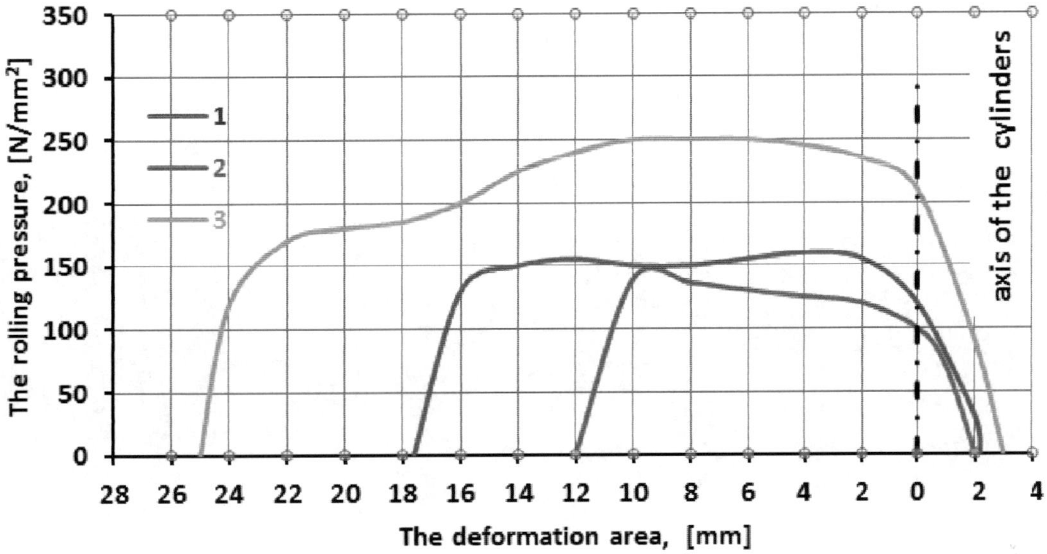

Figure 9. Variation of the pressure in the rolling contact arc length of aluminum, $h_0 = 12\,mm$ samples between same diameter cylinders $\frac{D_s}{D_i} = \frac{170}{170}[mm]$, applying various reductions

1-oscillogram number 6, ε=13,9[%]
2-oscillogram number 2, ε=32,5[%]
3-oscillogram number 4, ε=49,5[%]

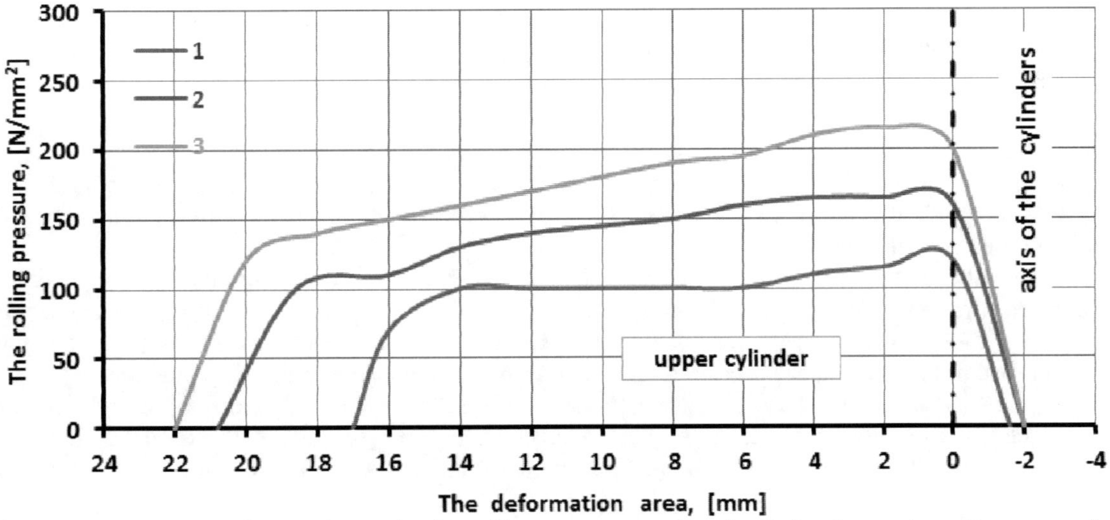

Figure 10. Variation of the pressure in the rolling contact arc length of aluminum, $h_0 = 12\,mm$ samples between same diameter cylinders $\frac{D_s}{D_i} = \frac{140}{170}[mm]$, applying various reductions, upper cylinder

1-oscillogram number 362, ε=19,5[%]
2-oscillogram number 353, ε=38,9[%]
3-oscillogram number 347, ε=45,8[%]

IOP Conf. Series: Materials Science and Engineering **106** (2016) 012019 doi:10.1088/1757-899X/106/1/012019

Thus, instead of the segments made with radii corresponding to the diameters of 170/170 mm, which were used for the symmetric process, we mounted the segments afferent to the ratios:

$$\frac{D_s}{D_i} = \frac{160}{180}; \frac{140}{200} [\,mm\,]$$

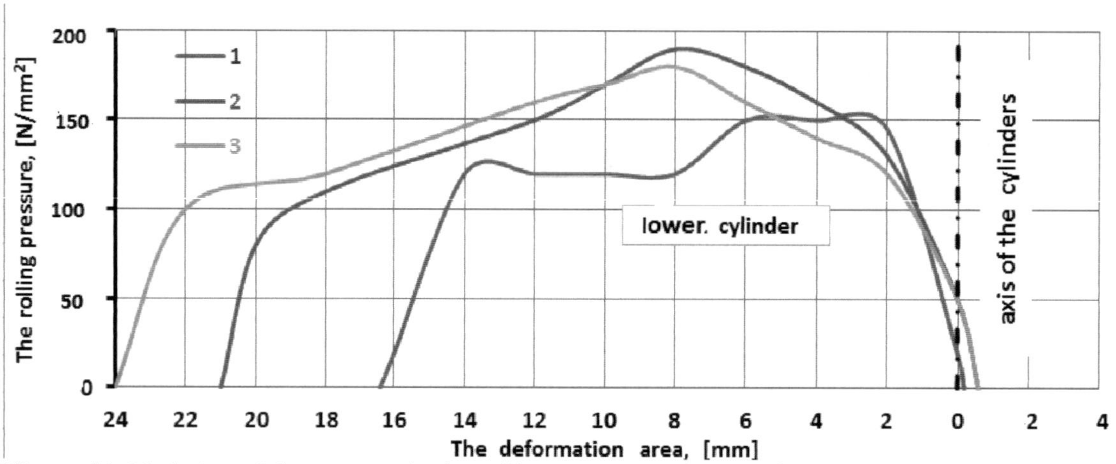

Figure 11. Variation of the pressure in the rolling contact arc length of aluminum, $h_0 = 12\,mm$ samples between same diameter cylinders $\frac{D_s}{D_i} = \frac{140}{170} [mm]$, applying various reductions, lower cylinder

1-oscillogram number 362, ε=19,5[%]
2-oscillogram number 353, ε=38,9[%]
3-oscillogram number 347, ε=45,8[%]

The data for the interim variant (160/180 [mm]) is not presented in this paper, because they just fully confirm the analysis made for the variant 140/200 [mm].

4. Results and conclusions

Through the conducted experiments, we obtained the laws of changing the average pressures, as well as the length of the contact arcs in the asymmetric process, which enabled us to eliminate the contradictions from the considered problem.

We have also clarified, by this research, the laws of qualitative transition from the symmetric to the asymmetric process, depending on the change in rolling conditions, thus confirming the Romanian researchers' idea, in particular the idea of professor Ilca, who stated that the symmetric process is just a special case of the asymmetric process which is widespread in the rolling industry practice.

- by using the experimental plant, we studied the dependency on reduction of the pressure and the actual lengths of contact arcs, at asymmetric rolling between rolls with unequal diameters.
- we found the dependency between the average pressure variation modality and the actual lengths of the contact arcs at the asymmetric longitudinal rolling.
- the experimental plant and equipment will continue to be used either for research, to find correlations between the technological and force parameters of the rolling process, or for operation, to find the strain degree of the technological equipment.
- we have also established that, depending on the thickness of the rolled strips and reduction, for a given ratio between the diameters of the work rolls, at asymmetric rolling, there is a critical value of the deformation degree up to which the pressure exerted by the roll with the larger diameter is high, and for values higher than this critical value the situation is reversed. The contact arc lengths variation is reverse in nature, i.e. the small lengths correspond to the maximum pressure values, respecting the equality of forces on both rolls.

IOP Conf. Series: Materials Science and Engineering **106** (2016) 012019 doi:10.1088/1757-899X/106/1/012019

- at the critical value of the deformation degree, for rolls with unequal diameters, the pressures become equal, as well as the lengths of the contact arcs. Therefore, the current opinions regarding the distribution of reductions between the rolls at asymmetric rolling are correct only for particular cases of rolling conditions.
- the locus of points corresponding to the value of critical reductions is a parabola that characterises the conditions of ideal symmetry (rolling pressure, length of the contact arcs) at rolling between rolls with unequal diameters.

References

[1] Alexa V 2002 *Contributions to research asymmetric rolling process*, Politehnica University of Timişoara, Romania, Doctoral Thesis

[2] Alexa V, Ratiu S and Kiss I 2005 Theoretical and experimental analysis of the rolling strain in the deformation area, *Metalurgia International* **5** 20-28

[3] Azbanbayev E, Isagulov A and Ashkeyev Z 2013 *Mathematical simulation of the process of rolling in the back taper rolls*, Metal 2013, Brno, Czech Republic, May 15-17

[4] Alexa V, Ratiu S and Kiss I 2003 Calculation program for the determination of force parameters in rolling, *Annals of the Faculty of Engineering–Hunedoara* **3** 149-154

[5] Alexa V 2004 Analysis of the lateral efforts in asymmetrical longitudinal rolling, *Annals of the Faculty of Engineering–Hunedoara* **2** 45-50

[6] Gudur P P, Salunkhe M A and Dixit U S 2008 A theoretical study on the application of asymmetric rolling for the estimation of friction,*International Journal of Mechanical Sciences* **50** 315-327

[7] Halloumi A, Desrayaud Ch, Bacroix B, Rauch E and Montheillet F 2012 A simple analytical model of asymmetric rolling, *Archives of metallurgy and materials* **57**(2) 425-425

Methods for recovering precious metals from industrial waste

L Canda[1], T Heput[1] and E Ardelean[1]

[1]Politehnica University of Timisoara, Department of Engineering and Management, 5 Revolution Street, 331128 Hunedoara, Romania

E-mail: letitia.canda@fih.upt.ro

Abstract. The accelerated rate of industrialization increases the demand for precious metals, while high quality natural resources are diminished quantitatively, with significant operating costs. Precious metals recovery can be successfully made from waste, considered to be secondary sources of raw material. In recent years, concerns and interest of researchers for more increasing efficient methods to recover these metals, taking into account the more severe environmental protection legislation.

Precious metals are used in a wide range of applications, both in electronic and communications equipment, spacecraft and jet aircraft engines and for mobile phones or catalytic converters. The most commonly recovered precious metals are: gold from jewellery and electronics, silver from X- ray films and photographic emulsions, industrial applications (catalysts, batteries, glass/mirrors), jewellery; platinum group metals from catalytic converters, catalysts for the refining of crude oil, industrial catalysts, nitric acid manufacturing plant, the carbon-based catalyst, e-waste.

An important aspect is the economic viability of recycling processes related to complex waste flows. Hydrometallurgical and pyrometallurgical routes are the most important ways of processing electrical and electronic equipment waste. The necessity of recovering precious metals has opened new opportunities for future research.

1. Introduction

Exponential growth of world's population and limited natural resources are causing new challenges. Pollution prevention, waste minimization, recycling and products at end of life (ELP) have become topics of interest. Sustainable development involves creating materials and products that can be more easily recycled through proper management. In terms of sustainability, recycling activities mean maximizing re-use, minimizing energy consumption, reducing costs and waste.

Increased pace of industrialization increases the demand for precious metals, while the high quality natural resources are disappearing (reducing the amount) with high operating costs. Alternative solutions lead to a need of increasing education regarding recovery of these metals. As a consequence of industrialization, solid and liquid waste generated from various technological processes or various consumer goods is increasingly high. These wastes may be organic or inorganic. Organic substances are, in general, biodegradable and environmentally friendly. Inorganic wastes contain toxic metallic and non-metallic elements with many serious negative consequences on ecosystems. Some of these non-ferrous metals have a mid deteriorating grade that lasts for a long time, having a cumulative toxic effect. Also, many of them are carcinogenic. The most significant solid waste materials contain non-ferrous metals such as gold, silver, nickel, molybdenum, copper, zinc, chromium and others, as potential heavy metals. Therefore, these residues would be considered secondary resources or artificial resources. Precious metals mainly originate from electric and electronic equipment, medical activities, spent catalysts, used batteries and ashes.

Non-ferrous metals, including aluminum, copper, lead, nickel, tin, zinc and others, are among the few materials that do not degrade and do not lose their chemical or physical properties during the recycling process. Consequently, nonferrous metals, and therefore precious metals, have the capacity to be recycled for an infinite number of times.

2. Overall view on the study

Currently, precious metals are used in a wide range of applications, not only for electronic and communications equipment, spacecraft and jet aircraft engines, but also for mobile phones or catalytic converters. The most commonly recovered precious metals are: gold from jewellery and electronics; silver from electronics, radiography, films and photographic emulsions, industrial applications (catalysts, batteries, glass/ mirror), jewellery; platinum group metals from catalytic converters, catalysts for refining crude oil, industrial catalysts, nitric acid manufacturing plants, carbon-based catalysts, e-waste. Technological progress must be seen in relationship with the protection of human health, ecosystems, having clean technologies at both levels: production and recycling, improvements of environmental legislation and improved methods in waste recycling as a source of secondary raw materials, avoiding emissions and contamination of water, air and soil. Another important aspect is the economic viability of recycling processes related to complex waste streams. In some situations the mixture of metals or the amount too small of these metals in mixtures makes them difficult to separate from waste. New technologies, such as spectroscopic analysis, facilitate and accelerate technical developments in this area.

Quality control waste and also establishing their chemical composition in terms of quality and quantity is critical in order to harness the most appropriate techniques for each type of waste. A proper diagnosis determines the profitability of the entire recycling process. This analysis can be made through spectroscopy which involves:

- chemical analysis of ferrous and non-ferrous materials by optical emission spectrometry and X-ray spectrometry;
- qualitative and quantitative analysis in various stages by X-ray diffraction.

X-rays can be used in analytical chemistry for the qualitative, quantitative and structure analysis. In the first category of methods, the analysis is based on X-rays emitted by the atoms of the sample. By determining the wavelength of the emitted radiation can be made a qualitative analysis and by determining their intensity, a quantitative analysis. This type of devices provides essential advantages over other traditional testing methods. Analysis can be done in seconds, and significantly reduces cost analysis, obtaining an accurate value of the precious metal concentration. Gathering concrete data is crucial for the recovery process. Traditional testing techniques take hours or even days.

Advantages of spectrometric analysis:

• The analysis is non-destructive
• No hazardous waste issued
• The results appear in seconds
• Sample preparation is not difficult [1].

Because of the relative rarity and of the high value per unit, precious metals continue to be recycled at a high recovery rate. US Geological Survey estimated that 240 tons of gold residues (new and old) were recycled in 2012 in the United States, which is more than the total domestic consumption of gold that was reported. In addition, Census Bureau data indicate that about 14,000 tons of remnants precious metals were exported in 2012 worth US $ 5.5 billion [2].

For an effective recycling on industrial scale several conditions must be met:

- waste involved in the recycling process to be accessible;
- an economic viability needs to exist;
- gathering mechanisms need to ensure availability of the product for recycling;
- the entry of the product into recycling chain has to ensure a safe storage of secondary waste;

- configuration of the recycling chain has to be optimized technical and organizational;
- enough capacity along the entire chain so that recycling to be complete [3].

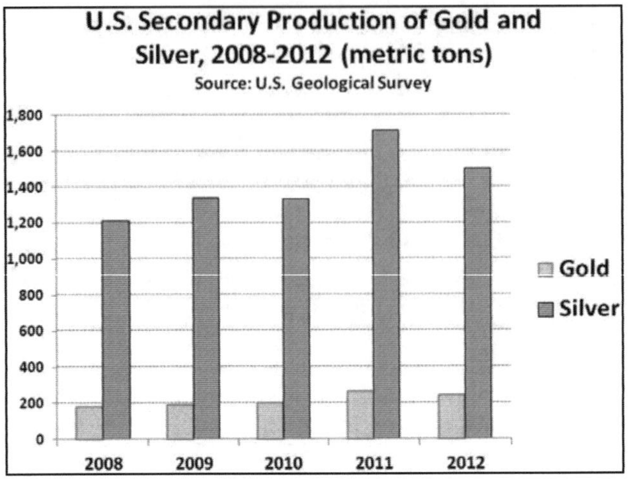

Figure 1. Secondary production of gold and silver between the years 2008-2012

In the last decade, many countries have issued regulatory laws regarding waste management. Underground storage or burning in incinerators is not allowed anymore without isolation of hazardous materials. Moreover, export of waste in underdeveloped countries is illegal according to international regulations [4].

The underground storage, the combustion with the flue gas released into the air endanger the environment, drinking water and lead to release of toxic gases into the atmosphere. Therefore, the recovery of precious metals is essential from the point of view of minimizing environmental pollution and of resource management. The incineration and mechanical biological treatment is by far the most important alternative of treatment.

By the year of 2020 it must be achieved the preparation for reuse and the recycling at least of paper, metal, plastic and glass from household waste and possibly originated from other sources, as far as these streams are similar to waste from the waste to a minimum of 50% from their total weight. The ways of treatment have to be designed to reduce the organic carbon content so as to meet the criteria of final disposal. Unlike simple storing, the new processing methods allow recover materials contained in solid waste containing precious metals. The production of such metals from primary resources is limited and the process is characterized through high energy consumption. Therefore, recycling can greatly contribute to an energy saving and reducing carbon dioxide emissions [5].

Precious metals play an important role in modern society, being inextricably linked to developments in technology and everyday life. Figure 2 presents the four pillars for environmental and energy by Tanaka [6]. Fractions of precious metals separated from waste during pre-processing can be further processed using hydrometallurgy, pyrometallurgy, electrometallurgy, biometallurgy and combinations of them [7].

Hydrometallurgical and pyrometallurgical processes are the most important ways of processing waste from electrical and electronic equipment. These routes can be followed by electrometric processes/electrochemical (eg electrowinning or electrolysis) for separation and recovery of a particular metal. Currently, studies on processing electronic waste through bio metallurgical ways are not confined to a simple lab work, an example may be the bioleaching of metals from e-waste. Pre-processing for electrical and electronic waste is not always necessary in pyrometallurgy. For example, complex electronic equipment such as mobile phones and MP3 players can be treated directly during the melting process [8].

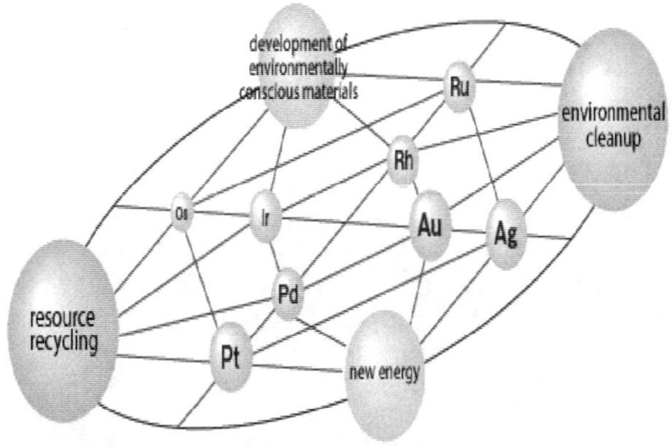

Figure 2. Tanaka Precious Metals Group's four pillars for its environmental and energy-related businesses [6]

However, for hydrometallurgical processes, pre-processing is required to separate fractions of metal from other fractions. This will increase the efficiency of each phase associated with hydrometallurgical processes. Each route has advantages and disadvantages that must be taken into account in selecting a suitable recycling process. Hydrometallurgical processes involve lower power consumption and lower cost compared to pyrometallurgical treatment, being also applicable in plants with relatively small capacities. The main problem of hydrometallurgical processes is the high consumption of water and/or chemicals that need further treatment. Lately hybrid methods were applied, which integrates chemical approach (more efficient) biological strategies (more eco compatible), thereby applying the benefits of both solubilization, chemical and biological.

3. Analyzes. Discussions

3.1. Gold recovery

The interest in gold recovery is due to its vast industrial applications, high market prices, but also for the value attributable during international political and economic crisis, as a limited resource. Experts believe that from the beginning of gold exploitation were extracted 160,000 tonnes [9]. Due to previous ineffective methods for gold recovery (an efficiency of only about 1 g of gold per 1 ton ore) the result was approximately 160 billion tons of tailings heavily contaminated, especially from the early 20th century, when for gold production the cyanide leaching process was the main method [10]. Gold is widely used in computer components. Motherboards and computer pins contain precious metals. Although computers and laptops contain a bigger amount of gold, precious metals are found in many other technological devices, from coffee makers to cars. Obvious deposits of gold from technological and household items can be fragmented. However, gold is too finely layered to be easily removed. For example, the amount of gold in a motherboard is one order of magnitude higher than the ore from which it is usually extracted. Given the growing demand for technological equipment in industrialized countries and in developing countries, is a priority and a real need the recycling of metals rather than extracting them from ores [11].

Gold leaching method with „aqua regia", a mixture of 3:1 between the hydrochloric acid and nitric acid, is a conventional one. The reaction mechanism is the following:

$$2HNO_3 + 6\ HCl \rightarrow 2\ NO + 4H_2O + 3Cl_2 \quad (1)$$

$$2Au + 11\ HCl + 3\ HNO_3 \rightarrow 2\ HAuCl_4 + 3NOCl + 6H_2O \quad (2)$$

Solubilization using chlorides is more difficult to apply because of its high corrosion level of solutions involved and because of its high toxicity. Leaching with thiosulphate is widely used in photography and pharmaceutical industry. Gold form a stable anion complex with thiosulfate [12]:

$$4\ Au+O_2+8\ S_2O_3^{2-}+2\ H_2O \rightarrow 4\ Au(S_2O_3)_2^{3-}+4\ OH^- \quad (3)$$

Gold is being dissolved slowly in thiosulfate, but good rates of leaching can be obtained if ammonia and copper are present. The problem is the large amount of reagent needed.

Another method is the gold recovery through leaching with thiourea:

$$Au+2\ CS(NH_2)_2 \rightarrow Au(CS(NH_2)_2)^{2+} + e^- \quad (4)$$

Gold recovery from various solutions can be made by cementation, adsorption, solvent extraction or ion exchange [13]. The high cost and high degree of environmental pollution, due to secondary waste generated in this process, open the path of research in the recovery of gold by enzymatic methods with various bio-adsorbents, such as bacteria, yeasts, fungi, actinomycetes, algae, biopolymers [14].

3.2. Platinum group metals recovery

The platinum group metals include platinum, palladium, rhodium, ruthenium, iridium and osmium. Secondary resources such as spent catalysts from motor vehicles and waste of electrical and electronic components, also represent important sources of recovery of these metals. They are also present in the anode sludge and can be separated from gold and silver through a variety of hydrometallurgical processes.

The main steps in the recovery of metals from the platinum group are:

• raw material pre-treatment, sampling and analysis;

• concentration and separation using pyrometallurgical and hydrometallurgical techniques such as chemical precipitation, chemical dissolution, extraction liquid/ liquid, osmium distillation, ion exchange and electrolytic processes;

• pyrolysis or reduction of chloro-metallic compounds to form the pure metallic sponge [15].

Some specific methods have been developed for the carbon-based catalyst, using incineration, prior the leaching step. „Aqua regia" is used on the commercial scale to recover Pt, Pd and Rh from spent catalysts and electronic scrap. However, the process suffers from an environmental viewpoint as it generates NOx, chlorine and acid fumes during leaching, due to the presence of HNO_3 in the leaching agent. Although platinum leaching rate by this method is high, the economy of the process needs further improvement [16].

Among solvent extraction techniques that have been referred to, a relevant example is that with 8-hydroxyquinoline or tri-butyl phosphate. The solvent for extraction requires substantial investment in factories, moreover solvents can be toxic. Electrochemical recovery of platinum group metals is feasible, but the recovery of thin coated metal deposits out of electrodes may limit application of it on an industrial scale [17].

Recovery becomes increasingly necessary both economically and environmentally. The recycling process helps to protect natural resources, also to set a limit for increasing prices caused by the growing demand in the automotive industry. A considerable yearly volume of scrapped auto catalytic converters makes the recovery of metals from the platinum group very profitable. The catalytic converter is a device used to reduce the toxicity of emissions from internal combustion engines. A catalytic converter provides an environment for a chemical reaction of combustion and the products are converted to less toxic substances [18].

The catalyst itself contains platinum group metals. Platinum is the most active catalyst and is widely used. It is not suitable for all applications because of unwanted additional reactions and/ or costs. Palladium and rhodium are two other precious metals used. Platinum and rhodium are used as a reduction catalyst, while platinum and palladium are used as oxidation catalyst. Cerium, iron, manganese and nickel are also used, although each has its own limitations. For example, nickel is not legal to be used in the European Union (due to its reaction with carbon monoxide), while copper can be used, but its use is illegal in North America due to the formation of dioxin [19].

Since 1981 3-way catalytic converters were used in vehicle emissions control systems in North America, Europe and Asia. 3-way catalytic converter has three simultaneous tasks:

1. The reduction of the nitrogen oxides which are converted to nitrogen and oxygen:

$$2NOx \rightarrow xO_2 + N_2 \tag{5}$$

2. The oxidation of the carbon monoxide which is converted to carbon dioxide:

$$2CO + O_2 \rightarrow 2CO_2 \tag{6}$$

3. The oxidation of unburned hydrocarbons (unburned or partially unburned fuel) which is converted to carbon dioxide and water [20]:

$$2C_xH_{2x} + 3xO_2 \rightarrow 3xCO_2 + 3xH_2O \tag{7}$$

The exact amount of platinum varies depending on the formula used. Catalytic converters for gasoline vehicles contain on average about 5 grams of platinum, while those for diesel vehicles can contain more than 15 g for high power machines. Researches in terms of optimization of methods for platinum group metals recovery have aroused a growing interest, as shown in Figure 3 [21].

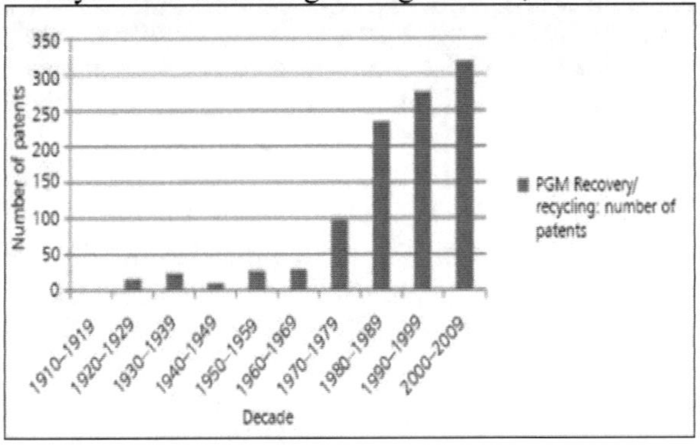

Figure 3. Number of patents for the platinum group metal recovery

3.3. Silver recovery

Recovery of silver from waste as solutions, such as those produced by conventional medical processing and industrial processing of X-ray films, photographic films and images has been practiced for over 100 years. However, the economic viability of the process has dramatically changed over the past decade. Silver is being used in infinite ways, from eye drops to lubricants for jet engines. It is used by each hospital, medical clinic, dentist, as well as by photographic processing departments, printers and by anyone performs photographic processing using a wet mixture. In Figure 4 is presented the forecast for global silver extraction.

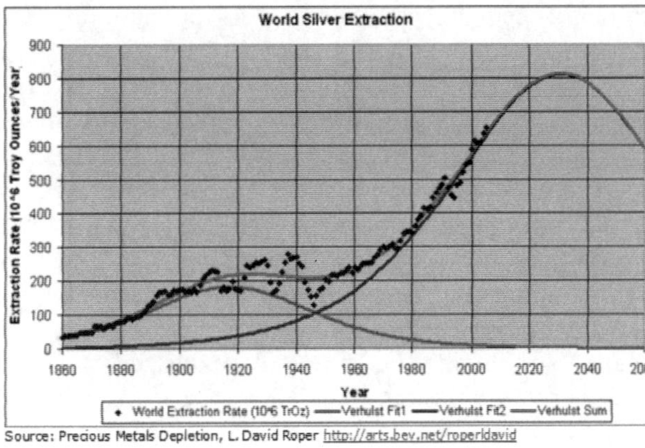

Figure 4. Forecasts for global silver extraction

Radiographic film is a polyester sheet coated on both sides with radioactive material, photosensitive. Every year around the world there are performed an estimated 2 billion radiographies, CT scans etc. Silver world production is currently not enough to meet the demand that is increasingly growing, respectively ~ 2-2.5% per year [22]. Traditionally, 94-98% of X-ray films are being used in healthcare services. Radiographies may contain 5-15 grams of silver per 1 kg of film, representing a significant secondary source of silver.

4. Approaches. Interpretations

Over the time, studies regarding the recovery of silver from radiographic films waste have shown two typical methods: through pyrometallurgical processes and through hydrometallurgical processes, more exactly by acid leaching. Pyrometallurgical processes for silver recovery require temperatures higher than 950° C (higher than 1742° F), with the destruction of the polymer substrate. Hydrometallurgical methods used for silver recycling were initially based on incorporating the silver in solution, in ionic form, and then through electrolysis in silver metal [23]. The silver recovery process depends on the leaching temperature of radiographic films in oxalic acid, in nitric acid or in sodium hydroxide solutions at boiling point, in order to separate the inorganic component out of the polymer substrate. For example, dissolution of silver in ammoniacal copper-thiosulphate system takes place according to the following reaction [24]:

$$Ag^0+5S_2O_3{}^{2-}{}_{(aq)}+Cu(NH_3)_4{}^{2+}{}_{(aq)} \rightarrow Ag\,(S_2O_3)_2{}^{3-}{}_{(aq)}+4NH_{3(aq)}+Cu\,(S_2O_3)_3{}^{5-}{}_{(aq)} \qquad (8)$$

In the next steps, metals are concentrated and purified through precipitation, cementing, extraction using solvents and ion exchange, with the purpose to be recovered after that in pure form using electrolytic methods and precipitation.

The method of electrolysis, in which a continuous electric current is applied between two electrodes, is capable of producing silver with purity higher than 98%. However, this method is used only for silver-rich effluents, because is unable to reduce silver concentration under 100 mg/ l, which is higher than the limit allowed for the environment of 5 mg Ag/ l [25].

Metallic replacement, also called cementation process, is based on the use of metals such as iron, zinc and copper, which are more active than silver to effectively recover of the effluent. Ions that belong to the more active metal are released into solution, while atoms that belong to the less active metals replace them in a solid state [6]. Impurities from active metals, such as Fe^{2+}, Zn^{2+} and Cu^{2+}, effluent and silver sludge require an expensive process for remediation. The process, generally, consists of two reactions: reduction of the metallic ion more active and oxidation of the less active metal. For an optimum performance of the process, it had been recommended a pH 5 to 7.6. Abdel-Aal and Farghaly have reported silver recovery of 98% at 90°C (at 194° F) over a period of 50 minutes retention time, using Zn metal powder 6% concentration of nitric acid. With a careful control, silver can be easily recovered through chemical precipitation. As precipitating agents were used: sodium sulfide, sodium dithionite, potassium boro hydride and 2, 4, 6-tri mercapto-s-triazine. It has been investigated the effect of ethylene glycol as a stabilizer to hydrogen peroxide in the process of silver recovering. Utilizing hydrogen peroxide with ethylene glycol leads to reducing the consumption of acid to less than 25%, resulting in an improvement of efficiency in recovery to 18.7%. The pH has also an influence on potassium boro hydride in silver recovery from radiographies and effluents. Increasing of the pH led to a reduced amount of potassium boro hydride required in order to achieve the desired purity and recovery [26].

Silver recovery from radiographies represents a long term solution, with a high potential in the future. Based on multiple studies and researches in this field, it seems that precious metals recovery is a promising solution in order to minimize the impact of industrial waste toward the environment. Scientists will conduct more thorough research on silver recovery from radiographies using leaching based on several types of acids and will analyze the relationship between the efficiency of result solutions according to various parameters: the leaching time, pH, the temperature and density of the paste. An example of technological flow for silver recovery from radiographic films is presented in Figure 5.

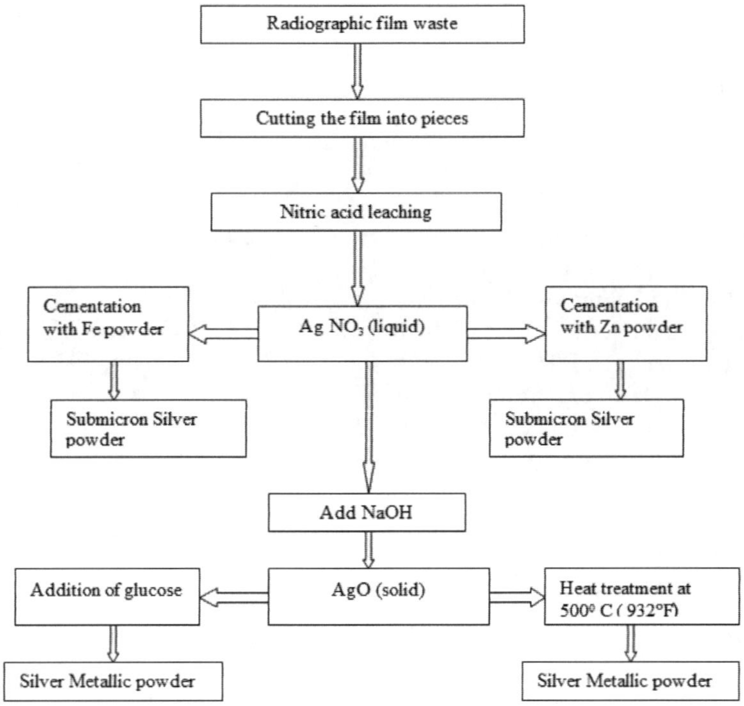

Figure 5.Technological flow in processing radiographic films and silver recovery [27]

5. Conclusions

Precious metals are not affected by the economic destabilization triggered by the financial crisis and this fact is confirmed by price-performance ratio achieved by the companies involved in the extraction of metals as a raw material. Sustainable development and increasingly severe legislation regarding the environment are making from the precious metals recovers, a possible approach in conserving natural resources and reducing waste without high costs. Optimizing recovery solutions opens up new paths in research field, considering growing demand, resources depletion and global need of development.

Precious metals resulted after the recovery processes have the same properties, even after multiple life cycles, therefore recycling allows:

- recovery of the valuable material without quality loss;
- energy saving in comparison with primary production;
- reducing resulted emissions and reducing greenhouse gas emissions;
- reduction of mining activities;
- reducing waste.

Industrialized countries have switched to a positive environmental policy, aimed at reducing technological risks, showing a great power of adjustment, through a culture of change and adopting long-term strategies. Finding new precious metal recovery technologies can balance the economic pressure with the need of preserving natural ecosystems and can have a positive global impact.

References

[1] ***http://ph.academicdirect.org/CAI_2006.pdf
[2] ***http://www.isri.org/recycling-industry/commodities-specifications/nonferrous scrap#.VASNZ6MkG9g

[3] ***http://www.technology.matthey.com/article/56/1/29-35/
[4] ***Puckett J, Byster L, Westervelt S, Gutierrez R, Davis S, Hussain A and Dutta M 2002 *Exporting Harm—The High-Tech Trashing of Asia*; The Basel Action Network (BAN) Silicon Valley Toxics Coalition (SVTC): Seattle, WA, USA
[5] Cui J and Forssberg E 2003 Mechanical recycling of waste electric and electronic equipment: A review, *J. Hazard Mater* **99** 243–263
[6] ***http://pro.tanaka.co.jp/en/features/environment/column05/
[7] ***http://www.academia.edu/6209804/Recovery_of_precious_metals_from_e-waste
[8] Anindya A 2012 *Minor Elements Distribution during the Smelting of WEEE with Copper Scrap*, RMIT University, Melbourne, Australia, Doctoral Thesis
[9] ***www.zeit.de
[10] ***http://www.united-commodity.com/en/Tailing-recycling.html
[11] ***Critical raw materials for the EU. In Report of the Ad-Hoc Working Group on Defining Critical Raw Materials; European Commission: Brussels, Belgium, 2010
[12] ***http://www.google.com/patents/US6344068
[13] ***Anand A, Jha M K, Kumar V and Sahu R 2013 *Recycling of Precious Metal Gold from Waste Electrical and Electronic Equipments (WEEE): A review*, Proceedings of the XIII International Seminar on Mineral Processing Technology (MPT)
[14] ***http://www.sciencedirect.com/science/article/pii/S0304386X11002659
[15] ***http://www.eurometaux.eu/DesktopModules/Bring2mind/DMX/Download.aspx?Command =Core_ Download&EntryId=7168&PortalId=0&TabId=57
[16] Jha M K, Lee J, Kim M, Jeong J, Kim B and Kumar V 2013 Hydrometallurgical recovery/recycling of platinum by the leaching of spent catalysts: A review, *Hydrometallurgy* **133** 23–32
[17] ***http://www.ncbi.nlm.nih.gov/pmc/articles/PMC106693/
[18] ***http://media.gm.com/media/ro/ro/opel/news.detail.html/content/Pages/news/ro/ro/2014/opel/05_10_25_ani_de_convertoare_catalitice.html
[19] ***http://en.wikipedia.org/wiki/Catalytic_converter
[20] ***http://www.tobecarbon.ro/catalizator.php
[21] ***http://www.technology.matthey.com/article/56/1/29-35/
[22] Bas A D, Yazici E Y and Deveci H 2012 Recovery of Silver from X-ray film processing effluents by hydrogen peroxide treatment, *Hydrometallurgy* **121-124** 22-27
[23] Marincovic J, Korac M, Kamperovic Z and Matic I 2006 Recycling of silver from exposed X-ray films, *Acta Metalurgica Slovaca* **12** 262-268
[24] Yazici E Y, Bas A D and Deveci H 2010 *Extraction of Silver from Waste X-ray Films by Thiosulphate Leaching*, Proceedings of the XIIth International Mineral Processing Symposium, Cappadocia-Nevsehir, Turkey
[25] Abdel-Aal E A and Farghaly F E 2007 Preparation of silver powders in micron size from used photographic films via leaching–cementation technique, *Powder Technology* **178** 51–55
[26] Aktas S, Morcali M H and Yucel O 2010 Silver recovery from waste radiographic films by cementation and reduction, *Canadian Metallurgical Quarterly*, available: maneyonline.com
[27] Masebinu S O and Muzenda E 2014 *Review of silver recovery techniques from radiographic effluents and X-ray film waste*, Proceeding of the World Congress on Engineering and Computer Science, Vol. II, WCECS 2014, 22-24 October, San Francisco, USA

International Conference on Applied Sciences 2015 (ICAS2015) IOP Publishing
IOP Conf. Series: Materials Science and Engineering **106** (2016) 012021 doi:10.1088/1757-899X/106/1/012021

The critical analysis of austenitic manganese steel T130Mn135 used for castings in the mining industry

A Josan[1], C Pinca Bretotean[1] and V Putan[1]

[1]Politehnica" University of Timisoara, Department of Engineering and Management, 5 Revolution Street, 331128 Hunedoara, Romania

E-mail: ana.josan@fih.upt.ro

Abstract. This paper presents the critical analysis of making technology of austenitic manganese steel *T130Mn135*, used for castings of the type *Mills hammer* at a Romanian foundry. Are analyzed 11 charges of steel for castings and is determined the diagram of the heat treatment. After the applying of the heat treatment results a single-phase structure, consisting of homogeneous austenite. For all the 11 charges is presented the variation of chemical composition.

1. Introduction

Austenitic manganese steels are part of the alloyed steel, to which the manganese is the main alloying element. The main particular physical property of high alloyed manganese steel (known as Hadfield steel) is its exceptional wear resistance (highest in comparison with all other known steel grades). Moreover, this steel, having an austenitic structure (is called austenitic manganese steel), is not magnetic, what determines also other uses for this steel. The austenitic structure of the steel is obtained after hardening at a temperature of 1000…1150 °C in water [1].

This steel is used to making castings that have a high wear resistance, combined with mechanical shock, but with no special conditions with regard to maintaining the exterior dimensions, of the type: Mill Hammer, jaw crusher, balls for mill, elements for tractors and tanks tracks, buckets for excavator etc., being used in the same time as the non-magnetic material [1], [2].

The mechanical properties of the Hadfield steel hardened are characterized by the following values [1], [3]:

- tensile strength: 800…1000 N/mm^2;
- flow limit: 225…400 N/mm^2;
- elongation: A_5 = 40…55 %;
- necking: z = 35…45 %;
- resilience: 20…30 daN/cm^2;
- Brinell Hardness: HB = 180…220.

Also, the austenitic structure of this steel determines the relatively low hardness (of approximately 200 HB) but however the Hadfield steel can not be mechanically processed with tools of common special steel (high speed steel-HSS). Only the carbide alloys (such as Widea) can process, to same extent (with low speed) this grade of steel. Therefore, *Hadfield steel parts are obtained only by casting* [1], [2], [3].

2. The analysis of technology for obtaining the parts of type *Mill Hammer*

The casting is the process of obtaining the manganese steel pieces. If the castings have internal hollows, these can be achieved only by cores, make separately, in core boxes. In this context, it is advantageous to obtain the pieces by casting, when is necessary to obtain the pieces with complicated shape, in conditions of certain easy and economical making technologies. Properties of the material have, often, a decisive influence on the making process of the casting [4], [5].

In order to perform the analysis of the causes leading to rejection of pieces of Mill Hammer type in industrial practice, was performed a study on a sample of rejected pieces, casting of steel T130Mn135 [5], [6]. The situation registered in the rejected castings is as follows (Figure 1):

- 56% of the castings presented crack, porosities, void holes;
- 24 % of the casings presented nonmetallic inclusions;
- 20% of the castings presented unhomogeneous structures.

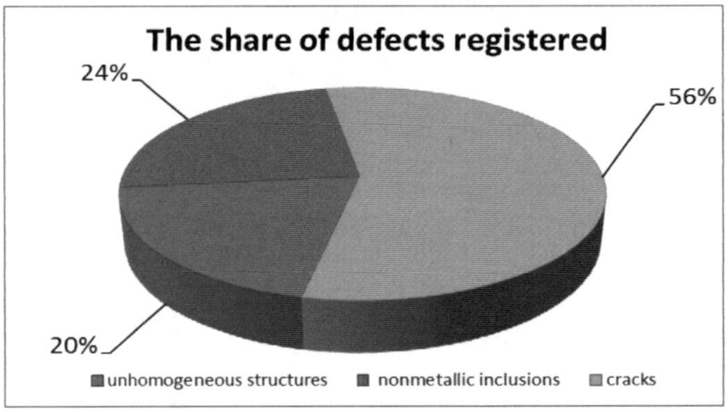

Figure 1. The share of defects registered

Following this study, were put in evidence causes for rejection of their [6], [7]:

- Failure to comply with prescribed parameters of thermal treatment, the castings studied presented a unhomogeneous austenitic structure, with carbides between austenite crystals;
- An example of structure by austenite, with carbides remaining undissolved at the grain-boundary, inclusions and micro-void holes is presented in Figure 2. This defect are due to non-compliance with the prescribed casting temperature and the oxidation of the steel during casting and thus forming MnO inclusions, which determines the occurrence of the hot cracking in the castings.

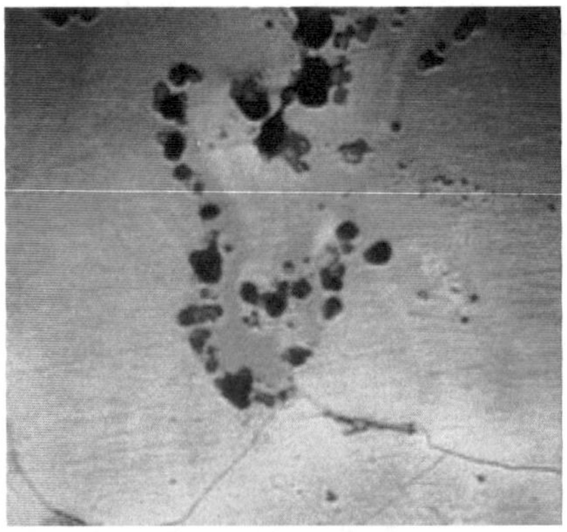

Figure 2. Austenitic structure 200X; Nital attack 2%

- Failure to comply with the casting parameters (temperature and speed casting), the castings presented a columnar primary structure with advanced transcrystallization who was not eliminated by the thermal treatment of hardening, causing the occurrence of the hot cracking in the castings; at occurrence of this serious defect of casting (hot and cold cracking) can contribute both the steel itself and the mould or casting process parameters and thermal treatment parameters of castings;
- The castings presented a large quantity of non-metallic inclusions, due to inadequate quality of the moulding mixture but also because the deoxidation process, which has been insufficiently completed.

Thus, a particulary important role is assigned to elaborated process of the liquid alloy, because it must to consider both the pouring properties of this alloy and moulding technology of piece to be obtaining. A particulary important role is assigned of the mould and positioning of core inside it.

3. The analysis of steel making technology (steel grade *T 130 Mn 135)*

In order to analyze regarding the establishment of measures that lead to improved physical and mechanical characteristics of manganese steel, were analyzed **11 charges** for the steel grade *T 130 Mn 135*. This steel was elaborated in order to obtain pieces of *Mill Hammer* type (Figure 3), used in mining, for coal grinding [2].

According to standard, for cast steel *T 130 Mn 135*, chemical composition must be within the following limits [3], [4]:

Table 1. Chemical composition of the steel *T 130 Mn 135*

Chemical composition, in %						
C	Mn	Si	P	S	Ni	O_2
1,25…1,4	12,5…14,5	0,3…1,0	max 0,11	max 0,05	max 0,8	max 1,2

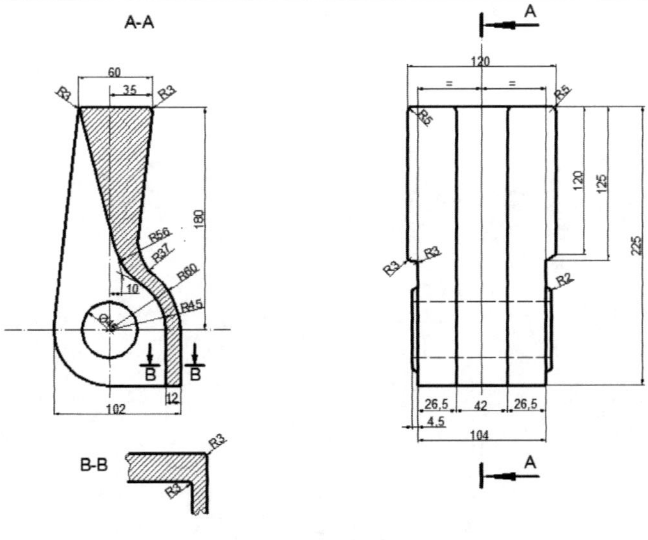

Figure 3. Drawing the piece *Mill Hammer*

The charges of steel were elaborated in electric arc furnace DSN-3 type, with basic lining, with a nominal capacity of 3500 kg. The length of time required to elaboration charges was 200…215 minute/charge.

The metal charge of the furnace consisted, for all charges studied, of the following materials [1], [4], [8]:

- pure scrap iron: 1000 kg;

- waste manganese steel: 2350 kg;
- ferro-manganese FeMn 75: 260 kg;
- ferro-manganrse FeMn 82: 70 kg;
- ferro-silicon FeSi 75: 20 kg.

In Table 2 are presented the chemical composition of the charges studied.

Table 2. The chemical composition of the charges studied

No.	Chemical composition, %							
	C	Mn	Si	S	P	Ni	Cr	Mn/C
Charge 1	1,34	14,37	0,36	0,02	0,07	0,33	0,54	10,72
Charge 2	1,38	13,60	0,20	0,01	0,18	0,20	0,61	9,85
Charge 3	1,45	12,61	1,08	0,01	0,09	0,11	0,85	8,83
Charge 4	1,28	11,62	0,55	0,03	0,08	0,55	0,51	8,68
Charge 5	1,23	12,4	0,42	0,04	0,09	0,62	0,52	10,08
Charge 6	1,32	12,82	0,48	0,01	0,09	0,32	0,54	9,71
Charge 7	1,42	13,2	1,02	0,04	0,10	0,59	0,92	9,29
Charge 8	1,33	14,36	0,35	0,02	0,07	0,32	0,53	10,79
Charge 9	1,35	13,59	0,20	0,02	0,07	0,19	0,60	10,06
Charge 10	1,27	11,61	0,54	0,02	0,07	0,53	0,50	9,14
Charge 11	1,43	12,59	1,06	0,02	0,06	0,09	0,83	8,80

Using the data registered in Table 2, was performed the critical analysis of the chemical composition of the 11 charges studied. Thus, conclusions were as follows:

- According to standard, the carbon content must be between 1,25…1,4% [3].

Of the diagram presented in Figure 4 follows that to charge No. 5, the carbon content is below the limit prescribed in standard, and to charge No.3 is registered a value above the maximum of carbon

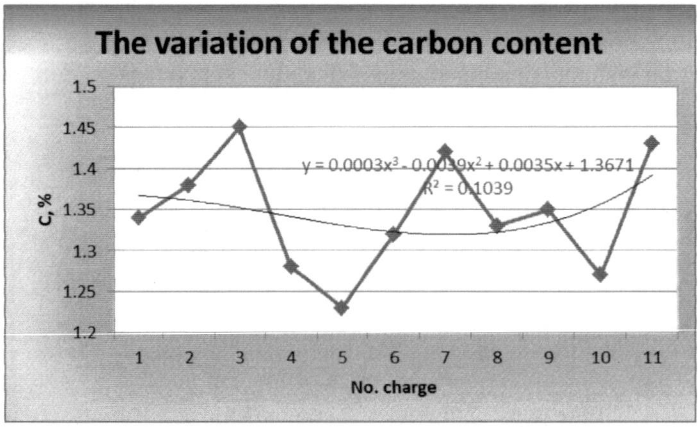

Figure 4. The carbon content of the charges elaborated

The increase over the acceptable limits of carbon content (1,4 % maximun), at a content of manganese in normal limits, has a negative influence of the quantity of carbides and their fineness in the basic mettalic mass, resulting in a decrease of the mechanical properties of the steel.

- According to standard, the manganese content must be between 12,5…14,5% [3].

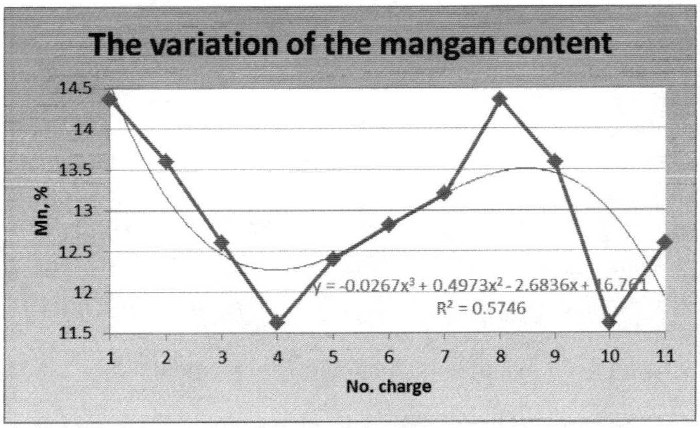

Figure 5. The manganese content of charges elaborated

Of the diagram presented in Figure 5 follows that only to charge No. 4 is recorded a manganese content slightly lower than the standardised value. This deviation causes a deterioration of physical and mechanical properties of steel.

- According to standard, the silicon content must be between 0,3...1,0% [3]. The diagram presented in Figure 6, indicates that, for charges 2 and 9, the silicon content is less than the admissible value, and for charges 3,7,11 is slightly than the maximum admissible value.

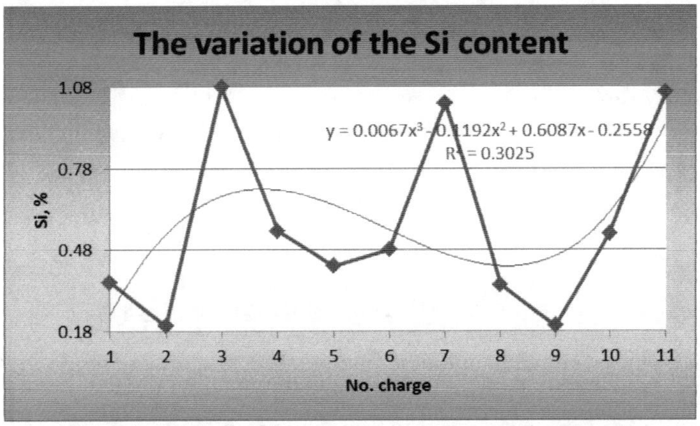

Figure 6. The silicon content of charges elaborated

- According to standard, the sulphur content must be max.0,05%, and the phosphor - max.0,11% [3].

Of the diagram presented in Figure 7, follows that to charge no.2 are registered an increase of the phosphor content above the admissible limits (max. 0,11%), which leads to an increase of britleness of the steel, simultaneously with the decrease its plasticity and toughness.

In all cases, the sulphur content is within the limits recommended, due to the high content of manganese in the steel.

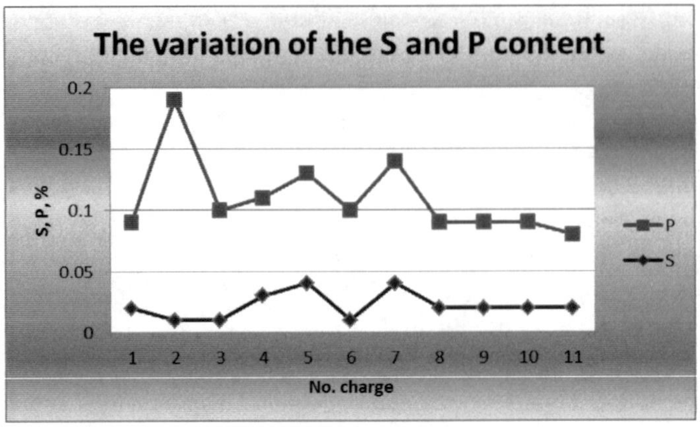

Figure 7. The sulphur and phosphor contents of charges elaborated

- According to standard, the nickel content must be max. 0,8% and the chrome content - max. 1,2% [3]. The diagram presented in Figure 8, indicate that the two elements correspond, in the case of all charges, with the limits recommended in the literature.

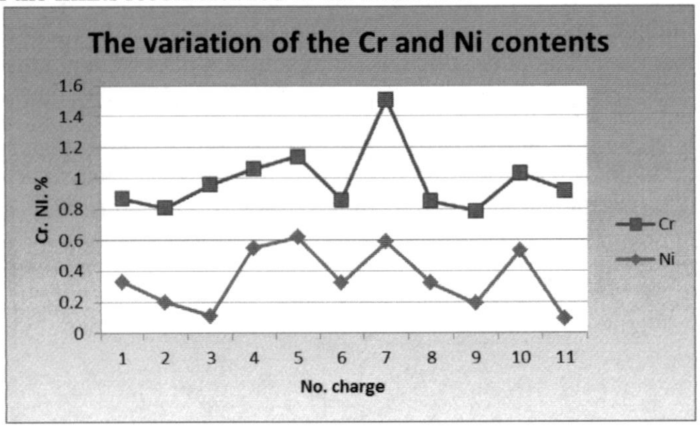

Figure 8. The chrome and nickel contents of charges elaborated

Following the analysis these results, is found that the best properties for austenitic manganese steel are obtained when the ratio $Mn/C \geq 10$ (when the quantity of carbon is the smallest possible, i.e. is towards the lower limit admissible and quantity of manganese is towards the upper limit admissible) [1], [3], [4], [8]. Figure 9 presents the report Mn/C for all 11 charges analyzed.

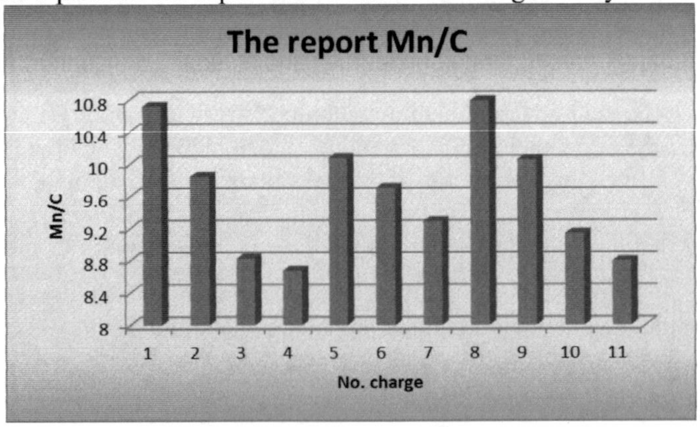

Figure 9. The report Mn/C of charges elaborated

Thus, can be mentioned that, in the case of charge no.1, the contents of elements are within the limits recommended for these steels. Is observed the report Mn/C = 10,72, which gives the castings the best physical and mechanical properties. Similary, also for charges 5,8,9 for which the report Mn/C > 10 (i.e 10,8; 10,79; respectively 10,06).

The parameters of the thermal treatment (heating temperature, heating time and maintaining, the speed of heating, the speed of cooling and medium cooling) must ensure the obtaining a uniform structure on the section of piece, obtaining a hardness which be within the recommended values and preserving the integrity of the metallic material (absence of cracks and internal and external microcracks) [1].

6. Conclusions

Obtaining the castings of the mangenese steel, without defects, requires to know the casting properties of this type of alloy. Then, must be establish the alloy elaborating technology and moulding-casting tehnology, taking into account of all factors that influence the quality of the castings.

Thus, to reduce losses and rejects, should take the following measures:

- At the steel making, to strictly observe the chemical composition in order to obtain the mechanical characteristics established by the standard and the beneficiary; it is necessary to give high priority to Mn and C content, so the ratio Mn/C $\geq$ 10;
- At the steel making should be carry out a advanced dezoxidation, due to the formation of the manganese oxide inclusions; to observe, if possible, the steel temperature to elaboration, the fluidity of the steel, the chemical composition;
- To increase the wear resistance of the castings, can be make a supplementary alloying with cromium (max. 2,5%) and molibdenum (max. 2%)
- At casting of steel into the mould to respect the parameters of the casting process, respectively the casting temperature (has the greatest influence on concerning the formation of hot crack), the casting speed and corectly pouring of the liquid alloy;

An important role in obtaining of quality castings is the correct positioning of the core into the mould [2], [5], [9], [10] (figures 10-12), so the mould to take over the stresses that appear during pouring and solidification of steel; thus, can be avoid the appearance of the cracks into the castings;

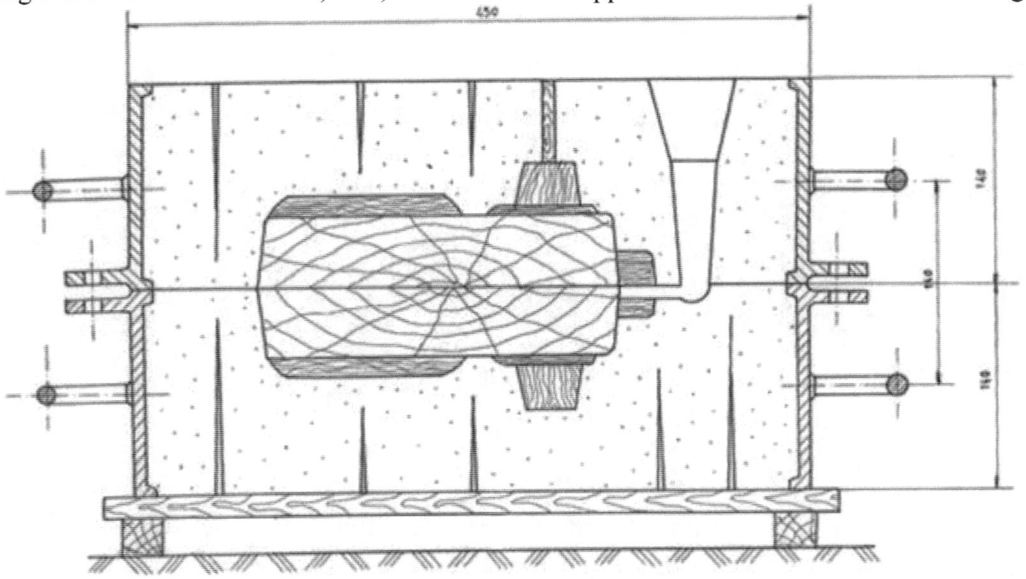

Figure 10. Making the mould parts

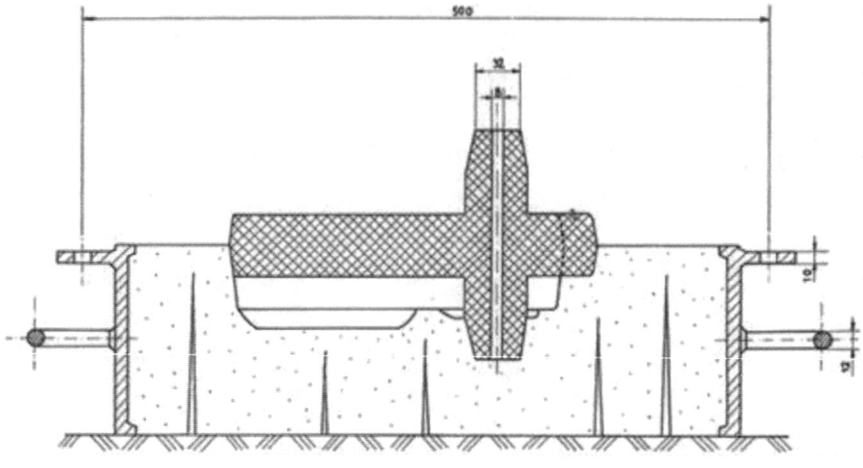

Figure 11. The positioning of the core in the lower mould part

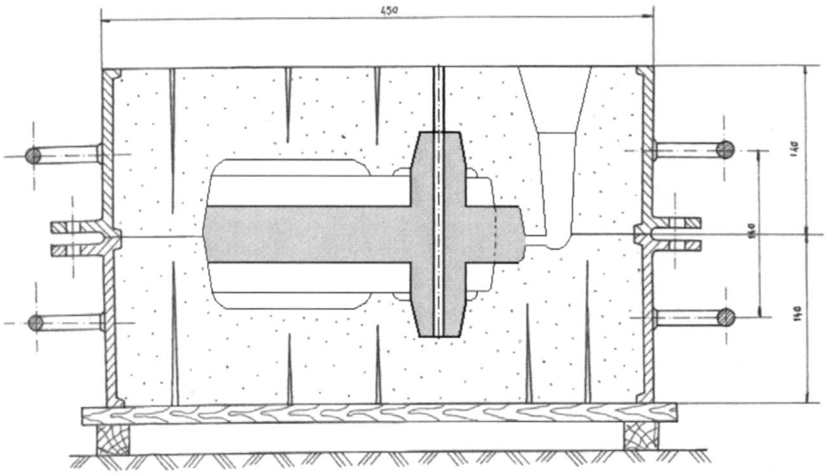

Figure 12. The mould reading of casting

- To respect the prescribed parameters of the hardening thermal treatment (Figure 13):

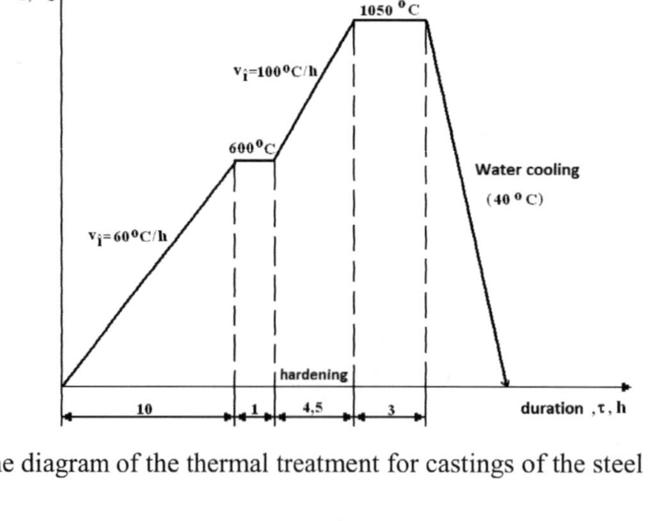

Figure 13. The diagram of the thermal treatment for castings of the steel *T 130 Mn 135*

References

[1] Sofroni L 1975 *The elaboration and pouring of alloys*, Publishing House Didactic and Pedagogical, Bucharest, Romania

[2] Josan A, Pinca Bretotean C and Pauca A 2012 *The Use of Regenerated Mold Mixtures in Manganese Steel Piece Casting* Proceedings of the 3-rd International Conference on Energy, Environment, Devices, Systems, Communications, computers (INEEE '12), Mathematical Modelling and Simulation in Applied Sciences, Rovaniemi, Finland, pp 162-164

[3] ***The Standard *STAS 3718-88 - Manganese austenitic steel castings. Grades and general technical requirements for quality ASTM A128 / A128M - 93(2012) - Standard Specification for Steel Castings, Austenitic Manganese*

[4] Stefanescu C and Cazacu I 1980 *Technologies of realization of the parts by casting*, Technical Publishing, Bucharest, Romania

[5] Josan A 2012 *Casting and moulding technologies of metal parts – the project guide-book*, Publishing Polytechnic, Timişoara, Romania

[6] Josan A 2010 *The analysis of casting defects recorded in the metallurgical enterprises*, International Symposium on Advanced Engineering & Applied Management - 40th Anniversary in Higher Education (1970-2010), Hunedoara, Romania (CD)

[7] ***1977 *International Atlas of casting defects* Technical Publishing, Bucureşti, Romania

[8] Socalici A, Popa E, Heput T and Vîlceanu L 2014 Contributions related to the control of steel ingot solidification, *IOP Conf. Ser.: Mater. Sci. Eng.* **57** 012004

[9] Josan A, Pinca Bretotean C, Ardelean E and Ardelean M 2015 Using lateral cores to casting of carbon steel parts, of drive wheel tyle in a metallurgical enterprise, *IOP Conf. Ser.: Mater. Sci. Eng.* **85** 012016

[10] Josan A and Pinca Bretotean C 2015 Using special additions to preparation of the moulding mixture for casting steel parts of drive wheel type, *IOP Conf. Ser.: Mater. Sci. Eng.* **85** 012017

Power variation during the functioning of nonlinear load

C Panoiu[1], R Rob[1] and M Panoiu[1]

[1]Politehnica University of Timisoara, Department of Electrical Engineering and Industrial Informatics, 5 Revolution Street, 331128 Hunedoara, Romania

E-mail: caius.panoiu@fih.upt.ro

Abstract. Present paper studies the variation of active, reactive, distorted and apparent powers measured during the functioning of an electro thermal installation with electromagnetic induction. In order to analyze these electrical parameters, measurement sets were accomplished using two methods. The first method is based on using a power quality analyzer and the second method consists in using a data acquisition system that contains an adapting interface and a data acquisition board connected to a computer.

1. Introduction

Distorted regime intensification into the power system implies important repercussions on the electrical parameters that characterize the functioning of the electric equipment that must work in sinusoidal regime. Some equipment is sensitive to voltage or current variation distortion and others are sensitive to the presence of some particular harmonic.

Circulation of the harmonic currents determines additional loss into the power network, increasing in this way the technological consumption. This fact leads to reducing the performances of the electrical machines, overvoltage, or electromagnetic interferences over telecommunication network. Analysing of the electrical parameters acquired during the functioning of nonlinear loads is very necessary for designing the power conditioning devices [1-3].

2. Description of electrical power analysis system

The electro thermal installation with electromagnetic induction that is studied in this paper is composed by an electronic converter CTC100K15 and a hardening inductor. The electrical characteristics are the followings: supplying voltage 3 x 400 V, frequency of the supplying voltage 50 Hz, rated current 27 A, control voltage 24 Vdc, consumed power at high frequency 15 kW, voltage at medium frequency 500 Vac. In order to study the variation of electrical powers, two methods were accomplished.

2.1. Electrical power analysis using CA 8334B

The first method consists in using of a power quality analyzer CA8334B connected to the point of common coupling on 0.4 kV. The analyzer is able to display in real time the electrical characteristics of the electric installation that is connected to. As well, parameters in transient and permanent regime and also current and voltage harmonics can be computed.

In Figure 1 the variation of active and reactive power is represented. As well, the variations of the total harmonic distortion of the phase voltage and phase currents are presented. The sampling period of the acquisition of active and reactive power is 1 second and the acquisition duration is 67 seconds.

The measurements were accomplished on the maximum power level. THD is computed using Fast Fourier Transformer of phase voltages and phase currents samples. In Figure 1, W represents the active power, VAR represents the reactive power, Athd and Vthd represent the total harmonic distortion of the phase currents and phase voltages. Studying the Figure 1, it can be noticed that in the hardening process of a steel piece, the variation of the active and reactive powers presents 3 stages.

The first stage is characterized by high values of power when the piece temperature is reaching Curie point. The duration of the first stage is depending of active power, i.e. it is decreasing with the increasing of the active power. In the second stage the piece temperature is increasing to hardening point, active and reactive powers being approximately constant and in the third stage the piece is extracted from the inductor and the powers is decreasing to zero.

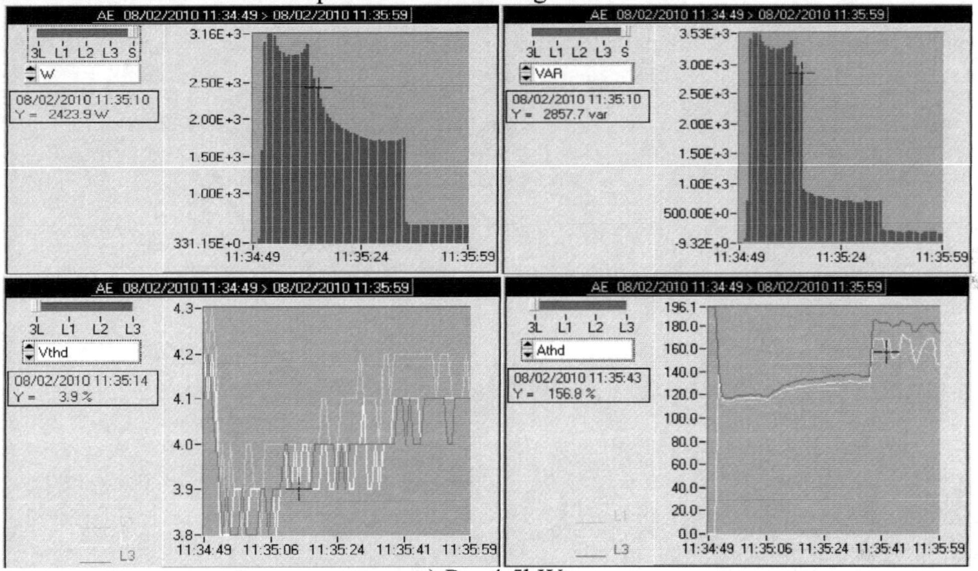

a) P = 4.5kW

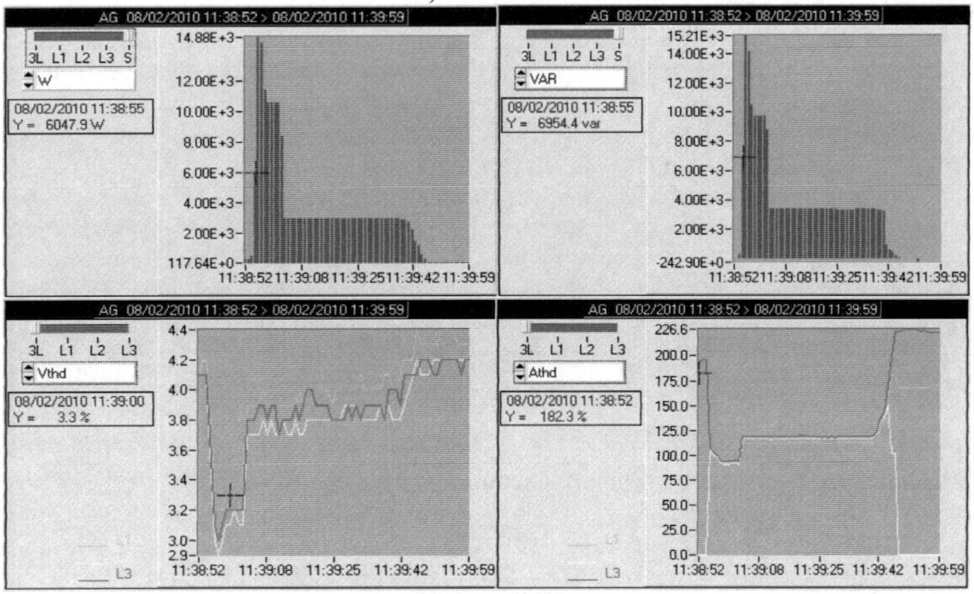

b) P = 15kW

Figure 1. Variation of active and reactive powers and THD for the phase voltages and currents (acquisition with CA8334B)

The same stages are also presented in variation of the total harmonic distortion of the phase currents where can be observed that THD is decreasing when the installation is in load operation. This can be explained by the increasing of the fundamental component in load operation.

A very important notice is referring to powers measurement possibility of CA8334B analyser. This analyzer indicates apparent power S, active power P and reactive power Q. This means that in situation in which the distorted power D is significant, when the analyser indicates a reactive power Q, this Q means $\sqrt{Q^2 + D^2}$. This fact leads to the wrong conclusion about the real value of the measured reactive power in situations in which the distorted power is significant irrespective to reactive power. The measurement results which are obtained using CA8334B analyser are presented in Table 1. These measurements results were made when the electro thermal installation works on 4.5 kW and 15 kW.

Table 1. Electric parameters measured with CA8334B analyser

Phase	L_1	L_2	L_3	L_1	L_2	L_3
		4.5 kW			15 kW	
U [V] line voltage	395.5	398.6	397	399.2	402.3	400.7
U_f [V] phase voltage	228.4	229.7	229.7	230	231.8	231.8
$THD_{voltage}$ [%]	4.3	4.2	4.2	3.9	3.9	3.9
I [A] phase current	7.7	6.6	6.8	31.1	31.1	31.1
A_{THD} (THD_I) [%]	118	118	118	105	106	105
P [W] active power		3194.15			14880	
$\sqrt{Q^2 + D^2}$ reactive+distorted power		3625.52			15210	
PF (cosϕ) [-] power factor	0.67	0.66	0.65	0.7	0.7	0.7

From Table 1 it can be observed that the phase currents and phase voltages are balanced and also it can be concluded that the power supplying system is balanced. Another conclusion is referring that when the active power is on the maximum level, the analyser indications referring to the reactive power are almost the same.

2.2. Electrical power analysis using the data acquisition system

Electrical power analyzing was also accomplished using a data acquisition system that contains an adapting interface and a data acquisition board connected to a computer [4-6]. The acquisition of voltage and current signals was made at the point of common coupling. The phase voltage uR(t), uS(t), uT(t) and phase currents iR(t), iS(t), iT(t) were acquired using the adapting interface which was designed in order to realize the galvanic isolation between the electro thermal installation and the data acquisition system and also to realize the compatibility of voltage levels.

In order to process the acquired data, two soft applications designed in LabVIEW 2011 were used: acquisition application and computing application. Using the acquisition application, the voltage and current samples are transformed into numeric data and stored into text documents. This information is loaded by the computing application which is able to compute the most important electrical parameters that characterize the functioning of the electro thermal installation. Using appropriate subroutines, 10 analogical input channels were configured.

Because the maximum sampling rate is 250 kS/s, the maximum sampling rate for a single channel is 25 kS/s.

The number of samples for a single period results from the relation:

$$N = f_S \cdot T = 25kS/s \cdot 0.02s = 500\,samples \tag{1}$$

IOP Conf. Series: Materials Science and Engineering **106** (2016) 012022 doi:10.1088/1757-899X/106/1/012022

The acquisition is developed on 10 seconds interval. The number of samples acquired on each channel is 250000. Using presented applications, 10 measurement sets were accomplished by increasing with 10% the power of electro thermal installation.

In Figure 2 the LabVIEW code of the computing application is presented. In Figure 3 the LabVIEW code for computing the electrical parameters and in figure 4 the front panel of the application is presented.

In Figure 5 the displaying parameters subroutine is depicted.

In Figures 6, 7 and 8 the variation of active, reactive, distorted and apparent powers and THD for phase currents and phase voltages when the electro thermal installation is functioning at 4.5 kW, 9 kW and 15 kW is presented.

The acquisition is developed on 10 seconds. Figures 6, 7 and 8 contain also the variation of phase voltage and currents on 0.02 seconds (power network period).

Some observations must be made. In figures that represent the variation form of the currents and voltages (on period of 0.02 seconds), phase 1 is red, phase 2 is green and phase 3 is blue.

In Figures 6, 7, 8, on 3 phase powers menu are represented the electric powers of the electro thermal installation: apparent power S (black), active power P (red), reactive power Q (green) and distorted power D (blue). For the other menus with variation on 10s, the variation for all three phases (black), phase 1 (red), phase 2 (green) and phase 3 (blue) is represented.

By analysing variations from Figures 6, 7, 8, observations concerning the measurements using data acquisition system can be made. Studying the variation on 0.02 seconds, can be observed that the phase currents are deeply distorted and the variation of phase voltage is almost sinusoidal.

The same conclusion can be made from the variation of THD of the phase currents (85% before reaching the Curie point and 110% after that) and phase voltages (2% before reaching the Curie point and 3% after that). Studying the power variation, can be concluded that the distorted power is significant, 12 kVAd, comparing with the reactive power of 2 kVAr.

However, the resulting component from the composition of distorted and reactive powers, $\sqrt{Q^2 + D^2}$, is comparable with the component measured with CA8334B analyser.

A very important conclusion is that a data acquisition system makes possible to measure the distorted power and CA8334B is not able to realize the distorted power acquisition.

In table 2 is synthesized the variation of the studied parameters increasing the power of the installation with 10%, using data acquisition system.

3. Conclusions

This paper proposes two different and accessible methods for following the time variation of the most important electric parameters that characterize the functioning of an electro thermal installation with electromagnetic induction.

The main conclusion is based on important possibility offered by data acquisition system. Using the samples acquired from phase voltages and phase currents, the present system is able to compute correctly the values of reactive and distorted powers. This opportunity leads to correct designing of the harmonic filters and the reactive power compensation system as well. On the other hand, the power quality analyser CA8334B offers correct information about the variation of apparent and active powers, THD for phase voltages and phase currents, even it is not able to separate the reactive from distorted powers.

From the powers variation can be observed that the distorted power is very significant comparing to the reactive power, so another conclusion is that the current harmonic spectrum contains harmonics with high amplitudes. Therefore, an important necessity is the power quality improvement at the point of common coupling using harmonic compensation devices. Analysing the information from Table 2, is observed a decreasing of THD for phase voltages (low values 2-3%) with the increasing the installation power.

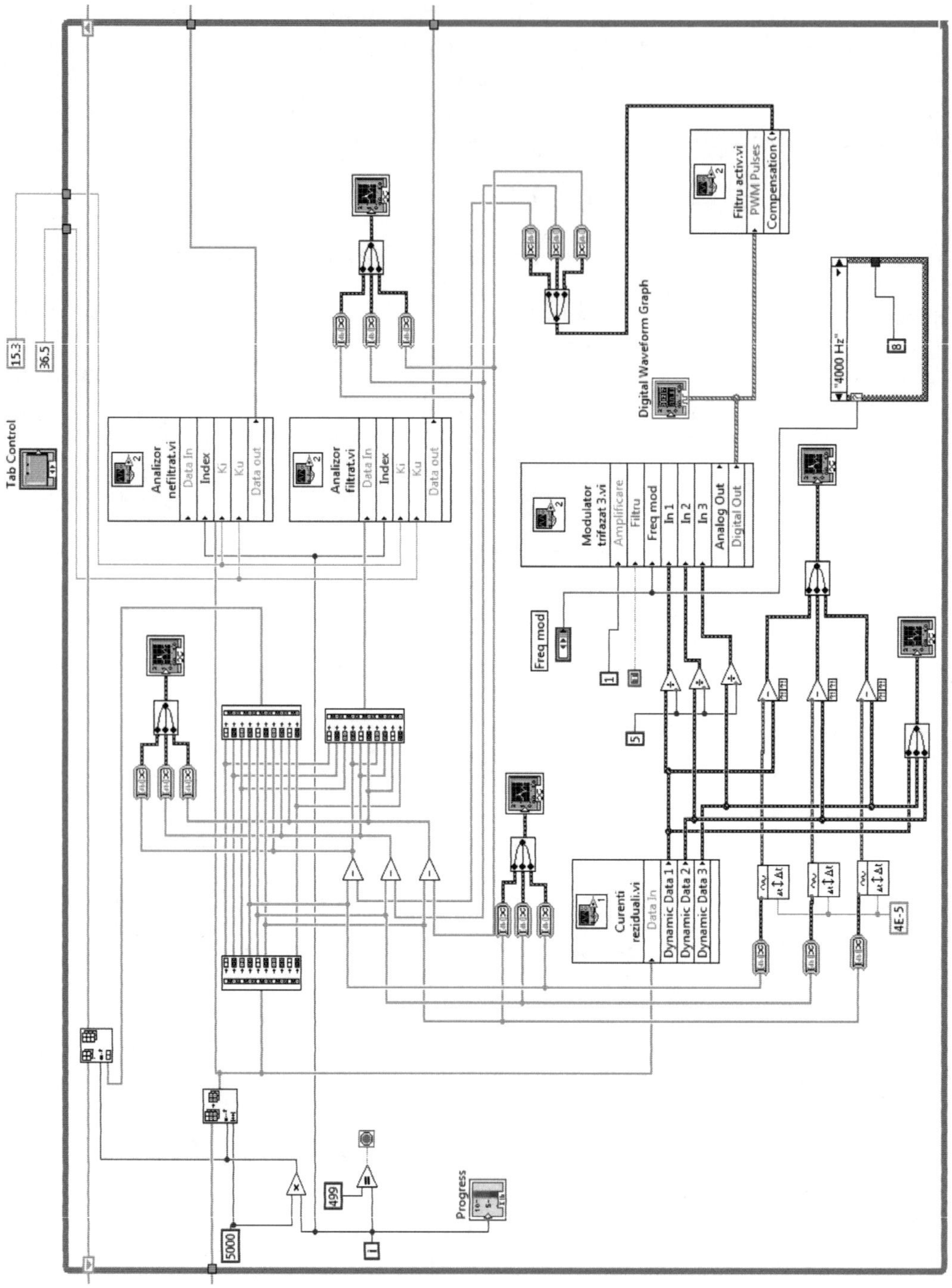

Figure 2. *Computing application* LabVIEW source code

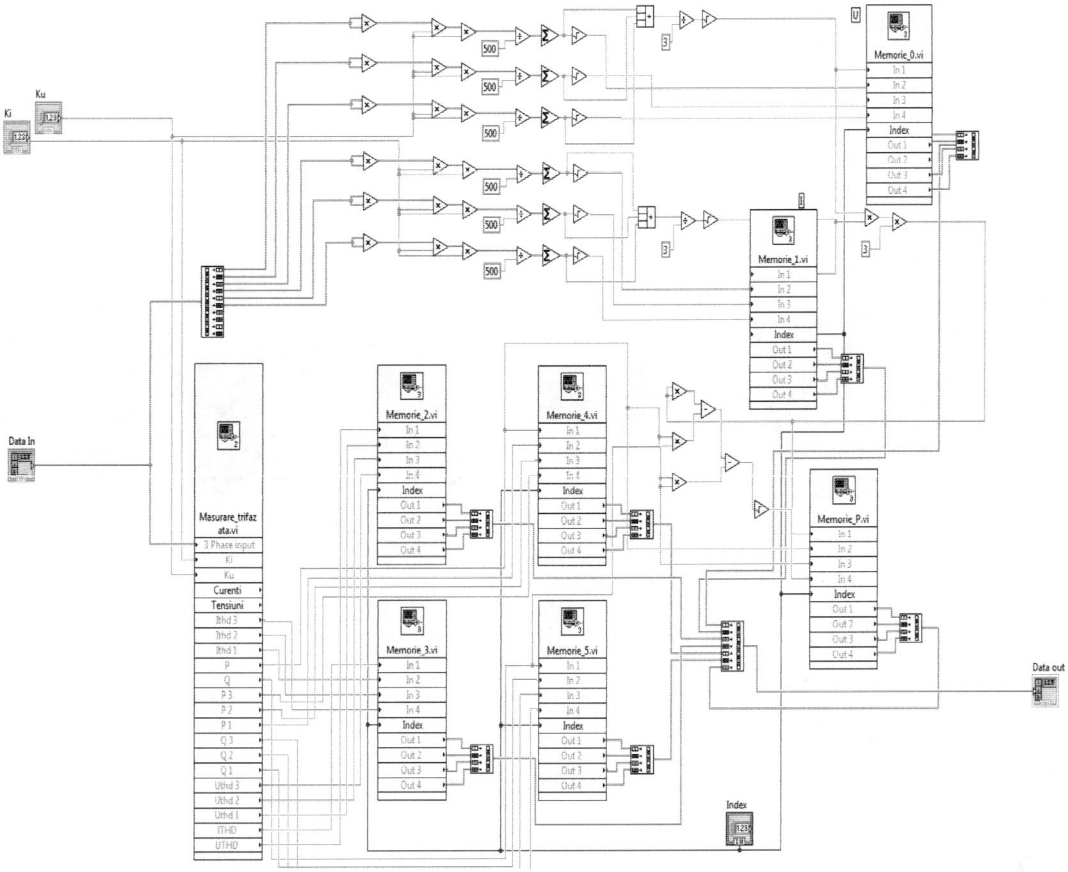

Figure 3. LabVIEW code for computing the electrical parameter
s.

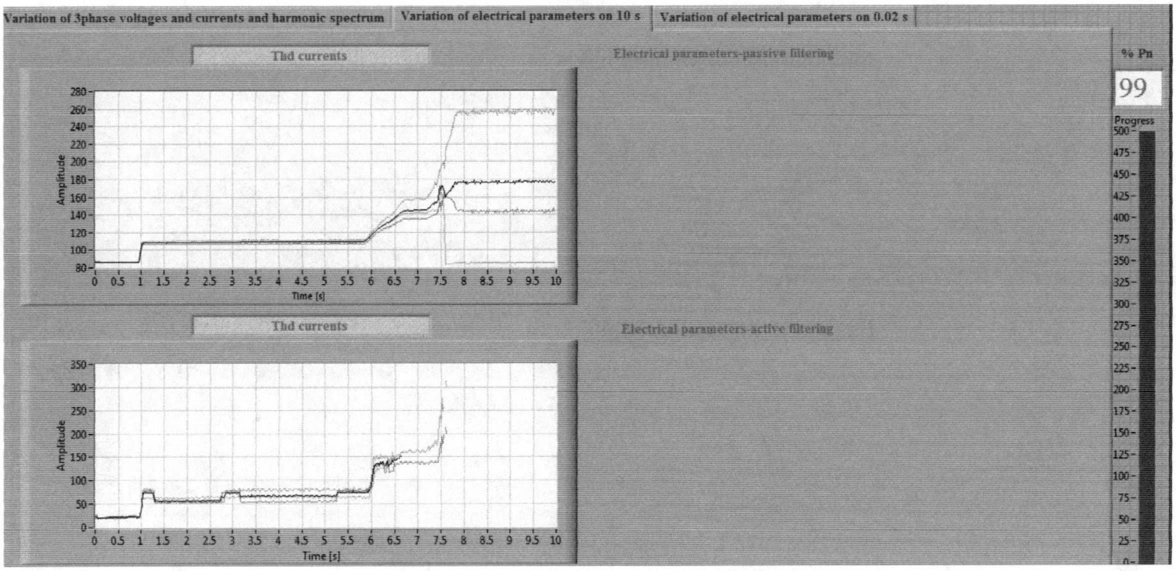

Figure 4. Front panel of the application

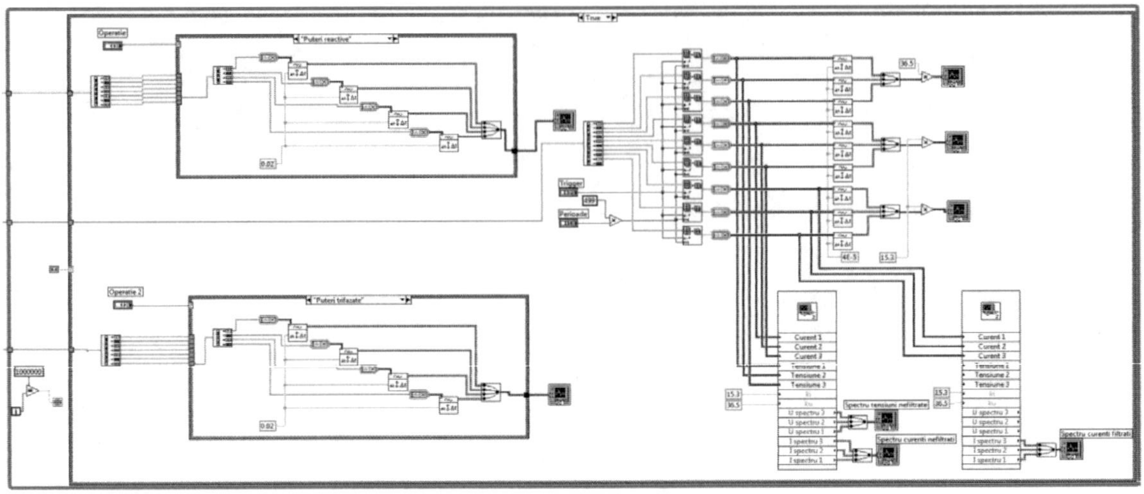

Figure 5. Displaying parameters subroutine

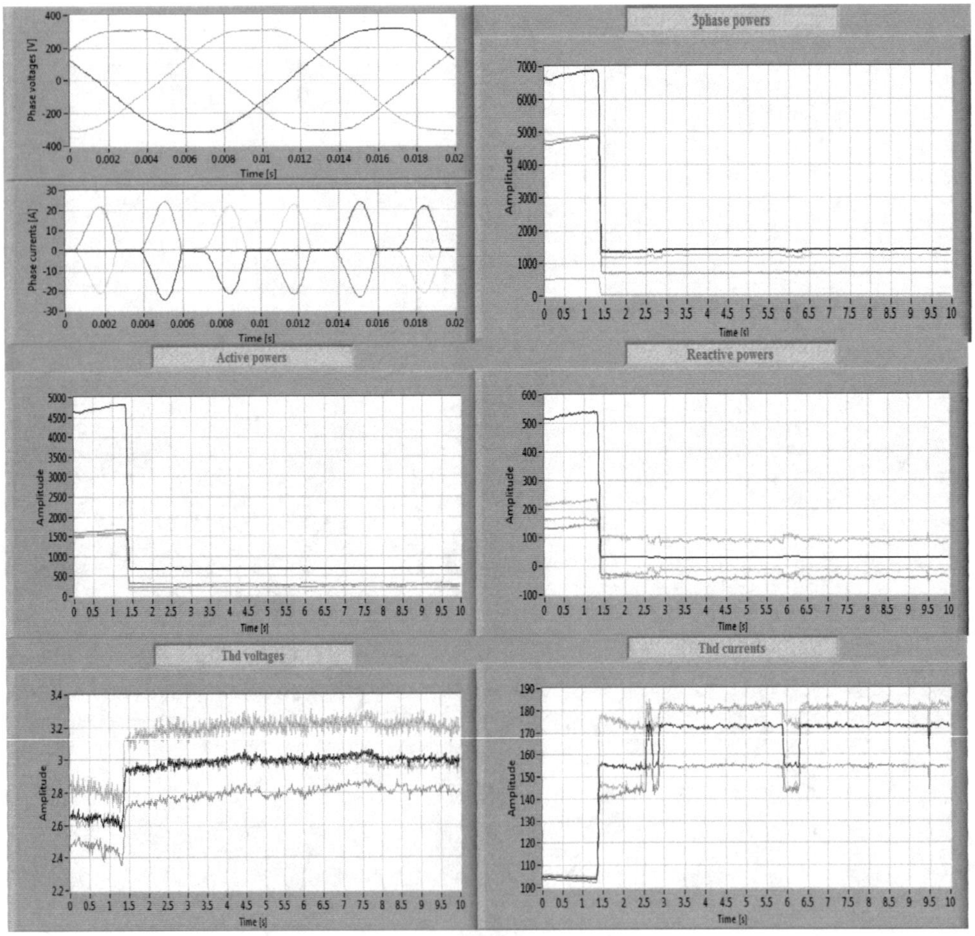

Figure 6. Variation of active, reactive, distorted and apparent powers, THD for phase voltages and currents, P = 4.5 kW

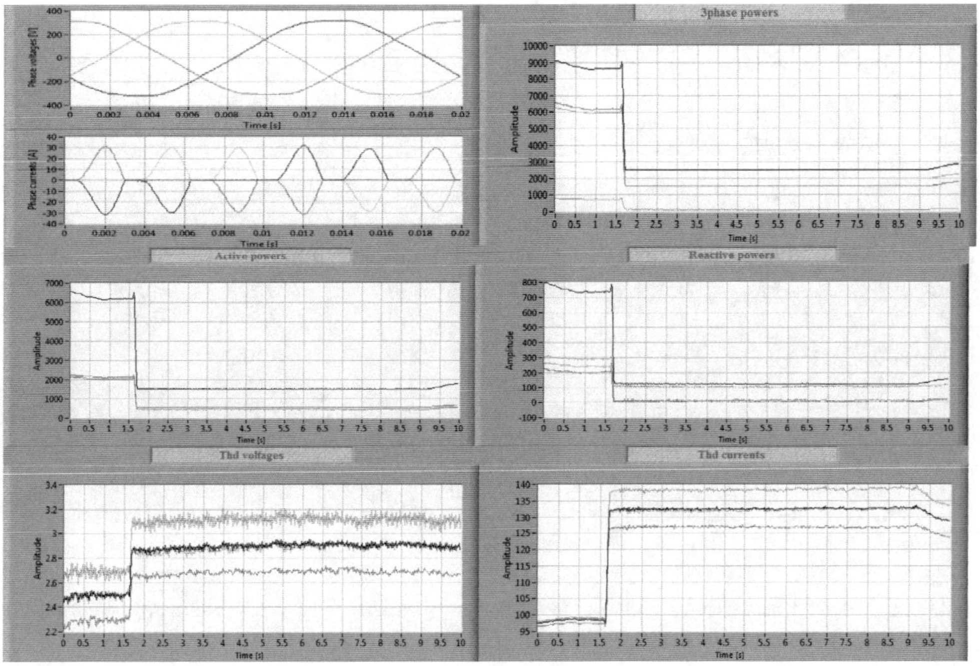

Figure 7. Variation of active, reactive, distorted and apparent powers, THD for phase voltages and currents, P = 9 kW

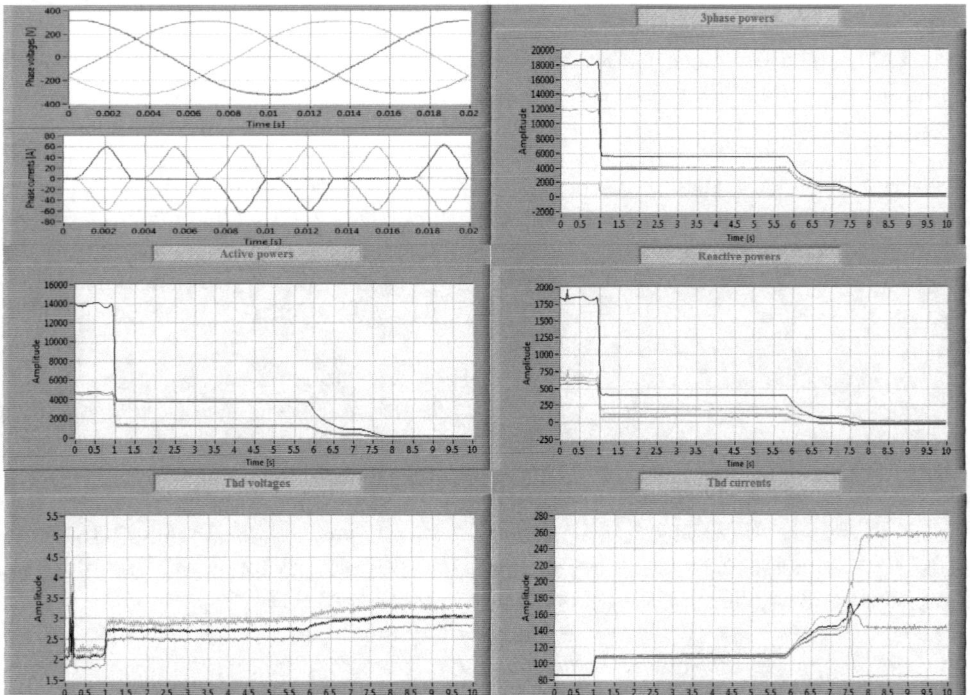

Figure 8. Variation of active, reactive, distorted and apparent powers, THD for phase voltages and currents, P = 15 kW

Table 2. Variation of electrical parameters using data acquisition system

Installation power [kW]	Phase voltage [V]	Phase current [A]	Active power [W]	Reactive power [VAr]	Distorted power [VAd]	Apparent power [VA]	THD voltages [%]	THD currents [%]
1.5	226.5	3.5	1300	180	1750	2187	2.95	140
3	227	6	2700	300	3100	4122	2.8	120
4.5	226.5	10	4700	500	4800	6737	2.65	105
6	226.6	10.5	4800	500	4900	6878	2.7	105
7.5	226.5	11	5500	750	5300	7675	2.65	100
9	226.5	13	6200	800	6000	8665	2.5	98
10.5	226.3	15.5	7500	1000	7000	10308	2.4	95
12	227	19	9500	1500	8500	12835	2.35	92
13.5	227	24	11500	1700	10000	15334	2.1	90
15	226.5	27	14000	2000	12000	18547	2	85

As well, THD for phase currents has a slow decreasing, but maintaining high values (85-140%).

Comparing the results obtained by both methods can be concluded that during the acquisition, the effective values of phase voltages are constant. The current values and the power factor are increasing with the increasing of the absorbed power of the installation. The total harmonic distortion of the voltage waveform is situated below standard limits (8%). The variation of phase current acquired at the point of common coupling is deeply distorted from sinusoidal shape, so THD can reach 140%. Following the results, the distortion of the current variation falls below the standard limits and the current harmonics must be compensated using appropriate devices [7-9].

References

[1] Tirian G O, Pinca-Bretotean C, Cristea D and Topor M 2012 Applications of fuzzy logic in continuous casting, *Transactions on Systems and Control* **5**(3) 133-142

[2] Iordan A E 2012 *Oriented object development of an interactive software designed for the study of A* method*, *AWERProcedia Information Technology & Computer Science* **2** 127-131

[3] Zamora I, Mazon, A J, Eguia P, Albizu I, Sagastabeitia K J and Fernández E 2003 *Simulation by MATLAB/Simulink of active filters for reducing THD created by industrial systems*, IEEE Bologna Power Tech. Conference, Bologna, Italy, June 23-26

[4] Panoiu C, Rob R, Panoiu M and Popa G 2011 Active power influence on the electrical parameters of an electro thermal installation with electromagnetic induction, *WSEAS Transactions on Power Systems* **6**(3) 63-72

[5] Rob R, Sora I, Panoiu C and Panoiu M 2010 *Using active filters in reducing of the total harmonic distortion on electro thermal installation with electromagnetic induction*, International Symposium of Power Electronics, Electrical Drives, Automation and Motion SPEEDAM 2010, Pisa, Italy, June 14-16, pp 233-238

[6] Iagar A, Popa G and Dinis C 2014 The influence of home nonlinear electric equipment operating modes on power quality, *WSEAS Transactions on Systems* **13** 357-367

[7] Akagi A, Watanabe E and Aredes M 2004 The P-Q theory for active filter control: some problems and solutions, *Revista Controle & Automacao* **15**(1) 78-84

[8] Iagar A 2011 Power quality problems generated by line frequency coreless induction furnaces, in *Power quality, harmonics analysis and real measurements data* (edited by Gregorio Romero Rey and Luisa Martinez Muneta), pp 235-260

[9] Rusu Anghel S, Tirian O, Osaci M and Dinis C 2009 *Algorithms for Raw Material Dosage Control for Cement Plants*, Proceedings of the 13th WSEAS International Conference on Systems, Rodos, Greece, July 22-24, pp 236-241

Study regarding the non-sinusoidal regime imposed by nonlinear loads

R Rob[1], C Panoiu[1] and M Panoiu[1]

[1]Politehnica University of Timisoara, Department of Electrical Engineering and Industrial Informatics, 5 Revolution Street, 331128 Hunedoara, Romania

E-mail: raluca.rob@fih.upt.ro

Abstract. Present paper represents a study concerning the variation of the phase voltages and phase currents acquired from the point of common coupling (PCC) during the functioning of an electro thermal installation with electromagnetic induction. The variation of the electrical parameters is followed on 0.02 seconds and also on 10 seconds using three methods. First method consists in modeling and simulation the electric scheme of the electro thermal installation. The second method uses a power quality analyzer and the third method is using a data acquisition system that contains an adapting interface and a data acquisition board connected to a computer. The variation of the phase voltages and phase currents and also their total harmonic distortions are presented. The electro thermal installation that is studied in this paper is a laboratory device. The technique implies reduce costs and the presented results are very useful in studying the distorting regime generated into the power network by the current harmonic sources.

1. Introduction

The presence of distorting regime into the power distribution leads to the increasing of power loss, to improper functioning of the communication systems or problems in functioning of the electronic equipment. The most unfavourable effects of the distorting regime are presented in the followings:
-overloading the capacitor banks used for power factor compensation;
-increasing the apparent power due to the distorting power;
-increasing the active power loss due to Joule – Lenz effect, determining overheating the electrical lines and the electric machines;
-improper functioning of the protection devices.

Analysing of the electrical parameters acquired during the functioning of nonlinear loads is very necessary for designing the power conditioning devices [1-2].

2. Description of the electro thermal installation

Analysing the electric parameters acquired during the functioning of nonlinear loads is very necessary for designing the power conditioning devices. A simplified scheme of the electrothermal installation is presented in Figure 1.

The electronic converter CTC 100K15 has the following characteristics [2-4]:
- supplying voltage 3 x 400 V, 50 Hz;
- rated current 27 A;
- control voltage 24V DC;

IOP Conf. Series: Materials Science and Engineering **106** (2016) 012023 doi:10.1088/1757-899X/106/1/012023

- consumed power at high frequency 15 kW;
- voltage value at medium frequency 500 V AC.

The hardening inductor is supplied from two high frequency toroidal transformers:

- T1: 660/500 V, 40 kVA, 70-100 kHz;
- T2: 150 kVA, 70-120 kHz with primary winding voltage of 500V and variable transformation ratio of 3:1, 4:1, 5:1, 6:1, 7:1, 8:1, 9:1 and 10:1. The electro thermal installation contains three coil shape inductors for hardening process.

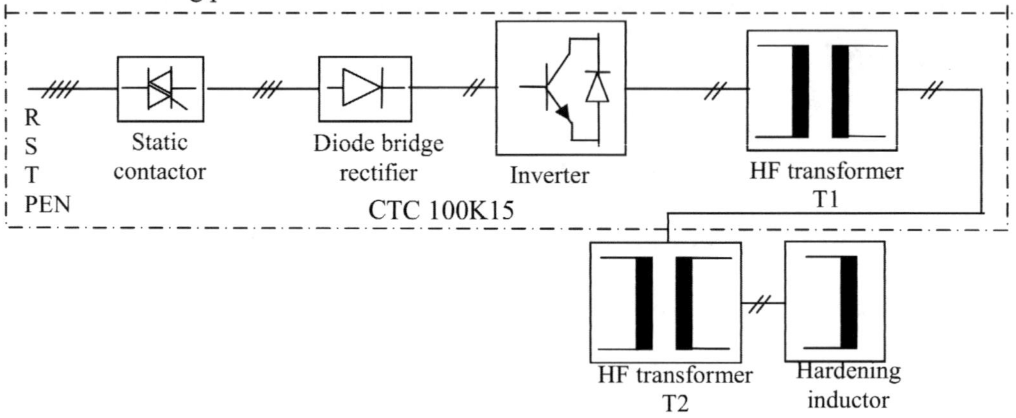

Figure 1. Electro thermal installation with electromagnetic induction

3. Methods used for studying the parameters variation

3.1. Modelling and simulation the electro thermal installation

In order to simulate the functioning of the electro thermal installation, PSCAD-EMTDC program was used. For simulating the power distribution a three phase voltage source was chosen: Un = 6 kV, f = 50 Hz. Power transformer has the following characteristics: Un1/Un2 = 6 / 0.4 kV and nominal power Sn = 400 kVA. The electric scheme presented in Figure 2 contains a thyristor static contactor controlled by a pulse control block. PSCAD model for static contactor command is also shown in Figure 2. The alternative voltage is rectified using a diode bridge rectified and a smoothing filter with L = 1mH, C = 300μF. The inverter consists in four IGBT transistors and the model of inverter command is presented in Figure 2. High frequency transformer T1 has the following characteristics: U1/U2=0.6/0.5 kV, ST1=40 kVA, f=100 kHz and T2 has the following characteristics: U1/U2=0.5/0.1 kV, ST2=150 kVA, f=100 kHz. Inductor is modeled using a RL series circuit R=0.25 Ω and L=1·10-7 H. The phase currents that must be studied are measured using the ammeters Ia, Ib, Ic from Figure 2. The variation of the phase currents measured at the PCC of the electro thermal installation with the power distribution is presented in Figure 3. The nominal power of the electro thermal installation is 15 kW.

From this figure can be observed that the distortions of the phase voltages can be neglected, but the variation of phase currents (THDI) is strongly distorted from the sinusoidal form, as in Table 1.

Table 1. PSCAD simulation for phase voltages, phase currents and THDI

Phase voltages (RMS) [V]	Phase currents (RMS) [A]	THDI [%]
233.3	35	78.8

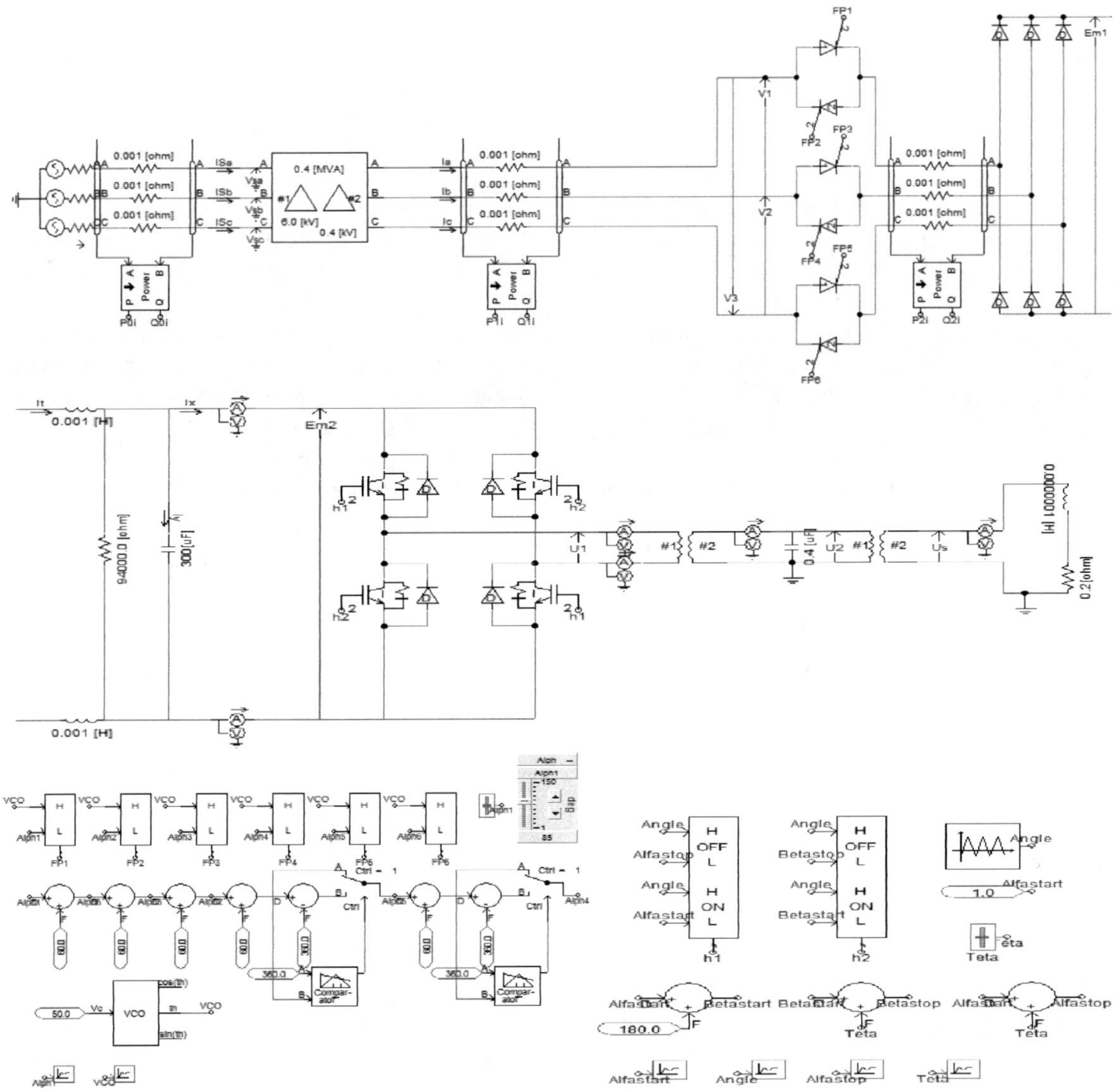

Figure 2. Electrical scheme of the electro thermal installation (PSCAD-EMTDC simulation)

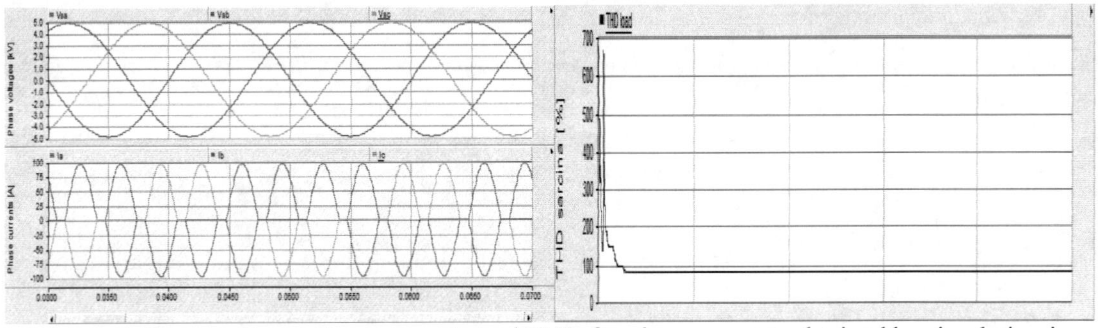

Figure 3. Phase voltages phase currents and THD for phase currents obtained by simulation in PSCAD (P = 15kW)

3.2. Data acquisition using CA8334B power analyser

CA8334B is a power quality analyser which is able to acquire in real time samples of phase voltages and phase current and also computes several electric parameters [5]. In this case the analyser has been connected to the PCC of the nonlinear load with the power distribution.

In Figure 4 the variation of phase voltages, phase currents and THD for phase currents is presented in situations when the active power is: no load operation, 4.5 kW, 9 kW, 15 kW. In Table 2 the values of these electrical parameters are related. Phase currents and phase voltages are processed when the acquisition period of the power analyzer is 0.1 ms. The variation of THD for phase currents is processed with the acquisition period of 1s and the following conclusions can be generated. The variation of THD presents 3 stages.

The first stage is characterized by high values of power when the piece temperature is reaching Curie point. The duration of the first stage is depending of active power; therefore it is decreasing with the increasing of the active power. In the second stage, the piece temperature is increasing to hardening point and in the third stage the piece is extracted from the inductor.

Table 2. Phase voltages, phase currents and THDI obtained with CA8334B

Absorbed power [kW]	Phase voltages (RMS) [V]	Phase currents (RMS) [A]	THDI [%]
No load	229.8	2.8	
4.5	228	6.9	130
9	228.2	10.3	126
15	229.8	29.6	120

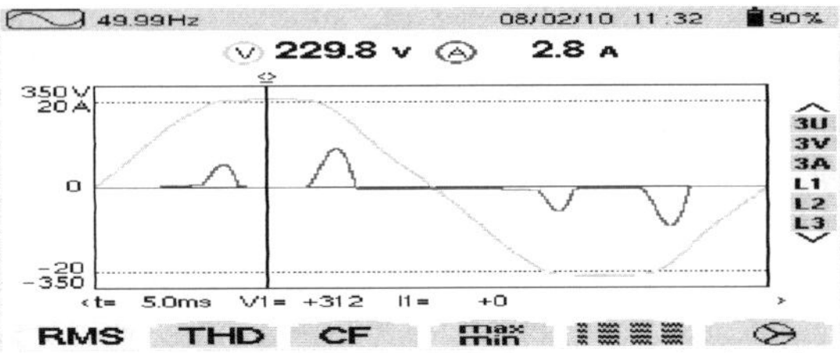

a) Phase currents and phase voltages acquired with CA8334B, no load operation

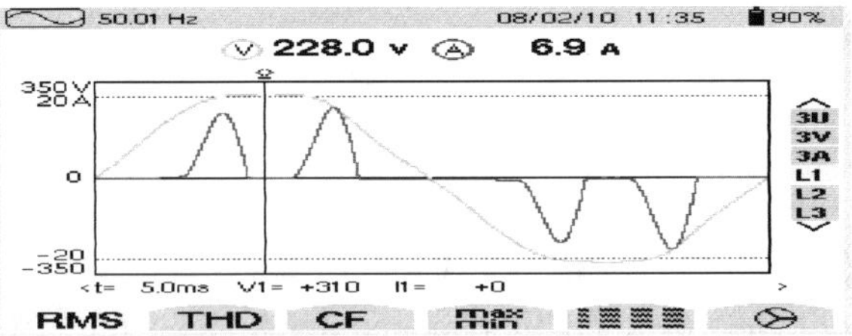

b) Phase currents and phase voltages acquired with CA8334B, P = 4.5kW

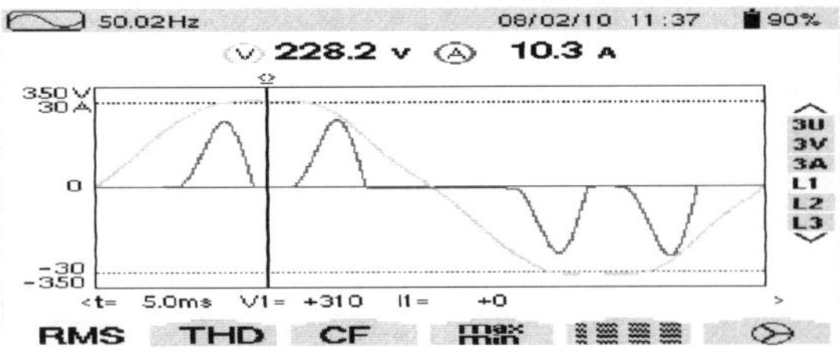

c) Phase currents and phase voltages acquired with with CA8334B, P = 9kW

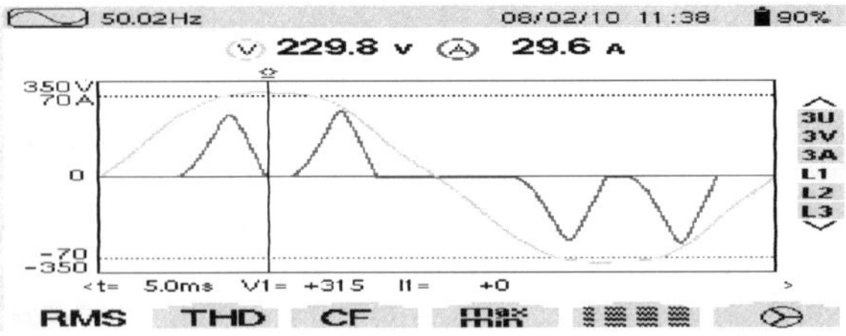

d) Phase currents and phase voltages acquired CA8334B, P = 15kW

Figure 4. Phase voltages, phase currents and THD for phase currents acquired with CA8334B

IOP Conf. Series: Materials Science and Engineering **106** (2016) 012023 doi:10.1088/1757-899X/106/1/012023

3.3. Description of data acquisition system

Data acquisition system also follows the acquisition of samples from phase voltages and phase currents at the point of common coupling. In order to accomplish the acquisition, an adapting block and a data acquisition board connected to a computer were used. The connection of data acquisition system to PCC is presented in Figure 5a.

The phase voltages and phase currents were acquired using the adapting interface which was designed in order to realize the galvanic isolation between the electro thermal installation and the data acquisition system and also to realize the compatibility of voltage levels.

The data acquisition board has the following characteristics:

-16 analogic inputs, with 250 kS/s and 16 bit resolution;
-2 analogic outputs, with 740 kS/s and 16 bit resolution;
-24 TTL digital in/out;
-1 digital trigger;
-Windows, Linux compatibility.

In order to process the acquired data, two software applications designed in LabVIEW 2011 were used: acquisition application and computing application.

Using the acquisition application, the voltage and current samples are transformed into numeric data and stored into text documents.

This information is loaded by the computing application which is able to compute the most important electrical parameters that characterize the functioning of the electro thermal installation. The algorithm of the computing application is presented in Figure 5b.

Using appropriate subroutines, 10 analogical input channels were configured. Because the maximum sampling rate is 250 kS/s, the maximum sampling rate for a single channel is 25 kS/s.

The number of samples for a single period results from the relation:

$$N = f_S \cdot T = 25kS / s \cdot 0.02s = 500\,samples \quad (1)$$

The acquisition is developed on 10 seconds interval. The number of samples acquired on each channel is 250000.

Using presented applications, 10 measurement sets were accomplished by increasing with 10% the power of electro thermal installation.

In Figure 6 the variation on 0.02 seconds and 10 seconds for phase voltages and phase currents is presented. As well, Figure 6 contains the variation on 0.02 seconds for THD for phase voltages and phase currents when the absorbed power of the power installation is 4.5 kW, 9 kW and 15 kW.

Some observations must be made. In figures that represents the variation form of the currents and voltages (on period of 0.02 seconds), phase 1 is red, phase 2 is green and phase 3 is blue.

For the other variations on 10 seconds, the colour signification is: black for all three phases, red for phase 1, green for phase 2 and blue for phase 3.

In Table 3 are synthesized the electrical parameters obtained by accomplishing 10 measurement sets using data acquisition system.

Studying the variation from Figure 6, it can be concluded that the total harmonic distortion for the phase currents reaches high values, comparable with the others obtained with the previous methods. These values are decreasing with the increasing of the absorbed power.

The total harmonic distortion for phase voltages is approximately constant during the variation of the absorbed power of the installation.

4. Conclusions

The electro thermal installation that is described in this paper represents a laboratory device. The electromagnetic pollution generated by its functioning deserves high importance. In order to study the variation of the electrical parameters at the point of common coupling with the power distribution during the functioning of the electro thermal installation, three methods were accomplished.

Table 3. Electrical parameters obtained with data acquisition system

Absorbed power [kW]	Phase voltages (RMS) [V]	Phase currents (RMS) [A]	THD voltages [%]	THD currents [%]
1.5	226.5	3.5	2.95	140
3	227	6	2.8	120
4.5	226.5	10	2.65	105
6	226.6	10.5	2.7	105
7.5	226.5	11	2.65	100
9	226.5	13	2.5	98
10.5	226.3	15.5	2.4	95
12	227	19	2.35	92
13.5	227	24	2.1	90
15	226.5	27	2	85

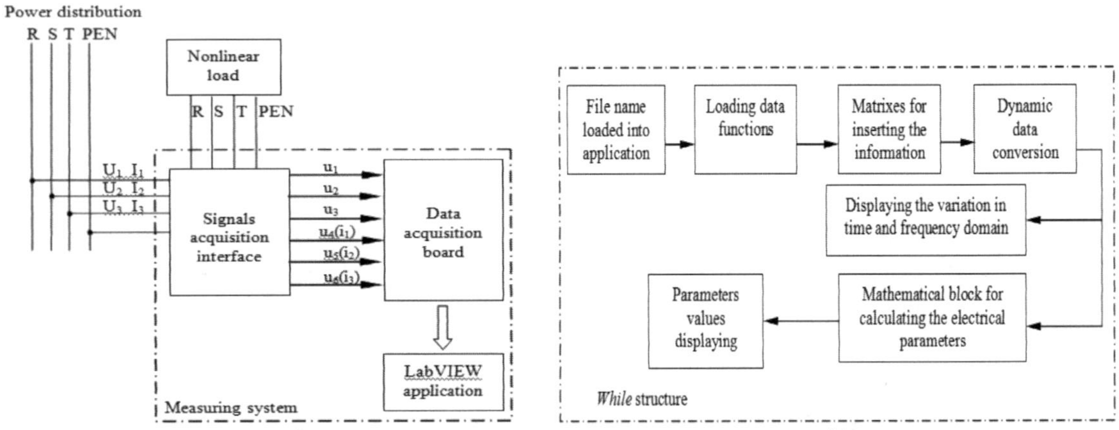

a) b)

Figure 5. a) Connection of data acquisition system at PCC, b) Computing application block scheme

Using PSCAD-EMTDC, simulations of the electric equipment's functioning can be accomplished. So, the command and control elements for the electro thermal installation were modelled. As can be seen from Figure 3, the variation of phase voltages, phase currents and THD for phase currents are presented. The voltage system is balanced and symmetrical. The phase voltages present a neglected distortion from the sinusoidal form, so the Figure 3 presents only the variation of the total harmonic distortion for the phase currents. By simulation the functioning of the electro thermal installation when the absorbed power is 15 kW, the THD for phase currents are 87.7 %, falling below standard limits of 8 %.

Acquisition with CA8334B power analyser represents a very practical method for studying in real time of the electrical parameters variation. From Figure 4 results a sinusoidal variation for phase voltages.

IOP Conf. Series: Materials Science and Engineering **106** (2016) 012023 doi:10.1088/1757-899X/106/1/012023

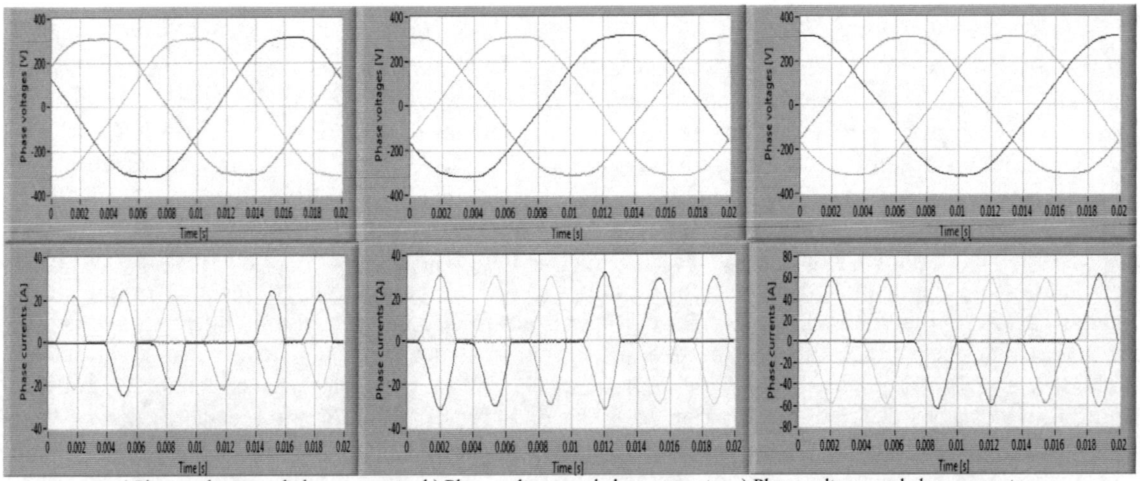

a) Phase voltages and phase currents (variation on 0.02s), P = 4.5kW b) Phase voltages and phase currents (variation on 0.02s), P = 9kW c) Phase voltages and phase currents (variation on 0.02s), P = 15kW

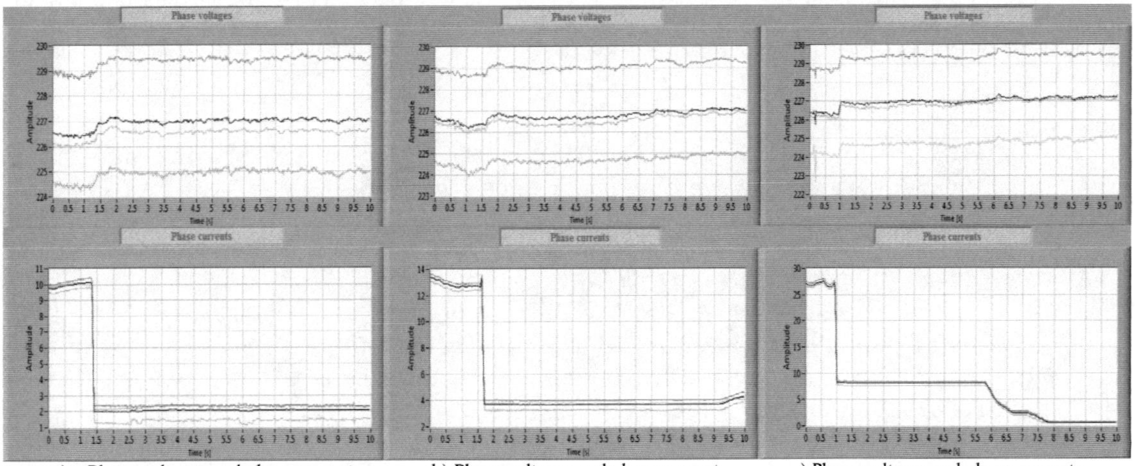

a) Phase voltages and phase currents (variation on 10s), P = 4.5kW b) Phase voltages and phase currents (variation on 10s), P = 9kW c) Phase voltages and phase currents (variation on 10s), P = 15kW

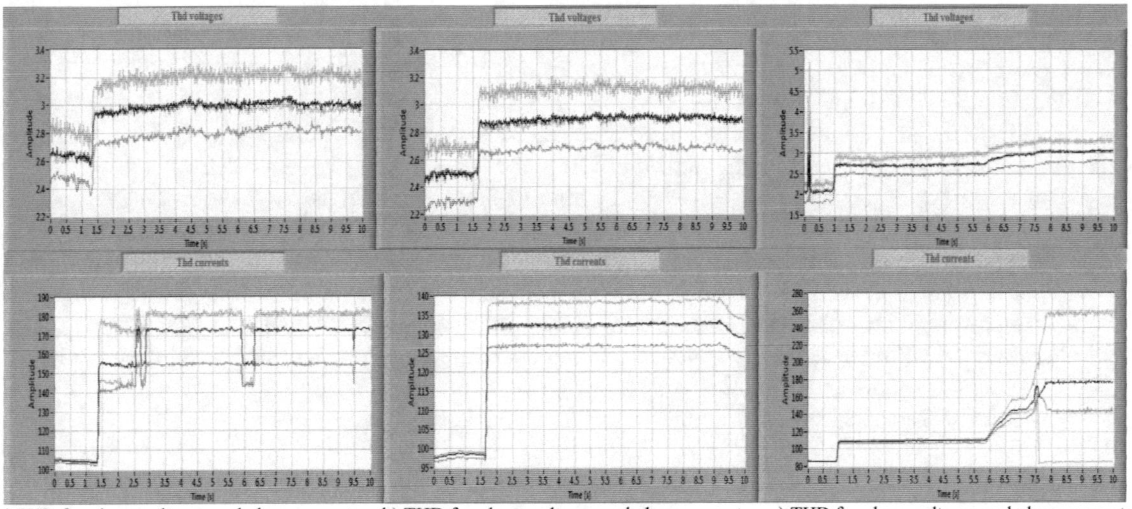

a) THD for phase voltages and phase currents (variation on 10s), P = 4.5kW b) THD for phase voltages and phase currents variation on 10s), P = 9kW c) THD for phase voltages and phase currents variation on 10s), P = 15kW

Figure 6. Phase voltages, phase currents and THD for phase voltages and phase currents acquired with data acquisition system

The phase currents are deeply distorted and this fact demonstrates that the electro thermal installation represents a current harmonic source. The RMS value of the phase currents is increasing with the increasing of the absorbed power. At 15 kW, RMS value of the phase currents is 29.6 A. The total harmonic distortion for phase currents exceeds 120 %. These values fall below standard limits of 8 %.

Data acquisition system which is described in the third method presents the advantage of permanently improvement concerning the soft applications that compute in real time the electrical parameters. The application displays the variation on 0.02 seconds and 10 seconds of the phase voltages, phase currents and THD for phase voltages and phase currents. Studying Figure 6, can be concluded that the voltages are balanced and symmetrical (the same conclusion from the second method) and THD for phase voltages are approximately constant and can be neglected. By studying Figure 6, we can observe that the variation on 10 seconds presents the same three stages described in the second method. Studying Table 3, the RMS values of the phase currents are increasing (3.5 – 27 A) and the THD for phase currents are decreasing (140 – 85 A) with the increasing of the absorbed power. Even in this situation, the values of THD for phase currents fall below standard limits. The most important conclusion is that the current harmonics must be compensated using appropriate devices [6-10].

References

[1] Akagi A, Watanabe E and Aredes M 2004 The P-Q theory for active filter control: some problems and solutions, *Revista Controle & Automacao* **15**(1) 78-84

[2] Panoiu C, Rob R, Panoiu M and Popa G 2011 Active power influence on the electrical parameters of an electro thermal installation with electromagnetic induction, *WSEAS Transactions on Power Systems* **6**(3) 63-72

[3] Rob R, Sora I, Panoiu C and Panoiu M 2010 *Using active filters in reducing of the total harmonic distortion on electro thermal installation with electromagnetic induction*, International Symposium of Power Electronics, Electrical Drives, Automation and Motion SPEEDAM 2010, Pisa, Italy, June 14-16, pp 233-238

[4] Panoiu C, Rob R, Panoiu M and Popa G 2010 *Measurements of the electrical parameters of an electrothermal installation with electromagnetic induction*, Proceedings of the 14th WSEAS International Conference on System, Corfu, Greece, July 22-25, pp 216-221

[5] Arnoux C *Three phase power quality analyzer CA 8332B, CA 8334B. User's manual*

[6] Fujita H and Akagi H 2009 The Unified Power Quality Conditioner: the Integration of Series and Shunt Active Filters, *IEEE Transactions on Power Electronics* **13**(2) 315-322

[7] Jacobina C de Rossiter Correa, M, Ferreira Pinheiro R, de Silva E and Nogueira Lima A 2010 Modeling and control of unbalanced three phase systems containing PWM converters, *IEEE Transactions on Industry Applications* **37**(6) 1807-1816

[8] Rusu Anghel S, Tirian O, Osaci M and Dinis C 2009 *Algorithms for Raw Material Dosage Control for Cement Plants*, Proceedings of the 13th WSEAS International Conference on Systems, Rodos, Greece, July 22-24, pp 236-241

[9] Iagar A, Popa G and Dinis C 2014 The influence of home nonlinear electric equipment operating modes on power quality, *WSEAS Transactions on Systems* **13** 357-367

[10] Iagar A 2011 Power quality problems generated by line frequency coreless induction furnaces, in *Power quality, harmonics analysis and real measurements data* (edited by Gregorio Romero Rey and Luisa Martinez Muneta), pp 235-260

Characterisation of the TiO_2 coatings deposited by plasma spraying

M L Benea[1] and L P Benea[2]

[1]Politehnica University of Timisoara, Department of Engineering and Management, Hunedoara, Romania
[2]The National Institute of Applied Sciences Lyon, France

E-mail: laura.benea@fih.upt.ro

Abstract. Plasma spraying of materials such as ceramics and non-metals, which have high melting points, has become a well-established commercial process. Such coatings are increasingly used in aerospace, automobile, textile, medical, printing and electrical industries to impart proprieties such as corrosion resistance, thermal resistance, wear resistance, etc.
One of the most important characteristics of thermal barrier coatings is the ability to undergo fast temperature changes without failing, the so called thermal shock resistance.
The formation of residual stresses in plasma sprayed ceramic and metallic coatings is a very complex process. Several factors, such as substrate material, substrate thickness, physical properties of both the substrate and the coating material, deposition rate, relative velocity of the plasma torch, etc. determine the final residual stress state of the coating at room temperature.
Our objective is to characterize the titanium oxide and aluminium oxide coatings deposited by plasma spraying in structural terms, the resistance to thermal shock and residual stresses.

1. Introduction

Ceramic thermal barrier coatings deposited by plasma spraying are used aerospace, automobile, textile, medical, printing and electrical industries to impart proprieties such as corrosion resistance, thermal resistance, wear resistance, etc. [1], [2].

The coating is built up particle by particle and it has a lamellar structure resulting from the flattening of particles on already solidified flattened particles of the substrate. Thus the thermomechanical properties of the coatings depend strongly on the effective contacts between the lamellar and the cracks network resulting from relaxation stresses (stresses induced during spraying or in service condition) [3].

Our objective is to characterize the TiO_2 and Al_2O_3 coatings deposited by plasma spraying in structural terms, the resistance to thermal shock and residual stresses.

IOP Conf. Series: Materials Science and Engineering **106** (2016) 012024 doi:10.1088/1757-899X/106/1/012024

2. Experimental part

2.1. Experimental conditions at the TiO2 coatings spraying in plasma

The substrateis a martensiticstainlesssteel (Z12CNDV12) whose chemical composition is presented in Table 1.

Table 1. Chemical composition of martensitic stainless steel Z12CNDV12

	C	Si	Mn	S	P	Cr	Ni	Mo	V	N2
Chemical composition	0,06	≤	0,50	≤	≤	11,0	2,00	1,5	0,25	0,02
	0,15	0,35	0,90	0,025	0,035	12,5	3,00	2,0	0,40	0,04

The powder used is Metco 102, with 99% TiO2 and particle sizes between 7.8 and 88 μm.

The coatings have been made using the GPPR-400 plasma generator.

The work has been done using the following parameters: intensity of the current at the generator: 500 A, voltage: 70 V, spraying distance: 70 mm, plasma gas flow: 36,6 l/h, coating thickness: 0,3 mm.

2.2. Thermal shock testing

For the coats made by using the Metco 102 powder on a martensitic stainless steel substrate, the thermal shock testing consist in: fast heating to 800°C or 500°C in 10s and immediate water cooling (600°C/s) without maintaining the maximum temperature.

2.3. The model of Takeuchi for calculation of residual stresses in plasma sprayed ceramic coatings

The formation of residual strains in plasma sprayed ceramic and metallic coatings is a very complex process. Several factors, such a substrate material, substrate thickness, physical properties of both the substrate and coating material, deposition rate, relative velocity of the plasma torch, etc., determine the final residual stress state of the coating.

Several authors present models concerning the origin of residual stresses in plasma sprayed ceramic and metallic coatings. Many of the models are merely qualitative and predict only the sign of the stress in the coating, compressive or tensile [4-6]

Takeuchi [7], Elsing [8-10] and Buckley-Golder [11], [12] present quantitative models for the calculation of residual stress. The model of Takeuchi takes primary cooling into account by introducing a strain term εe, representing strains in the coating caused by primary cooling. The Takeuchi model uses a uniform substrate temperature and assumes that the entire coating is deposited in one pass.

The stress in the coating at elevated substrate temperature is describe by the following equation:

$$\sigma_c(ed, x, T_s) = E_c \cdot \left[\left(\frac{E_c \cdot x + E_s \cdot es}{E_c \cdot ed + E_s \cdot es} \right)^{\varepsilon_e} (1+\varepsilon_e) - 1 \right] \tag{1}$$

Where:

- x: position in coating [m]
- E_c: Young's modulus coating [Pa]
- E_s: Young's modulus substrate [Pa]
- t_s: thickness of substrate [m]
- t_c: thickness of coating [m]
- ε_e: strain in coating due to primary cooling [-]
- T_s: substrate temperature [K].

According to Takeuchi, the strain εe, caused by primary cooling equals:

$$\varepsilon_e = \alpha_c (T_m - T_s)$$

Where:

- α_c: coefficient of thermal expansion coating

IOP Conf. Series: Materials Science and Engineering **106** (2016) 012024 doi:10.1088/1757-899X/106/1/012024

T_m: melting point coating material.

When the stress, caused by primary cooling, exceeds the yield of tensile strength of the coating material, ε_c has to be derived from the stress strain curve of that material.

After cooling down to room temperature, the stress in the coating is equal to:

$$\sigma_c(ed, x, T_s) = E_c\left[\frac{(1+\varepsilon_e)(\gamma+1)}{\beta+(\gamma+1)^{(1+\varepsilon_s)}-1}\left(\frac{E_c \cdot x}{E_s \cdot es}+1\right)^{\varepsilon_s}-1\right] \quad (2)$$

With:

$$\beta = \frac{1-\alpha_c(T_s - T_r)}{1-\alpha_s(T_s - T_r)} \qquad \gamma = \frac{E_c \cdot ed}{E_s \cdot es}$$

T_s: substrate temperature [K]

T_r: room temperature [K]

ε_s: coefficient of thermal expansion substrate [K^{-1}].

The determination of the crystalline structure of the covering layers was done by means of RX spectra obtained of a DRON-3 refractometer. Radiation Cu K_α was used at a voltage of 30 kV and a current of 30 mA. The detector movement rate was 1°/min.

3. Results and discussions

By comparing the RX spectra for the Metco 102 powder and for the coating obtained with this powder (Figure 1), it can be observed that the position of peaks are identical in both cases.

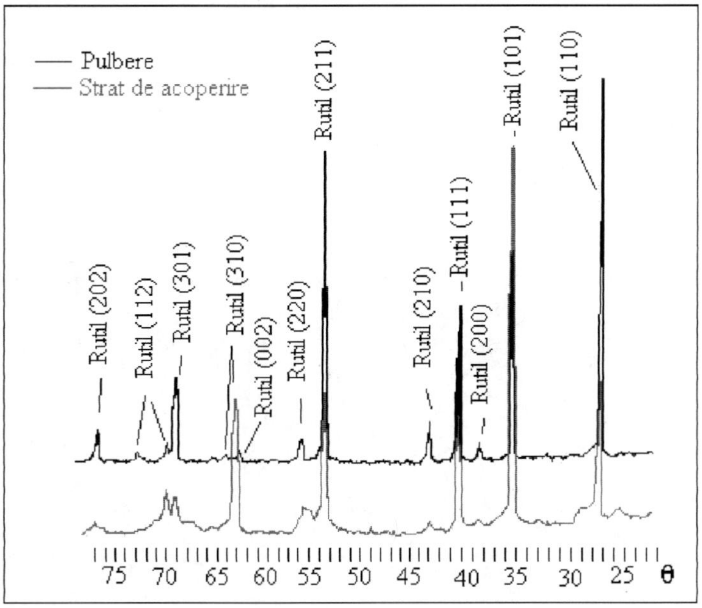

Figure 1. Comparison of the RX spectra obtained with the powder and covering layer achieved with Metco 102 (TiO_2)

3.1. Thermal shock determination

Some results of the experimental determination are shown in Table 2.

Table 2. Results of the thermal shock determination

Sample ref.	The number of thermal cycles	The maximum temperature cycle [°C]	Cooling medium	Observation
48	20	800	water	The appearance of scorches
50	30	800	water	The appearance of scorches
54	25	800	water	The appearance of scorches
	Medium		59	
3	180	500	water	No cracks, No scorches

Figure 2 shows some images for the coatings made using Metco 102 powder, in a different moments of the determination. Samples were examined by optical microscopy using a stereo microscope.

As expected, it can be observed that TiO2 coatings has a reduced resistance to heat shock at 800°C (over 25 cycles).

The coat, in its original state, has a characteristic ceramic deposition, with a pronounced roughness without visible cracks to the naked eye or to the optical microscope at low zoom. The images in Figure 2 reveal the fact that a smooth network of cracks is formed after the first set of 22 set of cycles. The cracks enlargement is fast, showing in this way the extended cracks and the exfoliations detected on the sample number 54. The breaking of the surface is highlighted by the glossy appearance of the detachment surface limits. The observed cracks have the propagation way perpendicular to the longitudinal axis of the sample, highlighting the direction with maximum stretching tensions during the experiment. Also, the cracks orientation on the layer thickness is from the substrate towards the deposit surface determined by the tensions in the substrate during the experiment.

Because the treatment chose in the first step was considered excessively harsh for the ability of TiO2 to take thermal shocks, the experiment was resumed on a new set of samples covered in the same conditions at a maximum temperature of the thermal cycle of 5000C with a cooling speed of 6000C/s. After 180 cycles there were no signs of exfoliation. The network of cracks appears to form at a higher number of cycles (54) and extends slowly, increasing the number of cycles. The number of cracks expand significantly over 135 cycles. At the limit of 180 cycles the apparition of the first dislocation of the covering layer on the limits of the cracks can be notified.

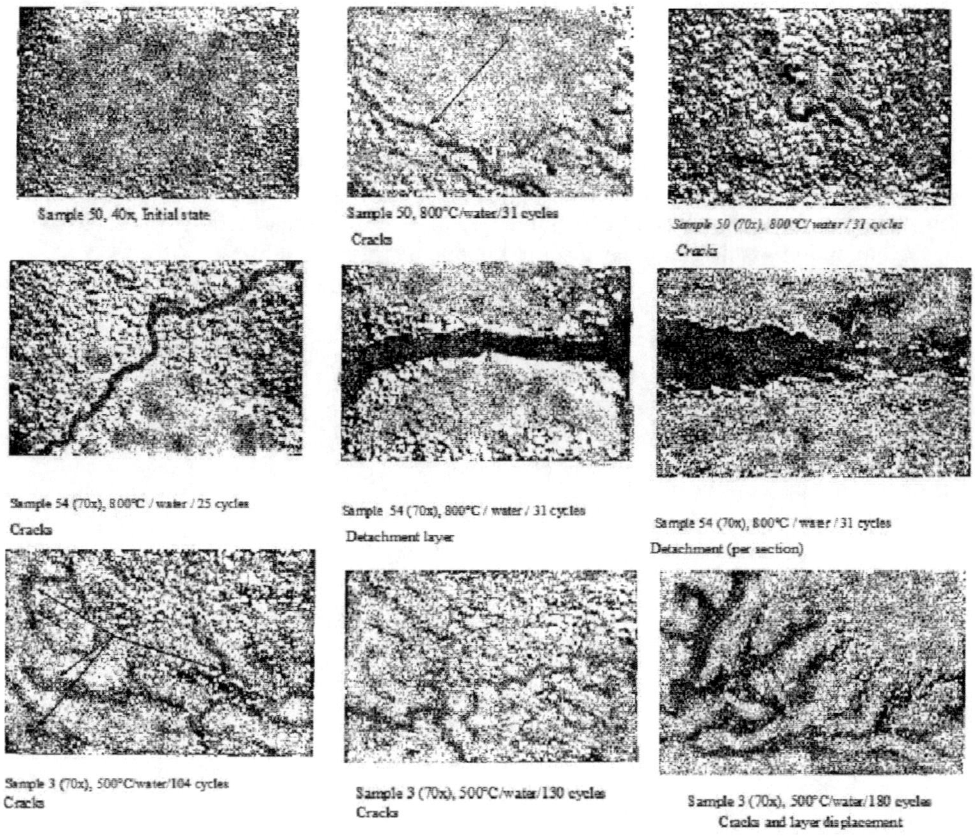

Figure 2. Images for the coats made by using the Metco 102 powder at the thermal shock

3.2. The determination of the strains in the coating layer using Takeuchi's method

The constants necessary for the determination of strains in the aluminium oxide and titanium oxide coating layers were obtained from the literature. They can be found in Table 3.

Table 3. The constants used for calculating the strains in the coating layers using Takeuchi's method

Constant	Symbol in the Matlab program	Al_2O_3 Z12CNDV12	TiO_2 Z12CNDV12
α_c	ac	$8\cdot10^{-6}$	$7{,}1\cdot10^{-6}$
α_s	as	$1{,}2\cdot10^{-5}$	$1{,}2\cdot10^{-5}$
e_d	ed	10^{-4}	10^{-4}
e_s	es	$1{,}5\cdot10^{-3}$	$1{,}5\cdot10^{-3}$
E_s	Es	$780\cdot10^{6}$	$780\cdot10^{6}$
E_c	Ec	$100\cdot10^{6}$	$70\cdot10^{6}$
T_m	Tm	2323	2033
T_s	Ts	423	423
T_r	Tr	293	293
x	x	$0\text{-}10^{-4}$	$0\text{-}10^{-4}$

IOP Conf. Series: Materials Science and Engineering **106** (2016) 012024 doi:10.1088/1757-899X/106/1/012024

For these calculations, the strain in the coating layer at high temperature described in equation (1) was noted with σ (sc in the Matlab program) and the the strain in the coating layer after cooling described in equation (2) was noted with σ_{max} (scmax in the Matlab program).

The data obtain was processed in Matlab, the strains σ and σmax which form in the coating layer obtained with aluminum and titanium oxide were represented in Figure 3 and Figure 4.

It can be observed that in the coating layers obtained with aluminium oxide, the strain after the primary cooling as well as the strain after cooling have higher values than in the case of the coatings obtained with titanium oxide, which can be explained by the fact that the coefficient of thermal expansion is higher for the aluminium oxide and thus closer to the one of the steel substrate.

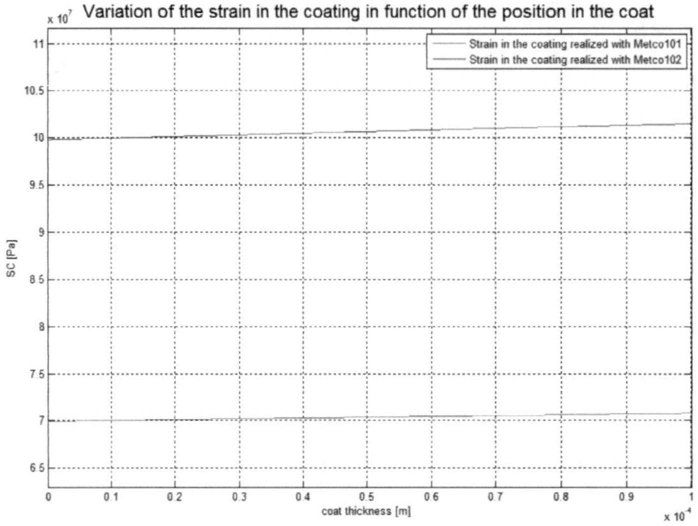

Figure 3. The variation of the strain σ in the aluminum and titanium oxide realized on a Z12CNDV12 stainless steel substrate

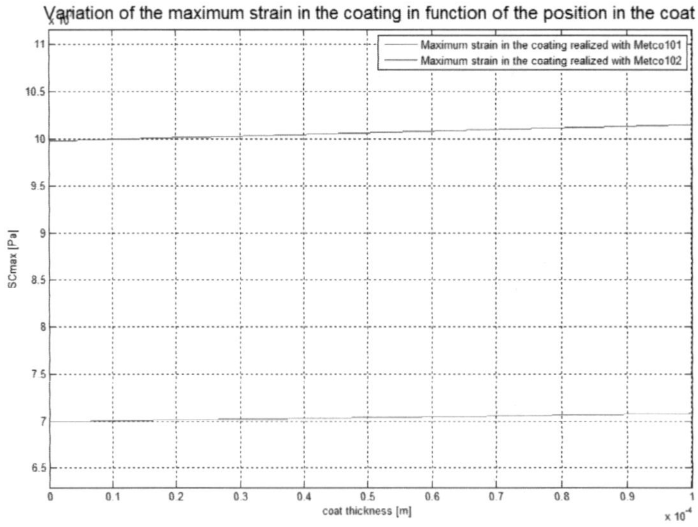

Figure 4. The variation of the maximum strain σ_{max} in the aluminum and titanium oxide realized on a Z12CNDV12 stainless steel substrate

4. Conclusion

Using plasma to achieve coatings is a complex process because it is influenced by many factors: generator parameters, characteristics of the powders used in the powder flow, spray distance, choosing the appropriate plasma gas, ensuring the substrate surface preparation, substrate characteristics, power, and so on

The spectra obtained from the Metco 102 powder and the coatings obtained from it, rutile characteristic peaks appear at the same values of θ, but they differ in their height and width.

Conclusions that results from analysing the results of determination the resistance to thermal shock of the TiO_2 layers are:

- The TiO_2 samples have a low resistance at thermal shock at 800°C (25 cycles);
- It has already formed a network of cracks after the first set of 22 cycles performed at 800° C;
- The growth of cracks is rapid, revealing the extended cracks and the exfoliation uncovered on the sample number 54;
- The breaking surface is highlighted by the glossy limits of the detachment surface;
- The cracks that have been seen have the sense of propagation perpendicular to the longitudinal axis of the sample attempted at 800 °C, highlighting the maximum stretch tension during the experiment;
- The orientation of the cracks on the coating layer thickness tested at 800 °C is from the substrate to the surface the deposit, this is determined by the tension of substrate during the experiment;
- Particularly in the first stage of the experiment is considered over harsh for TiO_2's capacity to take heat shock;
- At a maximum temperature of thermal cycle of 500 °C and a cooling of 600 °C/s after running a number 180 cycles, no samples presented peeling;
- The network of cracks is formed from a larger number of cycles and expands slowly when increasing the number of cycles;
- The maximum temperature of thermal cycle of 500 °C, the cracks are not extending significantly at more than 135 cycles;
- The apparition of the first displacement of coverage layer on the edge of the cracks is notified at the limit of the 180 cycles with the maximum temperature of 5000 °C.

The calculation of stresses that occur in the coating layers using the method of Takeuchi, leads to the conclusion that in the coating realized using Al_2O_3 powder (Metco 101) the tensions are higher than in the deposit made with TiO_2 (Metco 102). This can be understood by knowing of the fact that Al_2O_3 has a thermal expansion coefficient greater than TiO_2 and at the same martensitic stainless steel substrate, the difference in thermal expansion coefficients between the layer and the substrate is greater in the case of the Al_2O_3 and this tensions in the coating are greater.

References

[1] Meetham G W and Van de Woorde M H 2012 *Materials for High Temperature Engineering Applications*, Springer Verlag, Berlin Heidelberg New York, Germany

[2] Kvernes I, Lugscheider E and Norholm O 1988 *Potential of thermal and Wear Resistant Coatings in Combustion Engines*, 1st Plasma Technik Symposium, Lucerna, Elveţia, May 18-20, pp 41-55

[3] Bernard D, Vardelle M, Vardelle A and Fauchais P 1990 Influence of heat and momentum transfers between thermal plasma jet sand stabilized zirconia particulates on thermomechanical properties of theinsulting coatings, *Journal de Physique Colloque* **51**(C5) 331-341

[4] Clyne T W and Gill S C 1996 Residual stress in thermally sprayed coatings and their effect on interfacial adhesion - A review of recent work, *Journal of Thermal Spray Technology* **5**(4) 401-418

[5] Hseuh C H and Evans A G 1995 Residual stresses and cracking in metal/ceramic systems for microelectronics packaging, *J. Am. Ceram. Soc.* **68**(3) 120-127

[6] Gudge M, Rickerby D S, Kingswell R and Scott K T 1990 *Residual Stress in Plasma Metallic and Ceramic Coatings, Thermal Spray Research and Applications*, Proceedings of the 3rd National Thermal Spray Conference, Long Beach, California, USA, May 20-25, pp 331-385

[7] Takeuchi S, Ito M and Takeda K 1990 Modelling of Residual Stress in Plasma Sprayed Coatings: The Effect of Substrate Temperature, *Surf. Coat. Technol.* **43-44**(1) 426-435

[8] Elsing R, Knotek O and Balting U 1990 Calculation of Residual Thermal Stress in Plasma Sprayed Coatings, *Surf. Coat.Technol.* **43-44**(1) 416-425

[9] Elsing R, Knotek O and Balting U 1990 The Influence of Physical Proprerties and Spraying Parameters of the Creation of Residual Thermal Stresses during the Spraying Process, *Surf. And Coat. Technol.* **41** 147-156

[10] Knotek O, Elsing R and Balting U 1988 On the Influence of ThermoPhysical Data and Spraying Parameters on the Temperature Curves, *Surf. Coat.Technol.* **36** 99-110

[11] Eckold G, Buckley-Golder I M and Scott K T 1987 *A Theoretical Analysis of Residual Stresses in Flame Sprayed Brittle Materials*, Harwell Report, HAR 5123

[12] Rickerby D S, Eckold G, Scott K T and Buckley-Golder I M 1987 The Interrelationship between Internal Stress, Procesing Parameter sand Microstructure of Physically Vapour Deposited and Thermally Sprayed Coatings, *Thin Solid Films* **154** 125-141

Mathematical modelling in Matlab of the experimental results shows the electrochemical potential difference - temperature of the WC coatings immersed in a NaCl solution

M L Benea[1] and O D Benea[2]

[1]Politehnica University of Timisoara, Department of Engineering and Management, Hunedoara, Romania
[2]The National Institute of Applied Sciences: Department of Informatics, Lyon, France

E-mail: laura.benea@fih.upt.ro

Abstract. The method used for purchasing the corrosion behaviour the WC coatings deposited by plasma spraying, on a martensitic stainless steel substrate consists in measuring the electrochemical potential of the coating, respectively that of the substrate, immersed in a NaCl solution as corrosive agent. The mathematical processing of the obtained experimental results in Matlab allowed us to make some correlations between the electrochemical potential of the coating and the solution temperature is very well described by some curves having equations obtained by interpolation order 4.

1. Introduction

The simplest way of an interpolation consists in knowing the points $\{xi\epsilon[a,b], i=1,2,...,n\}$, called the interpolation points and the points $\{y_i, i=1, 2,...,n\}$ given by an interpolation function F(x) that belongs to a known class that needs to satisfy the relations $F(x_i)=y_i$, $i=1,2,...,n$. In this case, for a value $\xi \neq x_i$, $i=1, 2... n$, we can determine the value $\eta=F(\xi)$. If $\xi \in [a, b]$ the problem is called interpolation [1].

The mathematic interpolation can also be used with success for the interpolation of experimental data. In this case we used the data gathered using the potentiostatic method for the electrochemical resistance of the coating formed in plasma [2], [3].

Our preoccupations to find a parameter to reflect in a sensitive and easy to quantify way the corrosive action of the corrosion agent upon the coating led to the electrochemical potentials of the substrate and the coating immersed into the NaCl solution. Generally, in order to ensure the anti-corrosive action, there are preferred the coating/substrate couples in which the electrochemical potential of the coating is higher than that of the substrate [4].

The WC – Co coatings are designed to protect machines parts (roll cylinders, turbine shuffles, parts of diesel engines) during the combined wear and chemical corrosion stresses [5-7].

2. Experimental part

2.1. Experimental condition at the WC coatings spraying in plasma

The coated metallic substrate is the martensitic stainless steel Z12CNDV12. Before being coated, the surface of the substrate has been prepared by sanding with corundum powder.

The coatings have been made using a Metco 7MB equipment. The powder used is Metco 73, with 83% WC and 17% Co and particle size between 10 and 45 μm.

There have been working using the following parameters:

- Primary plasma gas: argon, 0.62 MPa pressure;
- Secondary plasma gas: helium, 0.62 MPa pressure;
- intensity of the current at the generator: 800A and 45÷55 V tension;
- spraying distance: 80 mm;
- (deposited) powder flow: 2.7 kg/h;
- coating thickness: 0.10 mm

2.2. Corrosion resistance determined by the potentiostatic method

The couple made of the coating (WC-Co) and the substrate (stainless steel) has been immersed in NaCl solutions having concentrations between 1 and 15%, at temperatures between 20 and 48˚C. In each case there has been determined the electrochemical potential of the coating and substrate, using as reference electrode the calomel electrode.

The difference between the electrochemical potential of the coating and that of the substrate represents a measure of the protection degree the coating provides. The greater this difference, the better the protection of the metallic substrate. [4]

3. Results and discussion

The results of the experimental determinations are shown in Table 1.

The standard deviation for the parabolic interpolations has been calculated with the relation (1).

$$= {}_4 = \sqrt{\frac{1}{15}\sum_{i=1}^{15}\left(y_i - a\cdot x_i^4 - b\cdot x_i^3 - c\cdot x_i^2 - d\cdot x_i - e\right)} \quad (1)$$

where a, b, c and d are the coefficients in the obtained regression equations, and y_i, x_i are pairs made of the values of the electrochemical potential differences between the coating and the metallic substrate and the concentration of the NaCl solution, for the same temperature.

Table 1. The values of the electrochemical potential differences [mV] between the coating and the substrate, when varying the temperature and the concentration of the NaCl solution

C [%]	The electrochemical potential difference [mV] for different temperatures of the NaCl solution [˚C]													
	20	22	24	26	28	30	32	34	36	38	40	42	44	46
1	253	250	248	245	240	245	237	235	233	225	225	225	225	227
2	225	215	209	205	198	199	195	194	192	191	190	188	190	198
3	210	200	195	190	180	175	174	180	172	171	170	170	172	172
4	195	179	168	170	159	159	157	155	152	154	149	152	152	154
5	190	176	161	150	142	144	143	142	141	141	140	140	141	142
6	185	160	152	150	141	142	135	134	133	132	132	133	133	135
7	184	181	165	151	140	135	134	132	132	131	130	131	132	132
8	183	165	150	143	140	137	133	132	132	131	131	132	133	134
9	184	160	155	145	142	138	137	140	133	132	132	133	135	136
10	183	160	160	143	140	135	134	133	132	131	131	132	133	135
11	183	177	168	157	150	143	142	140	137	135	132	132	133	137
12	183	170	159	155	140	137	135	135	133	133	131	132	132	137
13	183	172	165	151	145	142	139	138	140	134	133	135	138	139
14	183	170	160	153	150	148	142	139	136	135	135	137	140	139
15	180	168	157	150	145	143	140	137	136	135	135	137	139	140

In order to obtain some mathematical correlations between the corrosion behavior – estimated by the electrochemical potential differences and the concentration of the NaCl solution (for each temperature), the experimental data have been processed with Matlab program.

For each concentration of the NaCl solution it has been established the dependence: electrochemical potential differences between the coating and the substrate, noted as deps (y), and the temperature of the NaCl solution (x), by interpolation order 4. Table 2 shows the results of this interpolation.

In Figures 1-14 there are graphically represented the variations of the electrochemical potential differences between the Metco 73 coating and the metallic substrate related to the concentrations of the NaCl solution, for constant temperature.

Table 2. The results of the 4th order interpolation of the variation of the electrochemical potential difference between the coating and the metallic substrate vs. the concentration of the NaCl solution, in isothermal conditions

Temperature of the NaCl solution [°C]	The dependence relation: electrochemical potential difference between the coating andsubstrate (y) and the concentration of the NaCl solution(x)	Correlation coefficient r	Standard deviation s
20	$y=0.0077305\cdot x^4-0.3526\cdot x^3+5.7991\cdot x^2-40.8716\cdot x+287.7816$	0.99892	0.92062
22	$y=0.0059597\cdot x^4-0.31722\cdot x^3+5.9078\cdot x^2-45.3062\cdot x+287.955$	0.96932	5.8117
24	$y=0.0056683\cdot x^4-0.34677\cdot x^3+6.8183\cdot x^2-2.4896\cdot x+293.2594$	0.98489	4.4212
26	$y=0.0025746\cdot x^4-0.22919\cdot x^3+5.4366\cdot x^2-7.7457\cdot x+285.8302$	0.99091	3.7052
28	$y=0.0099036\cdot x^4-0.47267\cdot x^3+8.1471\cdot x^2-8.8958\cdot x+290.5068$	0.99329	3.1822
30	$y=0.012105\cdot x^4-0.53799\cdot x^3+8.8511\cdot x^2-62.679\cdot x+297.1169$	0.99586	2.6753
32	$y=0.009306\cdot x^4-0.44414\cdot x^3+7.739\cdot x^2-57.2276\cdot x+285.6101$	0.99678	2.2682
34	$y=0.0065474\cdot x^4-0.35716\cdot x^3+6.8111\cdot x^2-3.5447\cdot x+281.4159$	0.99381	3.1659
36	$y=0.0083619\cdot x^4-0.41563\cdot x^3+7.4291\cdot x^2-5.7598\cdot x+280.4742$	0.99781	1.8473
38	$y=0.0058911\cdot x^4-0.31456\cdot x^3+6.0075\cdot x^2-8.0569\cdot x+267.1885$	0.99877	1.3242
40	$y=0.0092604\cdot x^4-0.42376\cdot x^3+7.186\cdot x^2-52.7556\cdot x+271.1275$	0.99884	1.2881
42	$y=0.009237\cdot x^4-0.41715\cdot x^3+7.0324\cdot x^2-51.572\cdot x+269.1572$	0.99841	1.4764
44	$y=0.0082825\cdot x^4-0.38485\cdot x^3+6.6865\cdot x^2-0.2447\cdot x+268.7306$	0.99727	1.9199
46	$y=0.0061802\cdot x^4-0.33071\cdot x^3+6.3076\cdot x^2-9.9541\cdot x+272.6797$	0.99876	1.3315

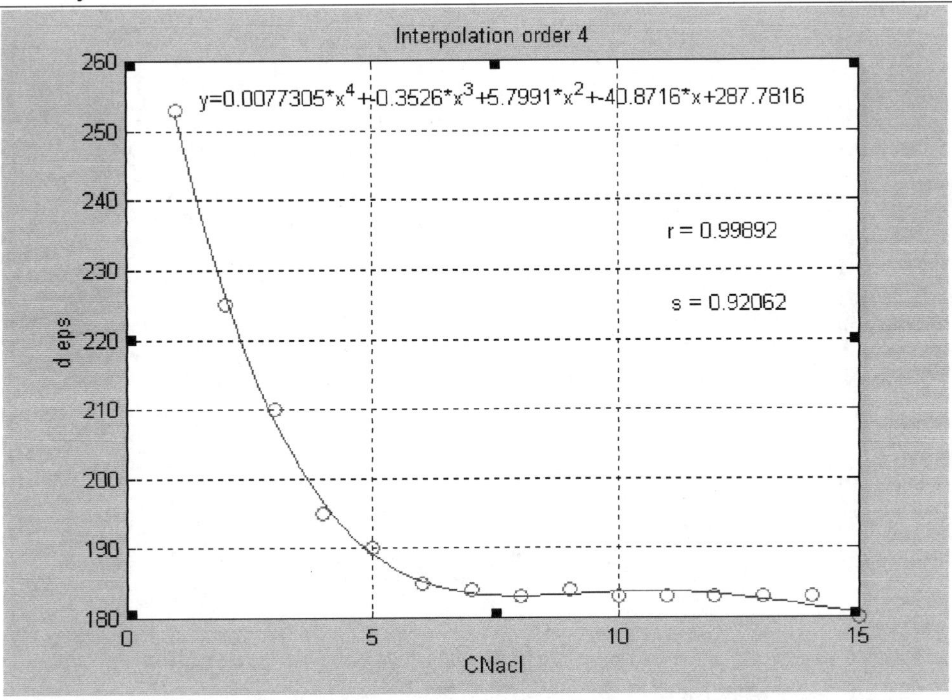

Figure 1. The variation of the potential difference deps [mV] depending on the concentration of the NaCl solution, for 20°C temperature

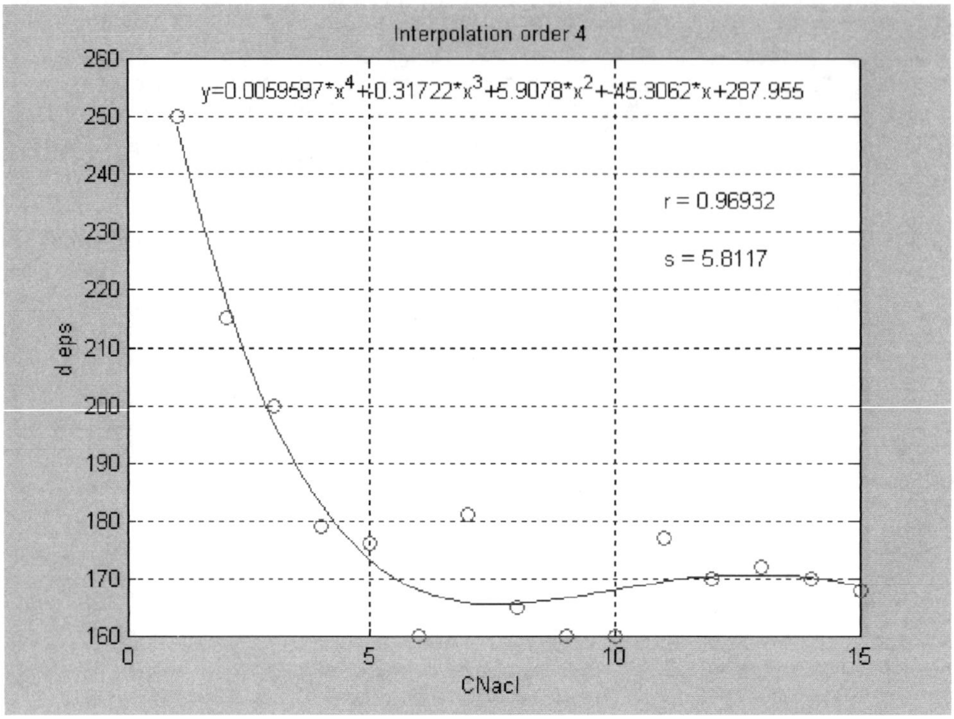

Figure 2. The variation of the potential difference deps [mV] depending on the concentration of the NaCl solution, for 22°C temperature

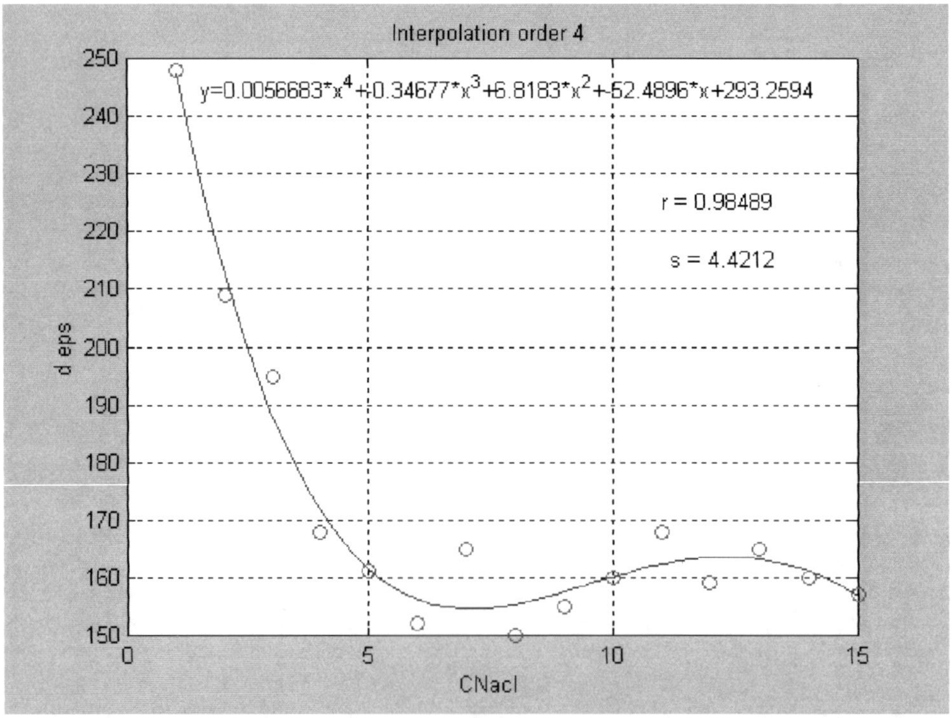

Figure 3. The variation of the potential difference deps [mV] depending on the concentration of the NaCl solution, for 24°C temperature

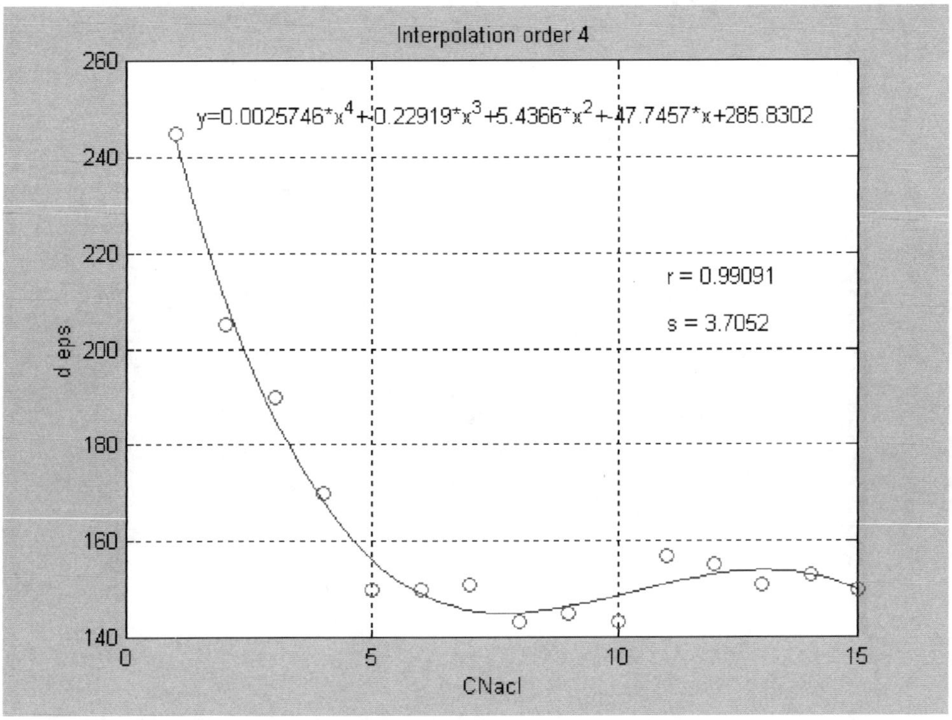

Figure 4. The variation of the potential difference deps [mV] depending on the concentration of the NaCl solution, for 26°C temperature

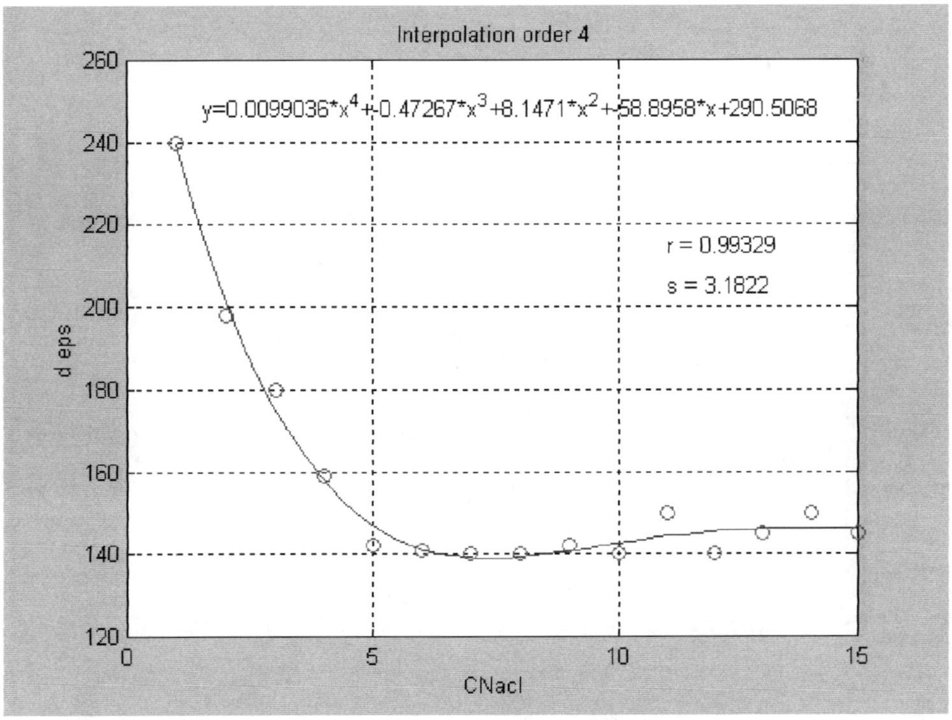

Figure 5. The variation of the potential difference deps [mV] depending on the concentration of the NaCl solution, for 28°C temperature

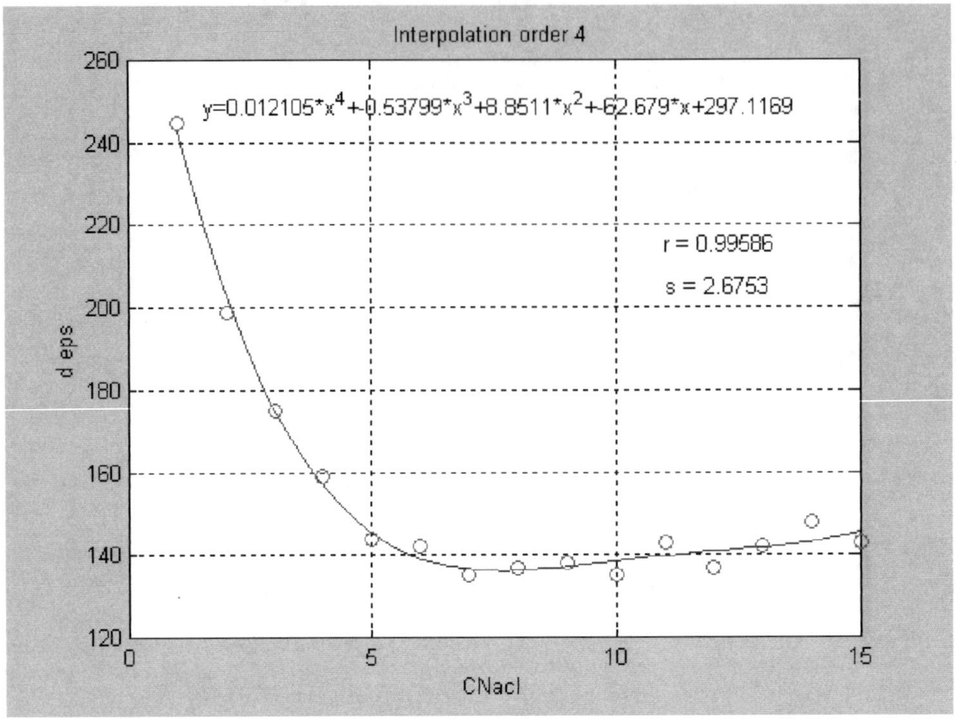

Figure 6. The variation of the potential difference deps [mV] depending on the concentration of the NaCl solution, for 30°C temperature

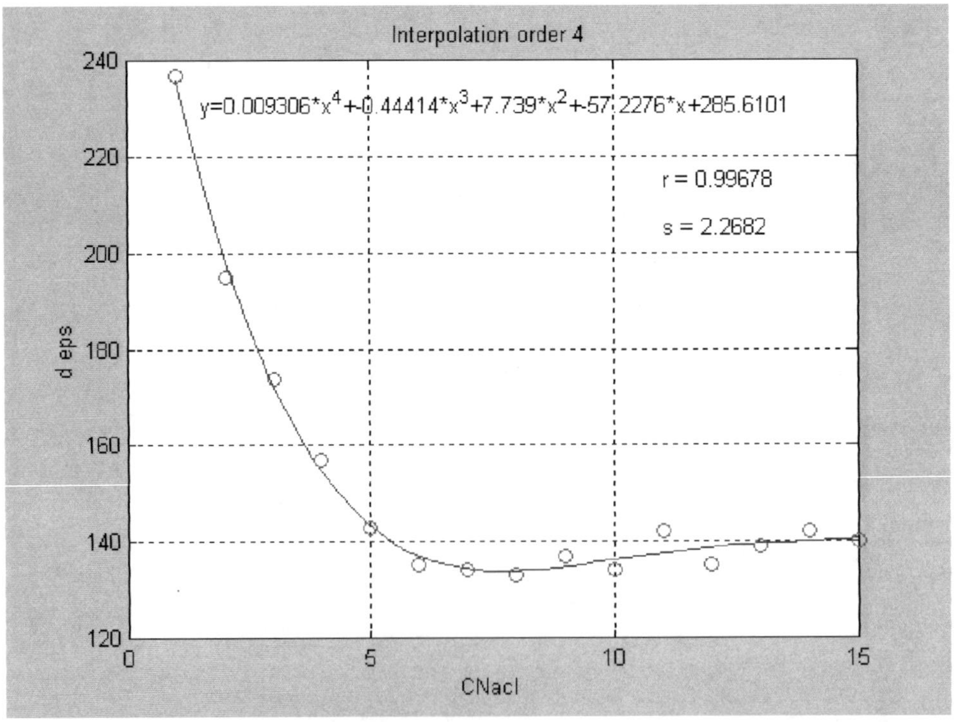

Figure 7. The variation of the potential difference deps [mV] depending on the concentration of the NaCl solution, for 32°C temperature

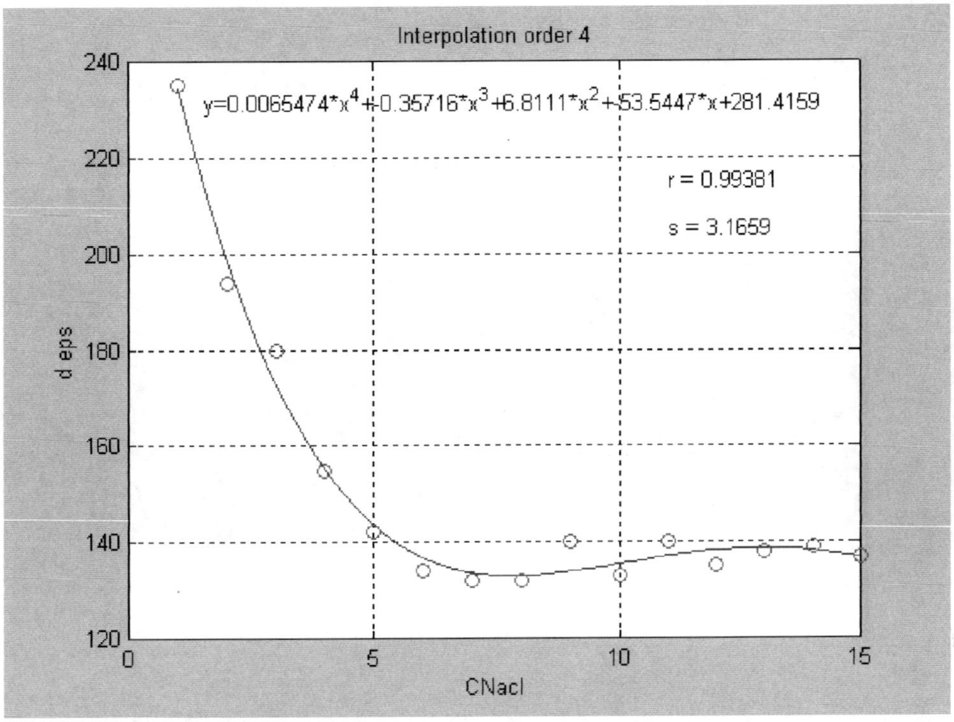

Figure 8. The variation of the potential difference deps [mV] depending on the concentration of the NaCl solution, for 34°C temperature

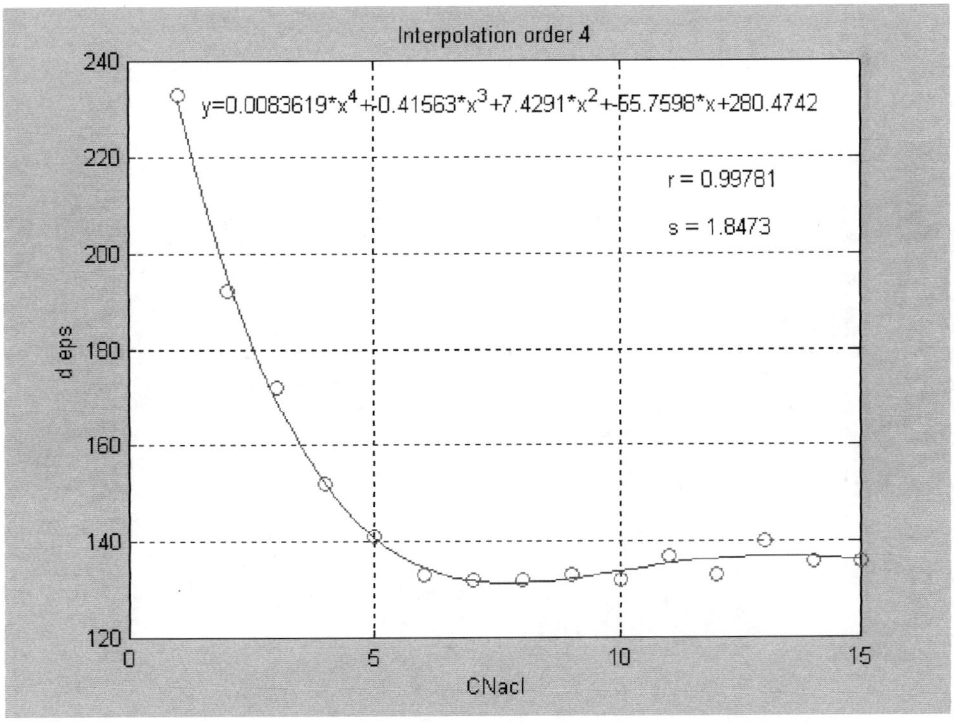

Figure 9. The variation of the potential difference deps [mV] depending on the concentration of the NaCl solution, for 36°C temperature

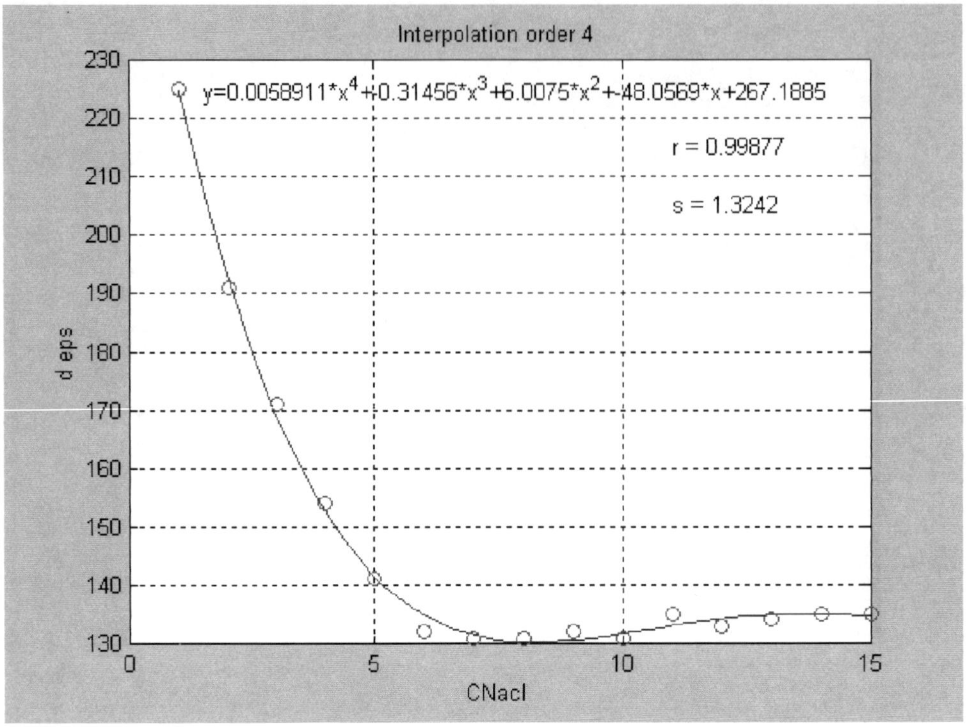

Figure 10. The variation of the potential difference deps [mV] depending on the concentration of the NaCl solution, for 38°C temperature

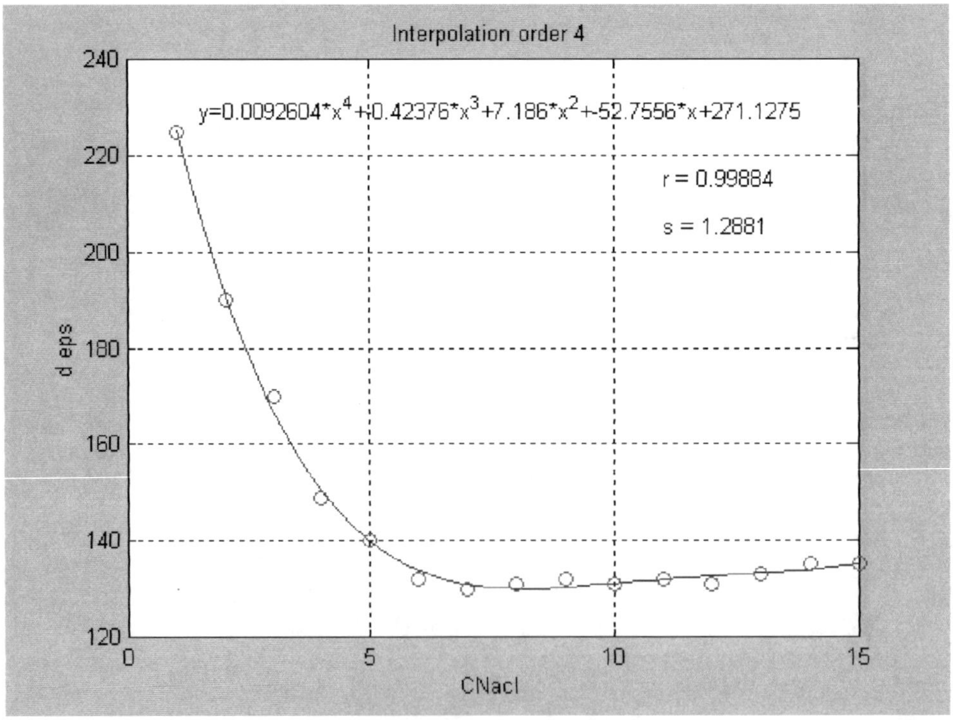

Figure 11. The variation of the potential difference deps [mV] depending on the concentration of the NaCl solution, for 40°C temperature

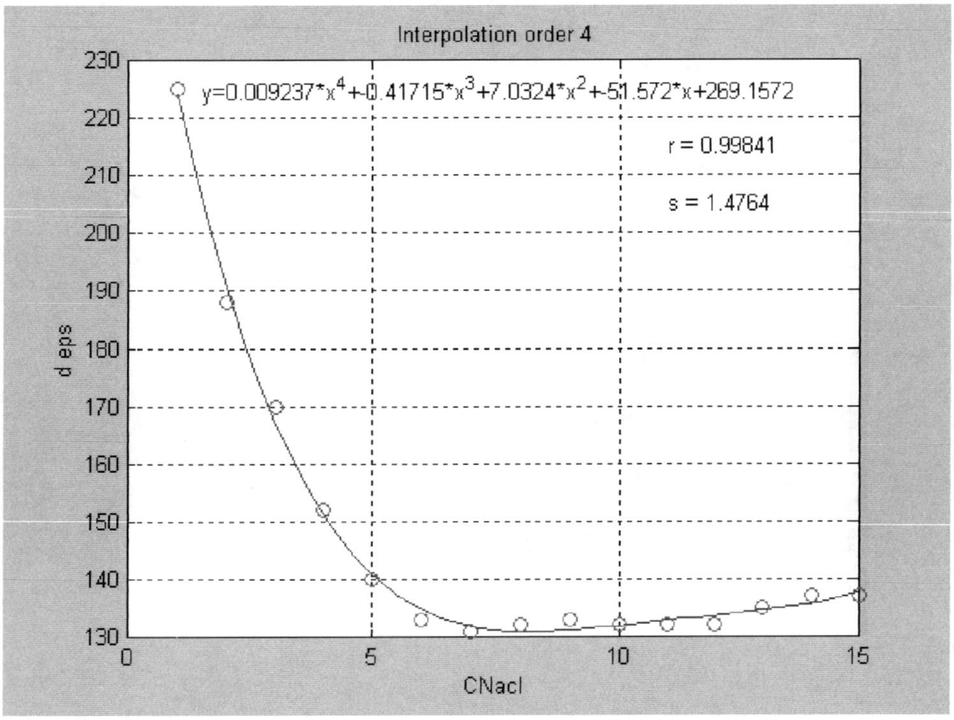

Figure 12. The variation of the potential difference deps [mV] depending on the concentration of the NaCl solution, for 42°C temperature

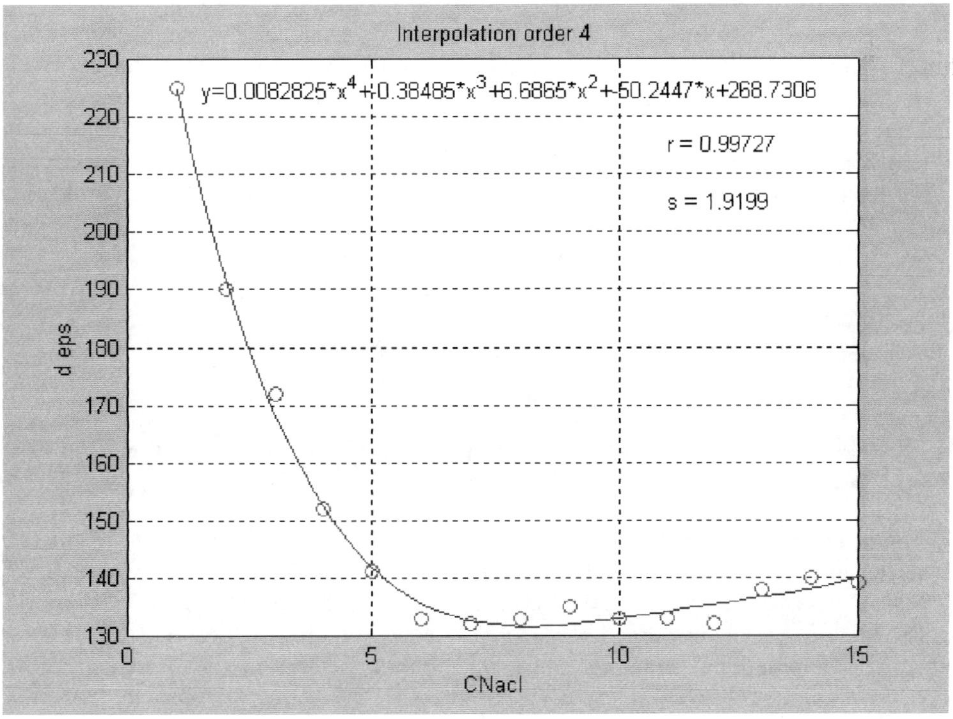

Figure 13. The variation of the potential difference deps [mV] depending on the concentration of the NaCl solution, for 44°C temperature

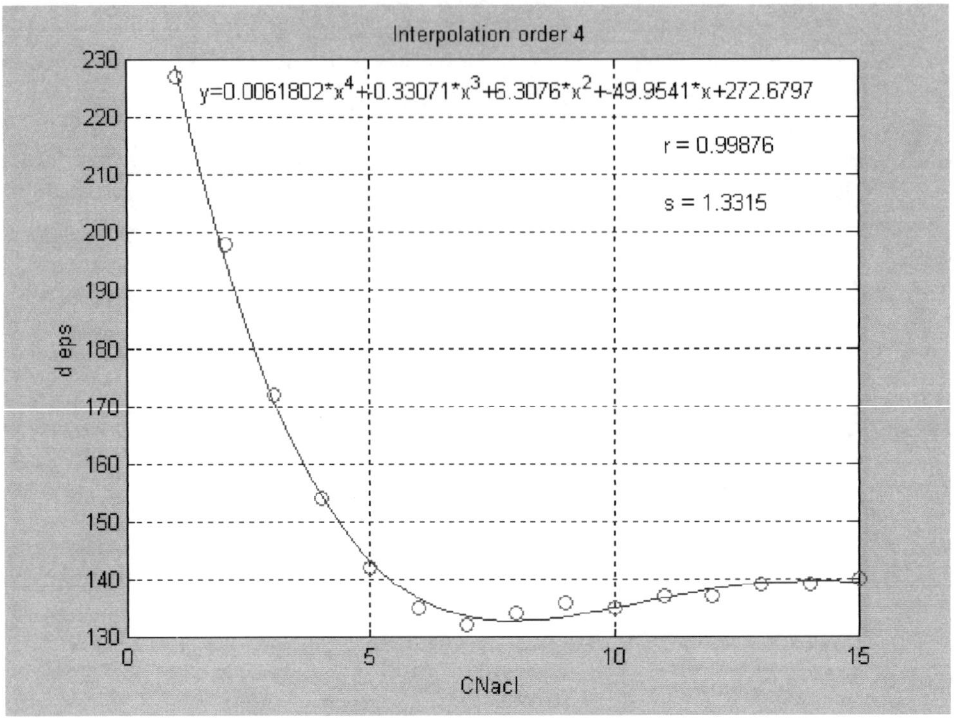

Figure 14. The variation of the potential difference deps [mV] depending on the concentration of the NaCl solution, for 46°C temperature

4. Conclusion

The obtained experimental results show that the coating/substrate electrochemical potential difference is a sensible parameter, which reflects the protection provided by the coating and may be used in quantifying coatings protection activity, in constant conditions concerning the nature of the couple coating/substrate and the spraying parameters.

The values of the potential differences between the coating and the substrate are decreasing when raising the concentration of the NaCl solution, which shows a decrease of the protection provided by the coating.

The mathematical processing of the experimental results shows that for all the of the temperature of NaCl solutions, the dependence between the electrochemical potential difference and the solutions concentrations is very well described by some curves having equations obtained by interpolation order 4 with the coefficients varying between 0.96932 and 0.99892.

References

[1] Maksay S 2001*Matematici speciale*, Sigma Publishing House, Deva, Romania (in Romanian)

[2] Benea M L, Maksay S, Lazău R I and Becherescu D 2003 Using the potentiostatic method in order to purchase the corrosion behavior of the WC coatings deposited by plasma spraying. I: The influence of the NaCl solution concentration, respectively of the temperature, upon the coating/substrate electrochemical potential difference, *Materials engineering* **4**(2) 127-140

[3] Benea M L, Maksay S, Lazău R I and Becherescu D 2003 Using the potentiostatic method in order to purchase the corrosion behavior of the WC coatings deposited by plasma spraying. II: The combined influence of NaCl solution concentration and temperature upon the coating/substrate electrochemical potential difference, *Materials engineering* **14**(2)141-155

[4] Benea M L and Benea L P 2015 Characterization of the WC coatings deposited by plasma spraying, *IOP Conf. Ser.: Mater. Sci. Eng.* **85** 012004

[5] Kasai S, Yanagisawa A, Ichihara A, Shimoyama Y, Ochiai K and Onishi H 1988 *An investigation of oxide cermet coatings for hearth roll aplicatin in a continous amelling fournace*, 1st Plasma Technik Symposium, Lucerne, Switzerland, May 18-20, pp 205-215

[6] Smith R W and Mutasim Z Z 1992 Reactive plasma spraying of wear – resistant coatings, *Journal of Thermal Spray Technology-ASM International* **1**(1) 57-63

[7] Kvernes I, Lugscheider E and Norholm O 1988 *Potential of thermal and wear resistant coatings in combustion engines*, 1st Plasma Technik Symposium, Lucerne, Switzerland, May 18-20, pp 41-53

IOP Conf. Series: Materials Science and Engineering **106** (2016) 012026 doi:10.1088/1757-899X/106/1/012026

Evaluating solar radiation on a tilted surfaces – a study case in Timis (Romania)

C Vasar[1], O Prostean[1] and G Prostean[2]

[1]Politehnica University of Timisoara, Department of Automation and Applied Informatics, Vasile Parvan Street, no. 1-2, 300223 Timisoara, Romania
[2]Politehnica University of Timisoara, Department of Management, Remus Street, no.14, 300191 Timisoara, Romania

E-mail: cristian.vasar@upt.ro

Abstract. In the last years the usage of solar energy has grown considerably in Romania, as well as in Europe, stimulated by various factors as government programs, green pricing policies, decreasing of photovoltaic components cost etc. Also, the rising demand of using Solar Energy Conversion Systems (SECS) is driven by the desire of individuals or companies to obtain energy from a clean renewable source. In many applications, remote consumers far from other energetic grids can use solar systems more cost-effectively than extending the grid to reach the location. Usually the solar energy is measured or forecast on horizontal surface, but in SECS there is needed the total solar radiation incident on the collector surface, that is oriented in a position that maximize the harvested energy. There are many models that convert the solar radiation from horizontal surface to a tilted surface, but they use empirical coefficients and the accuracy is influenced by different facts as geographical location or sky conditions. Such models were used considering measured values for solar radiation on horizontal plane, in the western part of Romania. Hourly values measured for global solar irradiation on the horizontal plane, diffuse solar irradiation on the horizontal plane and reflected solar irradiation by ground are used to compute the total solar radiation incident on different tilted surfaces. The calculated incident radiation is then compared with the real radiation measured on tilted surface in order to evaluate the performance of the considered conversion models.

1. Introduction

In the last years the usage of solar energy has grown considerable in Romania, as well as in entire Europe, stimulated by various factors as government programs, green pricing policies, decreasing of photovoltaic components cost etc.

In many applications, remote consumers that are located far from other energetic grids can use solar systems more cost-effectively than extending the grid to reach their distant location.

Also, the rising demand of using Solar Energy Conversion Systems (SECS) is driven by the desire of individuals or companies to obtain energy from a clean renewable source.

Usually the solar energy is measured or forecast on horizontal surface, but in SECS there is needed the total solar radiation incident on the collector surface, that is oriented in a position that maximize the harvested energy.

There are many models that convert the solar radiation from horizontal surface to a tilted surface, but they use empirical coefficients and the accuracy is influenced by different facts as geographical location or sky conditions. Such models were used considering measured values for solar radiation on horizontal plane, in the western part of Romania.

Hourly values measured for global solar irradiation on the horizontal plane, diffuse solar irradiation on the horizontal plane and reflected solar irradiation by ground are used to compute the total solar radiation incident on different tilted surfaces. The calculated incident radiation is then compared with the real radiation measured on tilted surface in order to evaluate the performance of the considered conversion models.

2. Solar energy components

In a point at the top of Earth's atmosphere, the beam of nearly parallel incident sunrays is referred to as extraterrestrial radiation (ETR). ETR fluctuates about 6.9 % during a year (from 1412.0 W/m2 to 1321.0 W /m2) due to the Earth's varying distance from the Sun.

The integration of the extraterrestrial spectrum over all wavelengths defines the solar constant GSC that represents the flux density of incoming solar radiation on a unitary surface perpendicular to the rays at the mean Sun–Earth distance. Since the Sun radiance varies [1], the solar constant does not remain steady over time. There is a variation of about ±1W /m2 around the mean solar constant during a typical Sun cycle of 11 years [2].

Based on data collected over 25 years from terrestrial to space observations, the actual best estimate of the average solar constant is GSC =1366.1 W /m2 [3].

When the solar radiation flux passes through the Earth's atmosphere, its spectral distribution is modified by absorption and scattering processes.

2.1. Solar energy components

Solar energy reach at photovoltaic panel through different ways as depicted in Figure 1. The most significant quantity of solar energy is provided by the direct radiation that came from the sun, especially on clear sky conditions. Due to various particles presented in earth atmosphere (such as air molecules, dust, fog or cloud elements, drops etc) occurs the diffuse radiation, which become significant in case of cloudy sky conditions. Also the ground surface, including buildings, water surface, vegetation etc, reflects solar radiation that can be collected by a tilted panel.

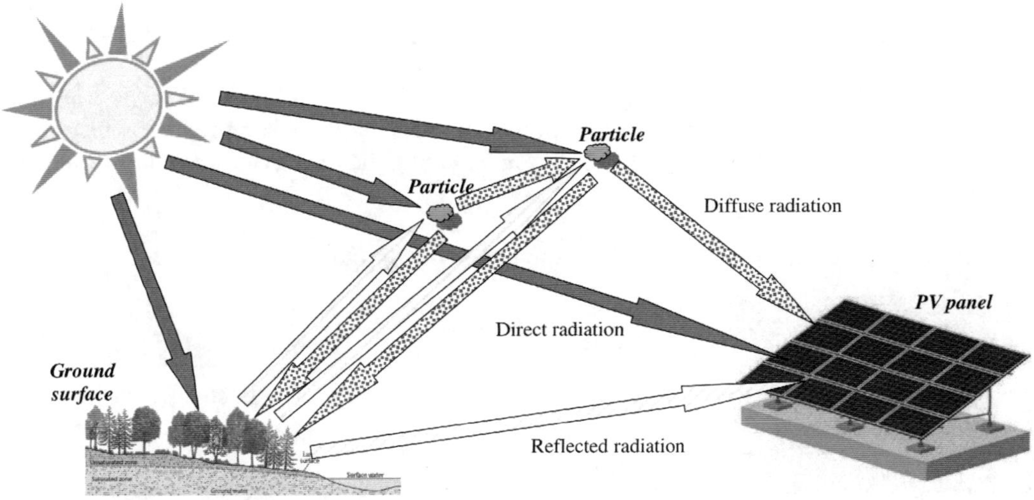

Figure 1. Solar radiation

In order to be able to analyze the solar radiation incident on a tilted surface is critical to know the surface orientation (azimuth angle and slope angle) as well as the exact geographical position (latitude and longitude) and astronomic time. Thus can be determined important parameters as the zenith angle, surface azimuth angle, elevation angle h and incidence angle (between sun direction and the normal of the tilted surface) – as depicted in Figure 2. [4]

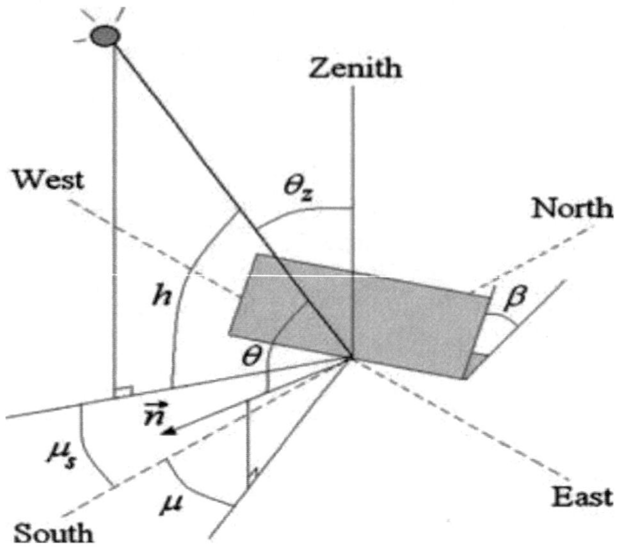

Figure 2. Tilted surface parameters

The solar radiation can be analyzed considering the following components:

- Direct beam irradiance (Gn) is the energy flux density (units: W/m2) of the solar radiation incoming from the solid angle subtended by the Sun's disk on a unitary surface perpendicular to the rays.
- Direct horizontal irradiance (Gb) differs from the direct beam irradiance in that it is measured on a flat horizontal plane.
- Diffuse irradiance (Gd) represents the energy flux density of the solar radiation incoming from the entire sky dome on a horizontal surface, excluding the direct beam coming from the Sun's disk.
- Global irradiance (G) is the sum of the direct horizontal and diffuse components.

$$G = G_b + G_d = G_n \cos\theta_z + G_d \quad (1)$$

- Reflected Irradiance (Gr) is the energy flux density of radiation reflected by the ground that is intercepted by the tilted surface.
- The total irradiance(Gt) received by a surface tilted with an angle to the horizontal plane is the sum of beam flux density, diffuse flux density, and the reflected flux density Gr of the solar radiation reflected from the ground.

$$G_t = G_n \cos\theta + R_d G_d + G_r \quad (2)$$

where : θ is the incidence angle,

R_d is the conversion coefficient taking into account the sky view factor and

Gr is the energy flux density of radiation reflected by the ground that is intercepted by the tilted surface.

Models for estimating global solar irradiance on tilted surfaces differ generally in their treatment of *Rd* which is considered the main potential source of errors.

2.2. Solar energy measurement
Solar radiation can be measured using dedicated equipments as pyrheliometers, pyranometers or less expensive equipments as sunshine recorders or even reference PV panels.

The pyrheliometer is used to measure the direct beam radiation, and must be always oriented directly to the sun (see Figure 3). The measurement accuracy is strongly influenced by the sensor characteristics, the performance of the sun tracking system and the dust that cover the lens. [5]

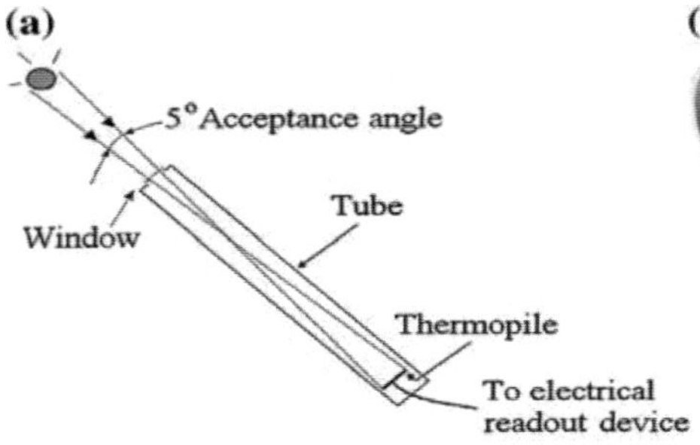

Figure 2. Pyrheliometer scheme

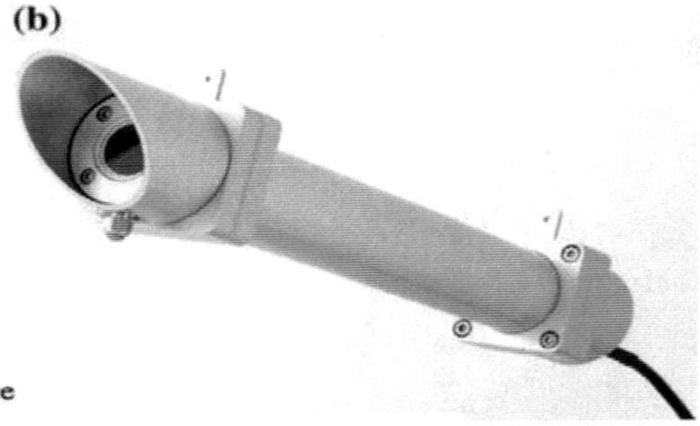

Figure 3b. Pyrheliometer

The pyranometer (Figure 4.a) is typically used for measuring the horizontal global radiation, but attaching a sun masking device (solid ring in Figure 4.b, or solid disk with active tracking system in Figure 4.c) can be used for measuring the diffuse radiation. [4][6]

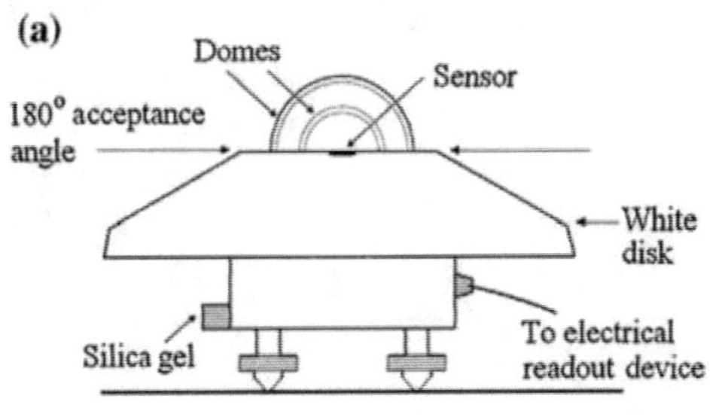

Figure 4a. Pyranometer (global radiation)

Figure 4b. Pyranometer with solid ring (difusse radiation)

Figure 4c. Pyranometer with solid disk (difusse radiation)

Measurement errors occur in case of dirty lens or improper positioned sun mask (solid ring must be adjusted everyday while solid disk must be adjusted continuously). The ring width or disk diameter also influences the measurement accuracy.

Using PV panels or other chipper equipments for solar radiation measurement will lead to a lower accuracy, which is inadequate for the solar radiation thorough analysis performed in this paper but usually is sufficient for controlling/monitoring the industrial solar energy conversion systems.

3. Solar energy on tilted surfaces

The beam radiation on a tilted surface as depicted in Figure 2 can be relative easy obtained from measured normal incident beam by applying simple geometrical relations between the horizontal and tilted surface.

$$G_b = G \cos\theta_z \tag{3}$$

Considering that the ground is an isotropic diffuse reflector is possible to determine the ground reflected diffuse radiation incident on a tilted surface, using the Liu and Jordan isotropic formula:

$$G_r = \rho G \sin^2\frac{\beta}{2} \tag{4}$$

where ρ is the albedo – ground reflectance.

Because the diffuse radiation arrives from all over the sky with different gradient, the computation of diffuse radiation on a tilted surface is not very simple, existing numerous models that try to correlate the diffuse radiation measured on horizontal surface with the diffuse radiation incident on a tilted surface. The large quantity of such conversion models proves the complexity of this task. [7]

The diffuse radiation $G_{d,\beta}$ incident on a surface with tilted angle β is:

$$G_{d,\beta} = G_d . \Psi_i \tag{5}$$

where Ψ_i denote a conversion function corresponding to the considered *i* model. [8]

In the performed case study there were considered the following models:

- Liu-Jordan model – assuming that diffuse radiation is isotropic

$$\Psi_{LJ} = \cos^2(\theta_z) \tag{6}$$

- Temps-Coulson model – assuming clear sky conditions

$$\Psi_{TC} = \cos^2\left(\frac{\beta}{2}\right)\left[1+\sin^3\left(\frac{\beta}{2}\right)\right]\left[1+\cos^2(\theta)\sin^3(\theta_z)\right] \tag{7}$$

- Hay model

$$\Psi_H = Fr_b + (1-F)\cos^2\left(\frac{\beta}{2}\right) \tag{8}$$

where F is Hay's sky-clarity factor defined as $F = \frac{G_n}{ETR}$ and $r_b = \max\left[0, \left(\frac{\cos\theta}{\sinh}\right)\right]$

- Klucher model – is a enhancement of the Temps-Coulson model

$$\Psi_k = \cos^2\left(\frac{\beta}{2}\right)\left[1+F'\sin^3\left(\frac{\beta}{2}\right)\right]\left[1+F'\cos^2\theta\sin^3\theta_z\right] \tag{9}$$

where $F' = 1 - \frac{G_d}{G}$ is Klucher's modulating function.

4. Case study

The considered study case was performed for a location from Timis County, in the western side of Romania. There were available data measured in several points from a small area for horizontal global radiation and radiation incident on two tilted surfaces one oriented to South and another oriented to South-West.

The available data measured in horizontal plan was averaged hourly and used for computation of the tilted surface incident radiation, using the four models presented in the above section. Best results were obtained using Klucher model (average error for one year values is less than 4% comparing with measured data on the tilted surface), respectively Liu-Jordan model (average error for one year is less than 6%).

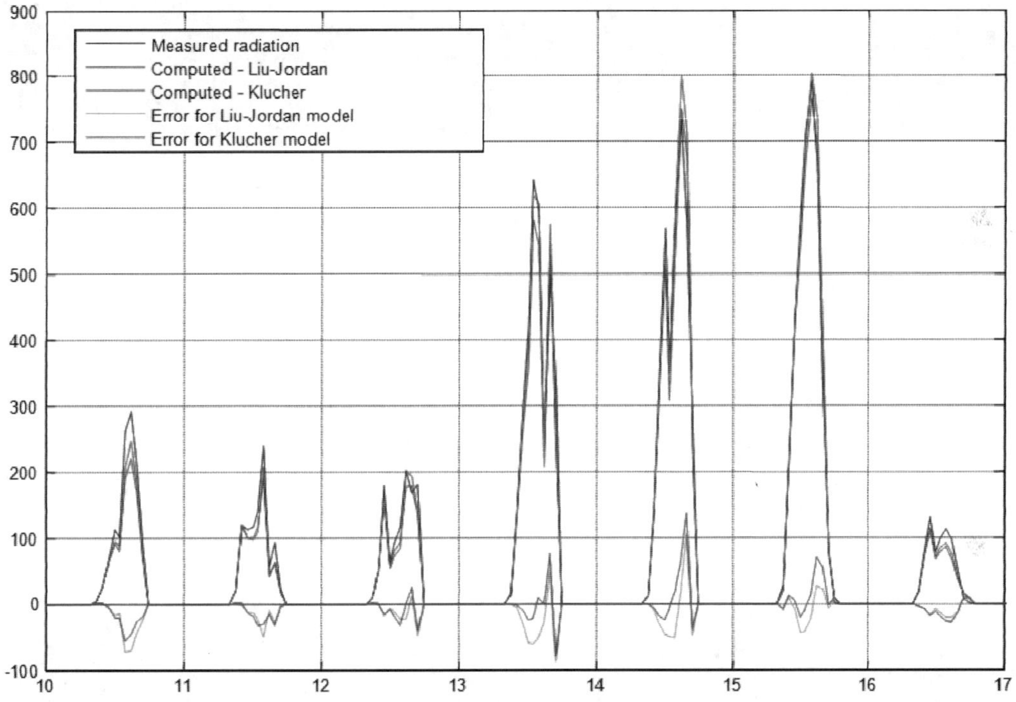

Figure 5. Solar radiation incident on a tilted surface (January 2012)

Other paragraphs are indented (BodytextIndented style). In Figure 5 there are presented the measured radiation (with blue solid line) on a tilted surface oriented to South-West, for a time period between 10.01.2012 and 17.01.2012 – there can be noticed both cloudy conditions (days 1, 2, 3 and 7) and clear sky conditions (days 4, 5 and 6). The computed values for the tilted surface radiation are depicted with green line (for Liu-Jordan model) and respectively red line (for Klucher model). The model error – difference between model output and the measured data – is depicted with cyan line (for Liu-Jordan model) and respectively magenta line (for Klucher model). It can be noticed a slightly better overall performance of the Klucher model. The abscise axis represents the day of the year and the ordinate axis represents the incidence solar radiation/error measured in W for a surface of unitary area.

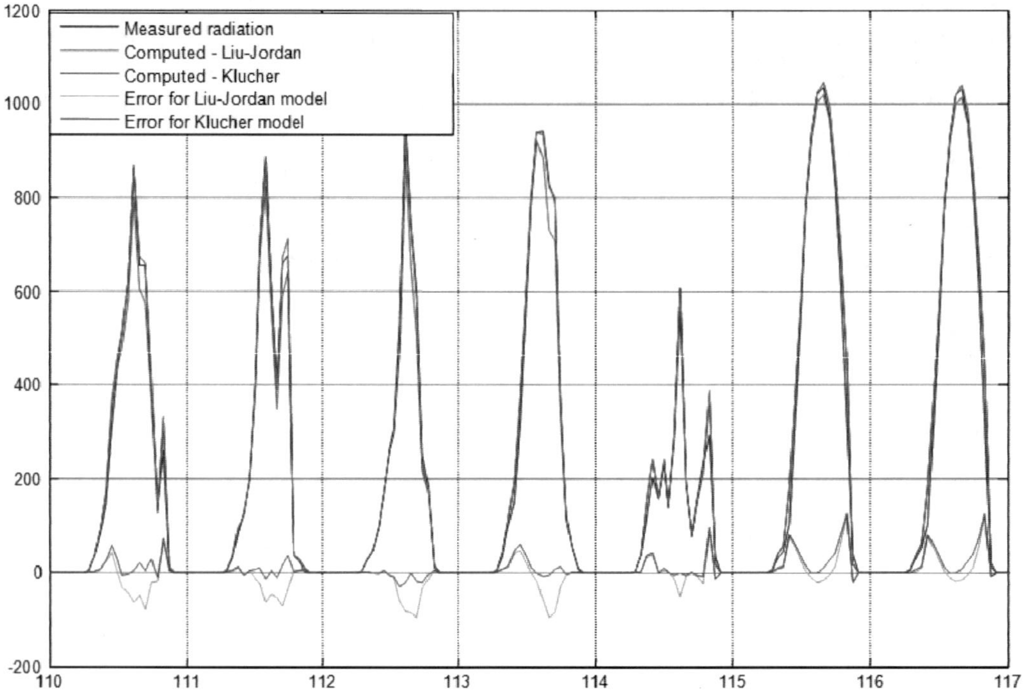

Figure 6. Solar radiation incident on a tilted surface (April 2012)

In Figure 6 there are presented the measured and computed radiation on a tilted surface oriented to South-West, for a time period between 19.04.2012 and 26.04.2012. Also in these spring sunny days the Klucher model was more accurate.

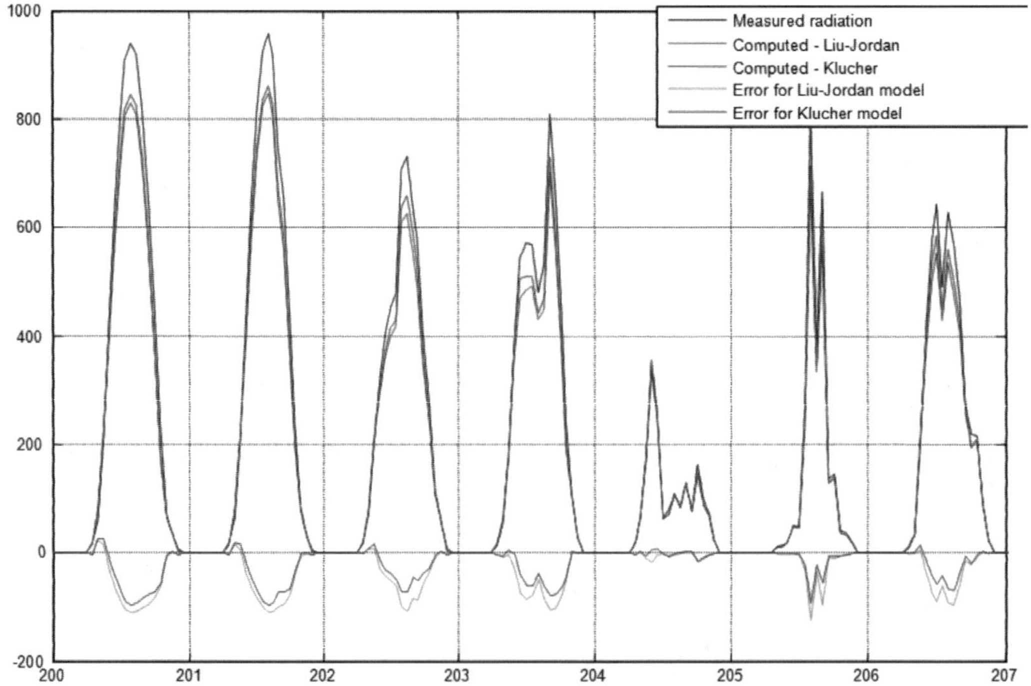

Figure 7. Solar radiation incident on a tilted surface (July 2012)

In Figure 7 is depicted the case of a tilted panel oriented to South, in summer days from 18.07.2012 to 25.07.2012. In this case the Klucher model performances are much better than the Liu-Jordan model.

5. Conclusions

Hourly values measured for global solar irradiation on the horizontal plane, diffuse solar irradiation on the horizontal plane and reflected solar irradiation by ground are used to compute the total solar radiation incident on different tilted surfaces.

The calculated incident radiation is then compared with the real radiation measured on tilted surface in order to evaluate the performance of the considered conversion models. Best results were obtained using the Klucher model.

These models can be used for evaluating the solar potential of a certain location before establish a solar park and also for optimal design of the planned SECS.

Acknowledgement

This paper was developed through the Partnerships in Priority Areas - PN II, with the support of ANCS, CNDI - UEFISCDI, Project no. 36/2012.

References

[1] Fröhlich C 1991 History of solar radiometry and the world radiation reference, *Metrologia* **28**111–115

[2] Gueymard C A and Myers D R 2008 *Solar radiation measurement: progress in radiometry for improved modeling*, Springer, Berlin, Germany

[3] Gueymard C A 2004 The sun's total and spectral irradiance for solar energy application and solar radiation models, *Solar Energy* **76** 423–453

[4] Paulescu M, Paulescu E, Gravila P and Badescu V 2013 *Weather Modeling and Forecasting of PV Systems Operation*, Springer-Verlag, London, UK

[5] ***2015 Hukseflux - Radiation measurement sensors. Available online http://www.huksefluxusa.com/radiation-measurement.php

[6] Paulescu M, Dughir C, Tulcan-Paulescu E, Lascu M, Gravila P and Jurca T 2010 Solar Radiation Modeling and Measurements in Timisoara, Romania: Data and Model Quality, *Environmental Engineering and Management Journal (EEMJ)* **9**(8) 1089

[7] Jakhrani A O, Samo S R, Rigit A R H and Kamboh S A 2013 Selection of Models for Calculation of Incident Solar Radiation on Tilted Surfaces, *World Applied Sciences Journal* **22** (9) 1334-1343

[8] Evseev E G and Kudish A I 2009 The assessment of different models to predict the global solar radiation on a surface tilted to the south, *Solar Energy* **83** 377–388

IOP Conf. Series: Materials Science and Engineering **106** (2016) 012027 doi:10.1088/1757-899X/106/1/012027

Study of the load variation for an electrical consumer

R C Vîlceanu[1], N Chiosa[2] and F D Şurianu[1]

[1]Politehnica University Timisoara, Faculty of Electrical and Power Engineering, Vasile Pârvan str., no. 1-2, 300223 Timisoara, Romania
[2]CNTEE Transelectrica SA- ST Timisoara, Romania

E-mail: radu_89_09@yahoo.com

Abstract. In this paper we study the load variation for an electrical consumer when it remains powered by an emergency power supply through an uninterruptable power supply due to a circuit failure and the way that the load varies when the main and backup power supply comes on. The study was conducted due to the fact that when the main power supply tripped, the emergency power supply, ensured by a diesel power generator with a time delay of 30 seconds, always tripped immediately after connecting, thus leaving the consumer powered by the uninterruptable power supply which has a 10 minutes limited supply.

1. Introduction

We conducted the study due to the fact that the electrical consumer [1] had to be powered at all time and when the main power supply tripped, the emergency power supply, ensured by a diesel power generator with a time delay of 30 seconds, always tripped immediately after connecting, thus leaving the consumer powered by the uninterruptable power supply which has a 10 minutes limited supply.

The simplified electrical scheme of the circuit is presented in Figure 1 in which we can see the point in which we connected with a Fluke 1760 Three-Phase Power Quality Recorder (shown in Figure 2).

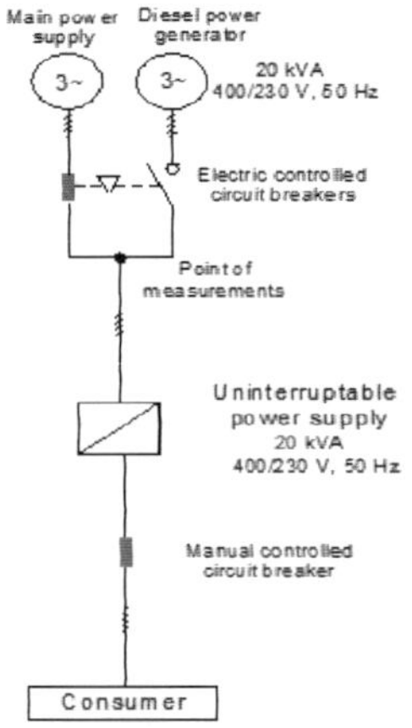

Figure 1. The simplified electrical scheme

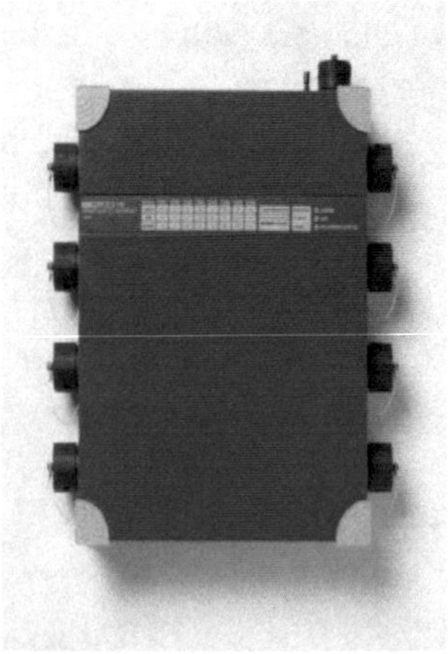

Figure 2. Fluke 1760 Three-Phase Power Quality Recorder

The Fluke 1760 recorder is designed for analysis of utility and industrial power distribution systems, in medium and low-voltage networks. It captures the most comprehensive details on user-selected parameters. The supplied PQ Analyze application software (shown in Figure 3) enables users to use the in depth measurement and produce sophisticated analysis and reports.

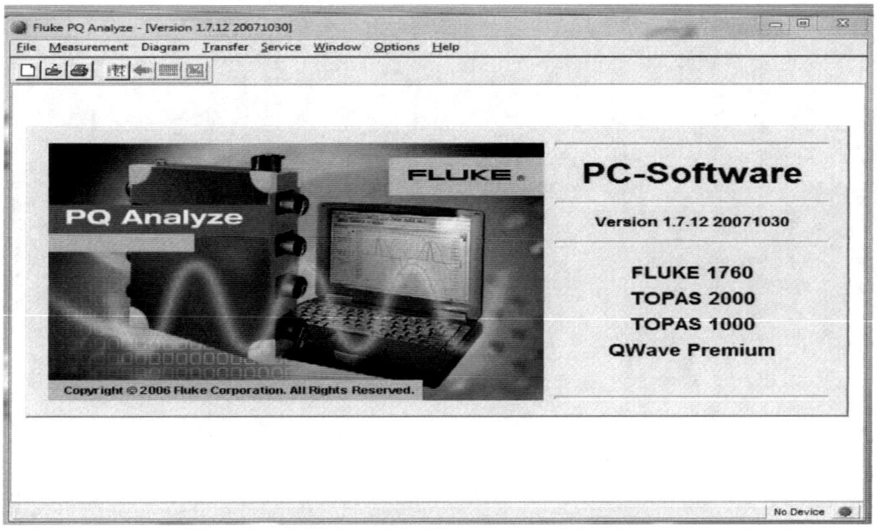

Figure 3. The PQ Analyze application software

2. Measurement tests done with the Fluke 1760 recording the data

To find out what the problem was and why did the electrical consumer trip when the emergency power supply connects, we conducted the following scenarios:

2.1. Measurements made with the electrical consumer connected at its average load

In this scenario we simulated a fault in the network and left the main power supply unpowered. After a time delay of 30 seconds, in which a PLC (programmable logic controller) device started the diesel power generator and monitored that the main power supply is still unpowered, the PLC device issues the connect command of the generator to power up the consumer [2]. Shortly after the generator connected, it tripped.

The Fluke 1760 recorded the following data:

- in Figure 4 we can see that when the diesel generator connects, the average 12 kVA load of the consumer is restored in three power spikes [3]. The second spike is the most relevant one and it appears approximately 1.25 seconds after the generator connects with a maximum value of 20.7 kVA, value which exceeds the generators nominal power and thus leads to the generator trip from the overload protection [4];

- in Figure 5 we can see the evolution of currents and frequency during the three power spikes. We can see when the diesel generator connects, the frequency increases to a maximum value of 51.6 Hz and during the power spikes it drops to a minimum value of 49.5 Hz. The nominal value of 50 Hz is reached only after the power spikes pay off.

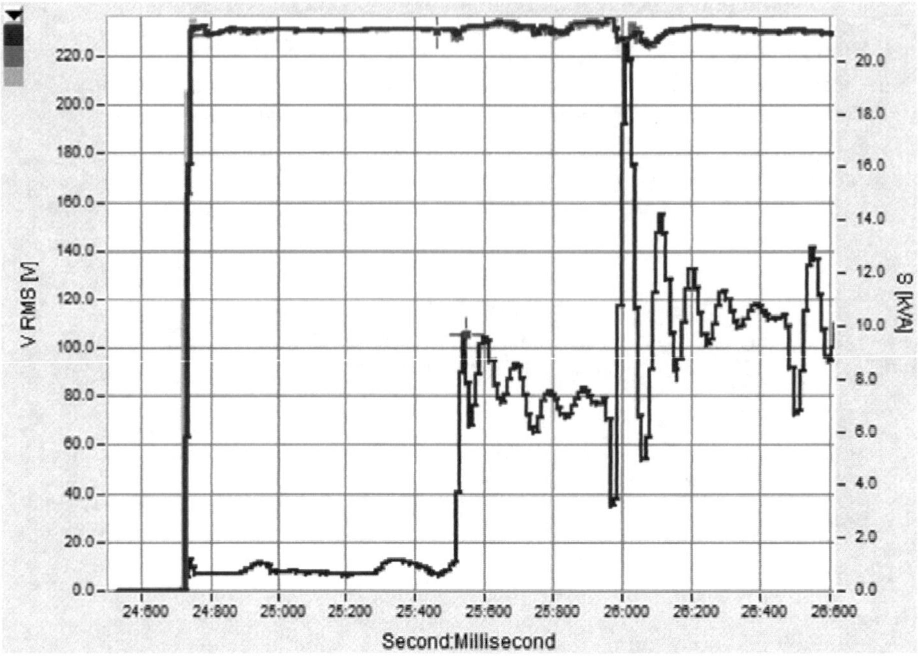

Figure 4. Graphic regarding the evolution of load and voltage when the diesel power generator connects with the consumer at its average load of 12 kVA

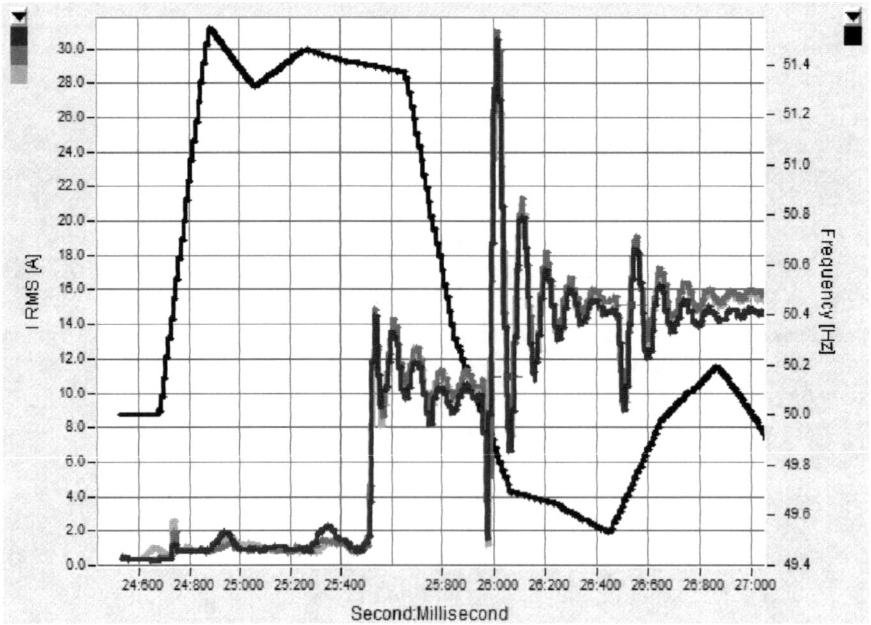

Figure 5. Graphic regarding the evolution of currents and frequency when the diesel power generator connects with the consumer at its average load of 12 kVA

2.2. Measurements made with the electrical consumer connected at a lower load

In this scenario we reduced the load of the consumer from 12 kVA to 6 kVA and left the main power supply unpowered. We conducted this scenario in order to see if the power spikes will appear again at a lower load or will the load restore uniformly.

The Fluke 1760 recorded the following data:

- in Figure 6 we can see that when the diesel generator connects, the 6 kVA load of the consumer is again restored in three power spikes. The second spike is the biggest one and it appears again approximately 1.25 seconds after the generator connects with a maximum value of 12 kVA;
- in Figure 7 we can see the evolution of currents and frequency during the three power spikes. We can see when the diesel generator connects, the frequency is stable very close to its nominal value of 50 Hz.

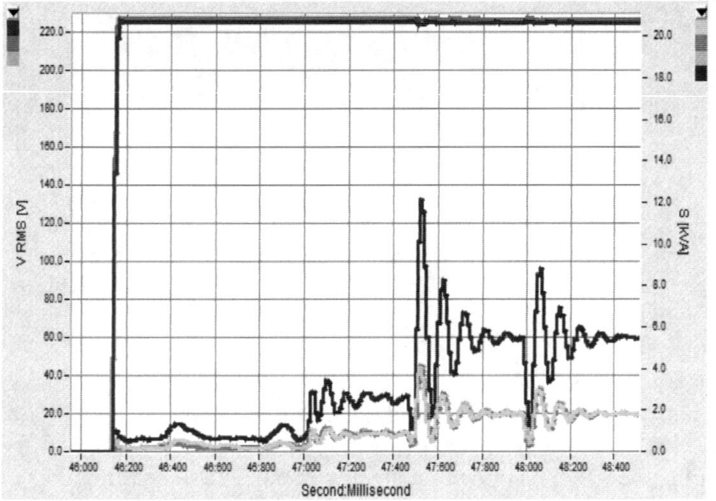

Figure 6. Graphic regarding the evolution of load and voltage when the diesel power generator connects with the consumer at a load of 6 kVA

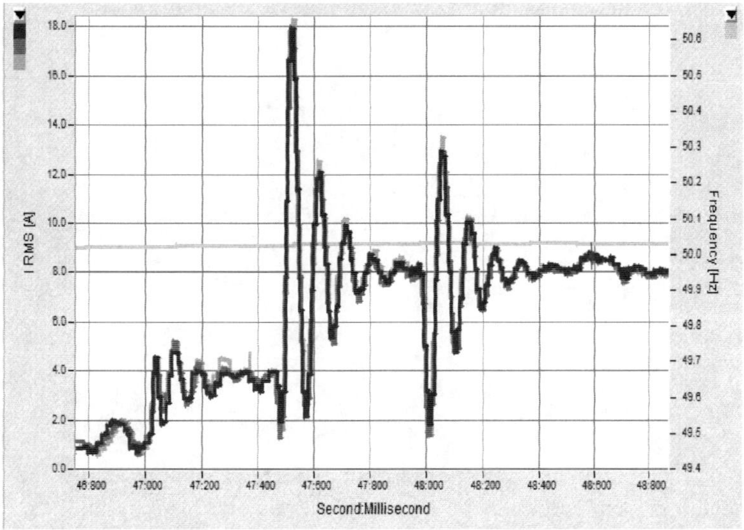

Figure 7. Graphic regarding the evolution of currents and frequency when the diesel power generator connects with the consumer at a load of 6 kVA

IOP Conf. Series: Materials Science and Engineering **106** (2016) 012027 doi:10.1088/1757-899X/106/1/012027

2.3. Measurements made with the electrical consumer connected directly to the power supplies

If this scenario we took the uninterruptable power supply out the power scheme thus leaving the electrical consumer connected directly to the two power supplies. The new power scheme is shown in Figure 8. The load of the consumer was reduced even more.

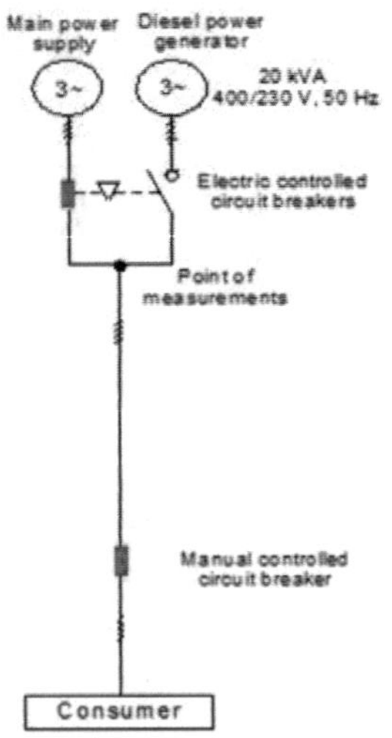

Figure 8. The electrical consumer connected directly to the power supplies

The Fluke 1760 recorded the following data:

- in Figure 9 we can see that when the diesel generator connects, the 3 kVA load of the consumer is restored uniformly. There are no more power spikes in the process;
- in Figure 10 we can see the evolution of currents and frequency during the load restore duration. We can see that when we switch the load from the main power supply to the diesel generator, there are no important spikes in currents and frequency.

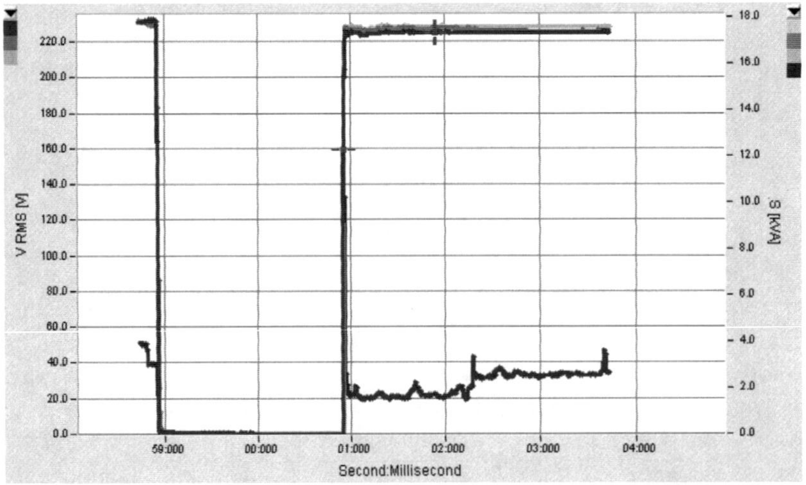

Figure 9. Graphic regarding the evolution of load and voltage when the diesel power generator connects with the uninterruptable power supply out of the power scheme

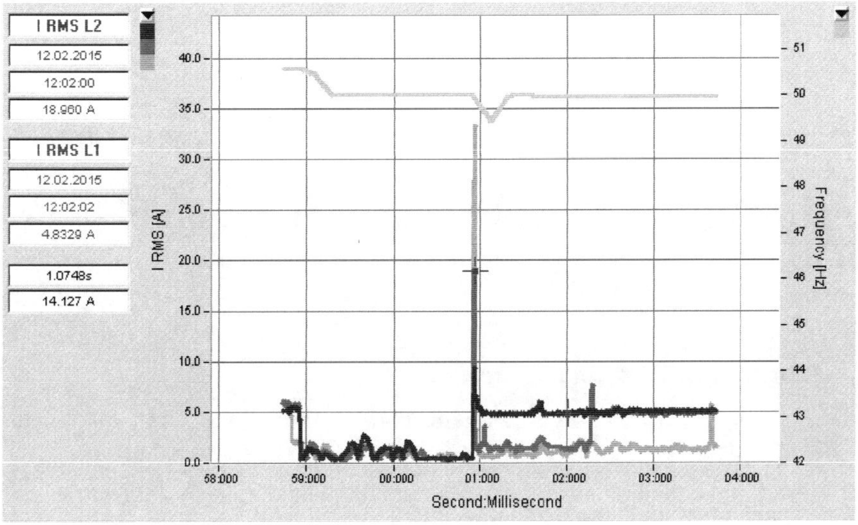

Figure 10. Graphic regarding the evolution of currents and frequency when the diesel power generator connects with the uninterruptable power supply out of the power scheme

3. Conclusions

Based on the results obtained from data processing and on the technical analysis of these data, we concluded the followings:

· when the diesel power generator takes a 12 kVA load, power spikes appear up to a maximum value of 20.7 kVA, value which exceeds the 20 kVA nominal power value of the generator and thus leading to the generator trip by the overload protection;

· after the Uninterruptable Power Supply was taken out of the scheme and the load reduced to 3 kVA, when the diesel power generator or main power supply connects, the load is uniformly distributed;

· the power spikes are caused by the Uninterruptable Power Supply and so it is recommended that when powering a n kVA consumer, to equip a 2 * n kVA nominal power value backup generator.

Acknowledgements

We would like to thank C.N.T.E.E. Transelectrica S.A., Romania for offering us the tools required for conducting this study. After we communicated our conclusions, the 20 kVA diesel generator was replaced with a 30 kVA one.

References

[1] Şurianu F D 2007 *Electric Power Consumers,* Orizonturi Universitare, Timisoara, Romania
[2] ***C.N.T.E.E. Transelectrica S.A. (private communication)
[3] Fanica V, Petru P and Ana P 2013 *Electric Power Quality,* SIER, Bucuresti, Romania
[4] Andrea P 2002 *Automation and protection of installations and power systems,* Orizonturi Universitare, Timisoara, Romania (in Romanian)

Issues regarding the modelling and simulation of hybrid micro grid systems

I Szeidert[1], I Filip[1] and O Prostean[1]

[1]Politehnica University of Timisoara, Automation and Applied Informatics Department, Vasile Parvan str., no. 2, 300223 Timisoara, Romania

E-mail: iosif.szeidert@aut.upt.ro

Abstract. The main followed objectives within control strategies dedicated to hybrid micro grid systems (wind/hydro/solar), that operate based on maximum power point tracking (MPPT) techniques are to improve the conversion systems efficiency and to maintain the quality of the produced electrical energy (the voltage and power factor control). One of the main goals of maximum power point tracking strategy is to achieve the harvesting of the maximal possible energy within a pre-set time period. In order to implement the control strategies for micro grid systems that operate at time variable parameter, there are usually required specific transducers (anemometer for wind speed measurement, optical rotational transducers, taco generators, etc.). In the technical literature there are presented several variants of the MPPT techniques, which are particularized at several applications (wind energy conversion systems, solar systems, hydro plants and micro grid hybrid systems). The maximum power point tracking implementations are mainly based on two-level architecture. The inferior level controls the primary variables, while the superior level represents the MPPT control structure. In the paper, authors present some micro grid structures proposed at Politehnica University Timisoara within the frame of a research grant. The paper is focused on the application of MPPT strategies on hybrid micro grid systems. There are presented several structures and control strategies and are highlighted their advantages and disadvantages, together with practical implementation guidelines.

1. Introduction

In technical literature, the micro grids are defined as aggregates/clusters of different generation systems that are interconnected together under a common control system/strategy. In those clusters there can be found solar/wind/hydro aggregates and/or other energy sources, storage devices and other additional equipment. In present, there are in progress numerous researches regarding the domain of hybrid micro grids [1], [2].

Also, at Politehnica University of Timisoara, there have been performed researches in afferent domains, since 1980 in domain of wind energy conversion systems. The authors of the present paper are one of the members of research grant that is focused on the hybrid conversion systems for renewable energies integrated into a micro-grid (research grant's acronym MICROREN).

The researchers are focused on the research, design and testing of flexible and intelligent micro grid, based on the integration of three types of renewable resources of energy: wind, solar and hydro. One of the main activities involves the study of the micro grid behavior under different operating conditions. In this order of idea there are required to be modeled and simulated all the component elements of different micro grid type structures.

In the paper there are presented issues on the above mentioned problematic, together with a proposed and analyzed control strategy of MPPT (maximum power point tracking) type.

2. Issues on modelling, simulation and control of micro grid elements/components

The increasing concerns of environmental issues require for the search for new and more sustainable electrical sources. Wind turbines, solar energy, hydro energy combined with fuel cells can be considered as possible answers for the environmental-friendly electrical energy production. Wind turbines are systems that include a rotor, a generator, turbine blades, and a drive or a coupling device. Mainly, those are equipped with squirrel cage induction generators, doubly fed (winded) induction generators or synchronous generators [3], [4], [5].

The authors present in Figure 1 a structure for a wind turbine emulation used in the research laboratory at the Politehnica University of Timisoara.

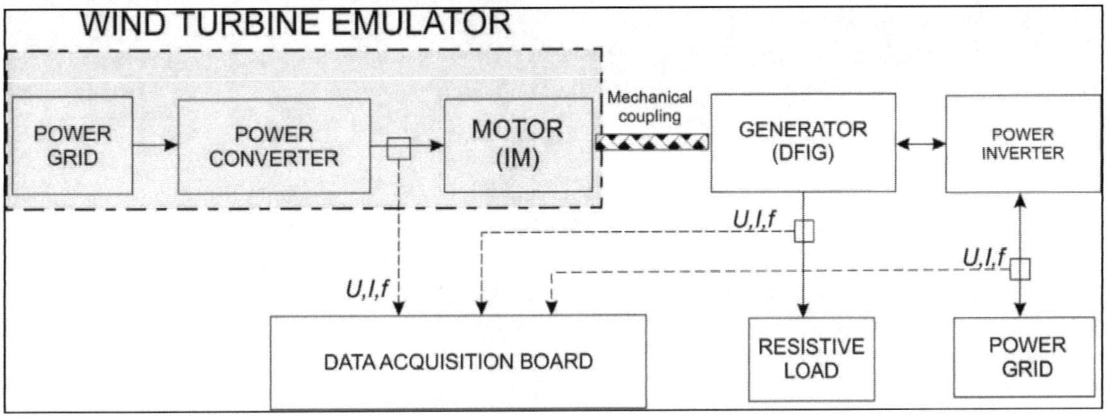

Figure 1. Schematics of the wind turbine emulator— experimental stand at Politehnica University Timisoara

The authors have performed the mathematical modeling and simulation of the wind energy conversion system in the Matlab environment. Some issues that must be considered in the case of the electrical machine modelling is to linearize the mathematical models in some operating regimes in order to reduce their order (as example in the case of the double winded induction machine the model has at least 7th order). The photovoltaic (PV) generation systems are also expected to increase significantly at worldwide level. PV's are an attractive source of renewable energy for distributed urban power generation due to their relatively small size and noiseless operation (versus wind turbines) [6], [7].

Photovoltaic aggregates present the major advantage that there can be easily added supplementary units. Photovoltaic systems convert the sunlight directly to electricity. The technology is quite well established and widely used for power supplies. The PV cells can be modeled as a diode in parallel with a constant current source and a shunt resistor (this being the most common model). They have been also modelled in Matlab-Simulink. The issues are similar in the case of hydro aggregates. Another important problematic in the case of hybrid micro grids is the establishing of the operation modes / regimes for the entire energy conversion system. Usually, electronic power converters are used to connect DG systems in parallel with the power network grid or other sources, but it may be useful for the power converters to operate in stand-alone mode, when the other sources become unavailable to supply certain loads. Electronic power converters connected to batteries or other storage devices will also need to be bidirectional to load/unload those de-vices/equipment. Main operating regimes that can be identified in the case of micro grids are usually: nominal grid connection mode and stand-alone modes.

The future evolution of the electrical energy system should certainly get a higher degree of decentralization concerning specially regarding the production and control issues. Therefore, the study through modelling and simulation of integrated micro grid systems completed with their validation on experimental setups should be of a high importance [8], [9], [10].

Regarding the control strategies there should be considered modern approaches such as the maximum power point tracking (MPPT) techniques. The authors proposed and implemented such structure as presented in Figure 2. The MPPT control strategy is placed on an upper hierarchical level fulfilling the control structure placed on the lower hierarchical level. So, the outputs of the MPPT control strategies are used as prescribed values for the control structure situated on the lower hierarchical level [11], [12].

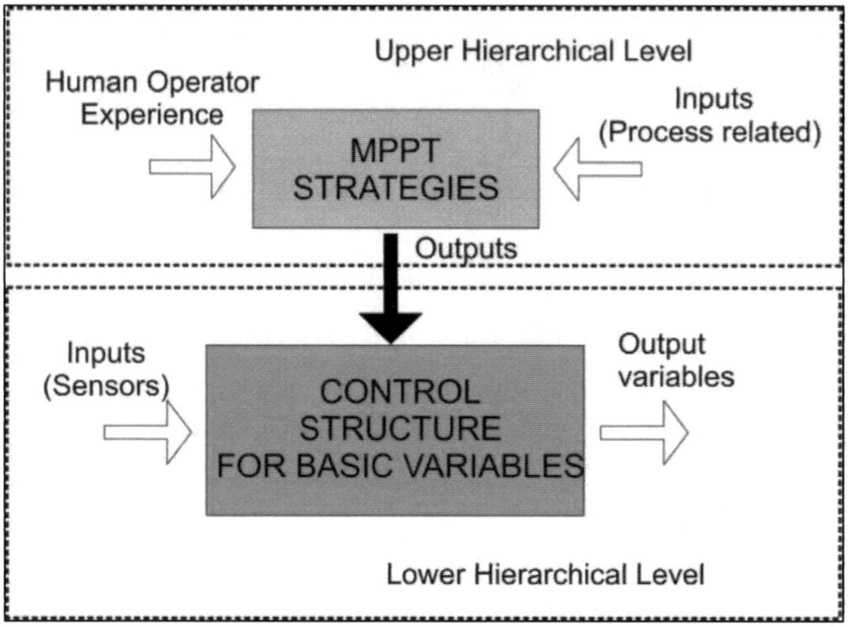

Figure 2. MPPT structure for hybrid micro grids (two hierarchical level diagram)

The proposed control strategy / structure was implemented and tested on laboratory setup. There was used dedicated hardware based on PLC (programmable logic controller). The solution considered the usage of a PLC from SIMATIC family and specific interfacing and connecting elements. The structure of the used equipment is consisted from: central unit PLC – SIMATIC S7, digital input module, digital output module, analog input module, analog output module, Ethernet communication module, other related accessories. The chosen architecture assures the local control of the laboratory setup (wind-solar hybrid system) and also the communication with the superior hierarchical supervisory level of the assembly's component elements. Through the mean of a SCADA system the entire micro grid is controlled and is assured a correct functioning of the assembly. The system is equipped with two data input/output devices constituted from a local touch panel and a remote SCADA system, the communication between them being assured via internet (through Ethernet modules). The programming language was STEP7 specific to the PLC from S7 family. There was assured the optimal functioning of the entire system so that the obtained electrical energy to be maximized in the given operating conditions.

The system control is performed by a power management unit. The hybrid system has three main nominal operating regimes:

- Regime 1: the wind turbine operates with power regulation and the PV is inactive;

- Regime 2: the wind turbine operates by MPPT and the PV completes the energy requirement, being integrated in the power loop;
- Regime 3: the both two renewable sources are operating by the mean of MPPT strategy.

3. Conclusions

In the paper, authors present some micro grid structures proposed at Politehnica University Timisoara within the frame of a research grant. There are presented issues regarding the modeling, simulation and control of small hybrid micro grids.

The main objectives to be achieved in the case of micro grid hybrid systems (wind/hydro/solar) that operate under the principle of maximum power point tracking (MPPT) is to increasing of energy conversion system's efficiency, meanwhile assuring that the delivered electrical energy into the power grid is maintained at required quality levels. The main goal of any MPPT strategy is to reach a maximum of harvested/converted energy in nominal operating regimes within a certain time period.

There are presented several structures and control strategies and are highlighted their advantages and disadvantages, together with practical implementation guidelines. There is considered the implementation using SCADA and PLC equipment. There are presented issues regarding practical implementation solution on laboratory rigs.

There can be must be considered, that the MPPT strategy is proper for micro grid systems, because those systems are complex and present high order mathematical models (in this cases there can be difficult to use classical control strategies).

4. Acknowledgement

This paper was developed through the Partnerships in Priority Areas - PN II, with the support of ANCS, CNDI—UEFISCDI, Project no. 36/2012.

References

[1] Venkateswarlu K and Krishna J 2012 Modeling and Simulation of Micro Grid System Based on Renewable Power Generation Units by using Multilevel Converter, *International Journal of Engineering Research & Technology* **1**(6)

[2] Deshmukh M K and Deshmukh S S 2008 Modeling of hybrid renewable energy systems, *Renewable and Sustainable Energy Reviews* **12** 235– 249

[3] Simoes G and Farret F 2004 *Renewable Energy Systems – Design and Analysis with Induction Generator*, CRC Press, Boca Raton, Florida, USA

[4] Budisan N, Prostean O, Boraci R, Szeidert I and Muller V 2010 *The Dual Induction Generator for Renewable Energy Conversion Systems. Experimental Results, Problems and Solutions*, IEEE International Joint Conferences on Computational Cybernetics and Technical Informatics (ICCC-CONTI 2010), Timisoara, Romania, May 27-29, pp 445-450

[5] Budisan N, Prostean O, Robu N and Filip I 2007 Revival by Automation of Induction Generator for Distributed Power Systems, in Romanian Academic Research, *Renewable Energy* **32**(9) 1484-1496

[6] Zhu Y and Tomsovic K 2002 Development of models for analyzing the load-following performance of microturbine and fuel cells, *Electric Power System Research* **62**(1) 1-11

[7] Nikkhajoei H and Iravani M R 2002 *Modeling and analysis of a micro-turbine generation system*, IEEE Power Engineering Society Summer Meeting, Chicago, USA, July 21-25, pp 167-169

[8] Valenciaga F and Puleston P F 2005 Supervisor Control for a Stand-Alone Hybrid Generation System Using Wind and Photovoltaic Energy, *IEEE Transactions on Energy Conversion* **20**(2) 398 - 405

[9] ***2009 European Commission website on Micro grids. [Online] http://www.microgrids.eu

[10] ***2006 European Smart Grids Technology Platform: Visions and Strategy for Europe's Electricity Networks of the Future. European Commission: Directorate-General for Research

[11] Lasseter B 2001 Micro grids [distributed power generation], *IEEE Power Engineering Society Winter Meeting* **1** 146–149
[12] Zoka Y, Sasaki H, Yorino N, Kawahara K and Liu C C 2004 *An interaction problem of distributed generators installed in a micro grid*, Proceedings of IEEE on Electric Utility Deregulation, Restructuring and Power Technologies Conference, Hong Kong, China, April 5-8, pp 795–799

IOP Conf. Series: Materials Science and Engineering **106** (2016) 012029 doi:10.1088/1757-899X/106/1/012029

Dual stator winding variable speed asynchronous generator: magnetic equivalent circuit with saturation, FEM analysis and experiments

L N Tutelea[1], N Muntean[1], S I Deaconu[2] and C D Cunţan[2]

[1]Politehnica University of Timisoara, Electrical Engineering Department, Vasile Parvan str., no. 1-2, 300223 Timisoara, Romania
[2]Politehnica University of Timisoara, Department of Electrical Engineering and Industrial Informatics, 5 Revolution Street, 331128 Hunedoara, Romania

E-mail: sorin.deaconu@fih.upt.ro

Abstract. The authors carried out a theoretical and experimental study of dual stator winding squirrel cage asynchronous generator (DSWA) behaviour in the presence of saturation regime (non-sinusoidal) due to the variable speed operation. The main aims are the determination of the relations of calculating the equivalent parameters of the machine windings, FEM validation of parameters and characteristics with free FEMM 4.2 computing software and the practice experimental tests for verifying them. Issue is limited to three phase range of double stator winding cage-asynchronous generator of small sized powers, the most currently used in the small adjustable speed wind or hydro power plants. The tests were carried out using three-phase asynchronous generator having rated power of 6 [kVA].

1. Introduction

Despite its simple and robust construction, the motion control for this type of generators should take into account the complexity of the dynamic model which is nonlinear and variable in time and that the physical parameters of the machine are not always known with great precision [1-3]. Usually, the electric machines are designed to be supplied in sinusoidal regime. The presence of higher harmonics will have as result the appearance of a deforming and saturation regime in the machine, with adverse effects in general in its operation [4], [5]. The appearance of the deforming regime in the machine is inevitable because any static frequency converter based on semiconductor technique produces voltages or currents, which contain, in addition to fundamental and harmonic, higher odd time harmonics.

The introduction of distributed generation through renewable sources of energy has opened a challenging area for power engineers. As these sources are intermittent in nature, variable speed electric generators are employed for harnessing electrical energy from these sources. However, power electronic control is required to connect these sources to the existing grid [6]. A wind turbine can be designed for a constant speed or variable speed operation. Variable speed wind turbines can produce 8% to 15% more energy output as compared to their constant speed counterparts, however, they necessitate power electronic converters to provide a fixed frequency and fixed voltage power to their loads. Most turbine manufacturers have opted for reduction gears between the low speed turbine rotor and the high speed three-phase generators. Direct drive configuration, where a generator is coupled to the rotor of a wind turbine directly, offers high reliability, low maintenance, and possibly low cost for certain turbines [6-11].

For small to medium power wind or hydro turbines, permanent magnet generators and squirrel cage asynchronous generators are often used because of their reliability and cost advantages. For certain high power wind or hydro turbines, effective power control can be achieved with double PWM (pulse width modulation) converters which provide a bidirectional power flow between the turbine generator and the utility grid [12-17].

2. Equivalent Circuit with Considering the Saturation

One of the main goals is the reduction of the necessary reactive excitation power by the arrangement of the load windings versus the excitation windings. Some arrangements of the windings can establish an internal positive load current reaction and thus can determine the desired effect. Such arrangements can be obtained with the excitation winding and load winding displaced one from another by a certain space angle α, where the rotor winding and the consumer impedance are also shown. Because of this complexity, a reconsideration of the main flux linkage saturation effect is called for in the design of the machine and in the development and practical implementation of speed/torque control algorithms. To avoid deep magnetic saturation in the stator and rotor cores and rotor and stator teeth, magnetic design methodologies have been suggested both for the dual stator-winding and brushless doubly fed asynchronous machines [9], [10].

In the literature there are known various mathematical models associated to asynchronous machines fed by static frequency and voltage converters.

In the case of magnetic saturation, the mmf of the iron U_{Hfe} is no longer negligible with respect to the mmf of the air-gap $U_{H\delta}$, a deformation of the magnetic curvature takes place, and the polar coverage factor depends on the saturation factor k_s [17]:

$$k_{st} = \frac{U_{H\delta} + U_{Hts} + U_{Htr}}{U_{H\delta}}, \qquad (1)$$

for the teeth saturation and a new proposal:

$$k_{s\alpha p} = \frac{U_{H\delta} + U_{Hts} + U_{Htr} - U_{Hys} - U_{Hyr}}{U_{H\delta}}, \qquad (2)$$

for the teeth and yoke saturation, where U_{Hts} and U_{Htr} represent the mmf in the stator and rotor teeth, and U_{Hys} and U_{Hyr} represent the mmf in the stator and rotor yoke.

In the references [17], [18] it was shown that if there is yokes saturation (both stator and rotor), this influences in the opposite way the distribution of the flux density in the air-gap. The teeth saturation leads to a bending of the flux density curvature (B_{1flat}), and the yoke saturation leads to a sharper peak ($B_{1\ peak}$) (Figure 1).

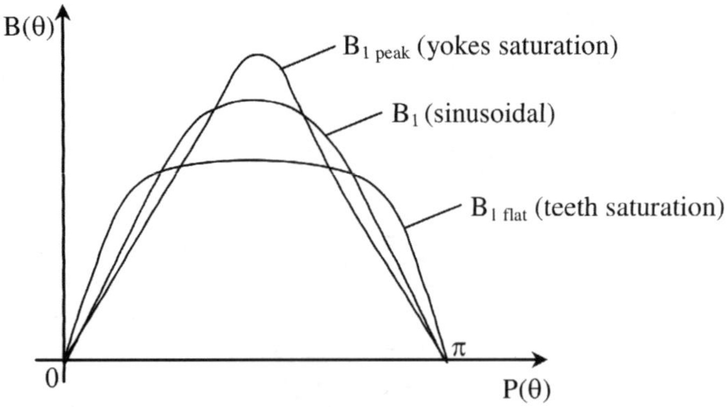

Figure 1. Sinusoidal, flat and peak air-gap flux density

If the level of teeth and yoke saturation is identical, then the distribution of the flux density stays sinusoidal (B_1).

Considering the air-gap distribution of the flux density sinusoidal, sometimes with the 3rd harmonic included, the yoke flux density will be computed in 4-5 points, and the corresponding mmf is calculated through digital integration using the Simpson formula:

$$\int H_y dl \approx \frac{D_x \cdot \pi}{2p} \cdot \frac{1}{c_x} \quad , \tag{3}$$

with

$$c_x = f(B_y) \ . \tag{4}$$

Usually, the magnetization curvatures H = f (B) are given by the producer up to a certain value of the flux density B (for instance 1, 8 - 2T). There are areas in which the flux density is higher than the maximum value obtained in the magnetization curvature very often when designing modern electrical machines, which require high torque densities with respect to the mass and the volume of the machine. A method of extrapolation of the magnetization curvature is proposed, and this method is based on the property of relative magnetic permeability of the magnetic material μ_r to tend to 1 and is also based on the property of differential permeability μ_d to tend to 0 when the flux density becomes infinity:

$$\mu_r = \frac{B}{\mu_0 \cdot H}, \tag{5}$$

$$\mu_d = \frac{dB}{\mu_0 \cdot dH}. \tag{6}$$

The continuity of both the permeability μ_r and μ_d is imposed, in the last points of the given magnetization curvature. The intensity of the magnetic field H_x in a random point may be written as:

$$H_x = H_{ref\,end} + \left[\left(B_x - B_{ref\,end} \right) + B_0 \left(e^{a_1 B_x} - e^{a_1 B_{ref\,end}} \right) \right] \cdot \frac{1}{\mu_0}, \tag{7}$$

where H_{refend} and B_{refend} are the intensity of the magnetic field, and the flux density respectively in the last point from the characteristic, and a_1 and B_0 are variables calculated so that the function is continuous and convergent. They have the expressions:

$$a_1 = \ln[(1-\mu_{d1})(1-\mu_{d2})] \cdot \frac{1}{x_1 - x_2} \quad , \tag{8}$$

$$x_1 = \frac{B_{n-2} + B_{n-1}}{2} \quad , \tag{9}$$

$$x_2 = \frac{B_{n-1} + B_n}{2} \quad , \tag{10}$$

$$\mu_d = \mu_0 \cdot \frac{dH_{ref}}{dB_{ref}}, \tag{11}$$

$$\mu_{d1} = \mu_0 \cdot \frac{H_{ref\,n-1} - H_{ref\,n-2}}{B_{ref\,n-1} - B_{ref\,n-2}} \quad , \tag{12}$$

$$B_0 = (\mu_{d2} - 1) \cdot e^{-a_1 x_1} \cdot \frac{1}{a_1} \ . \tag{13}$$

In Figure 2 the dependence between the flux density and intensity of the magnetic field (extended magnetization curvature) is presented.

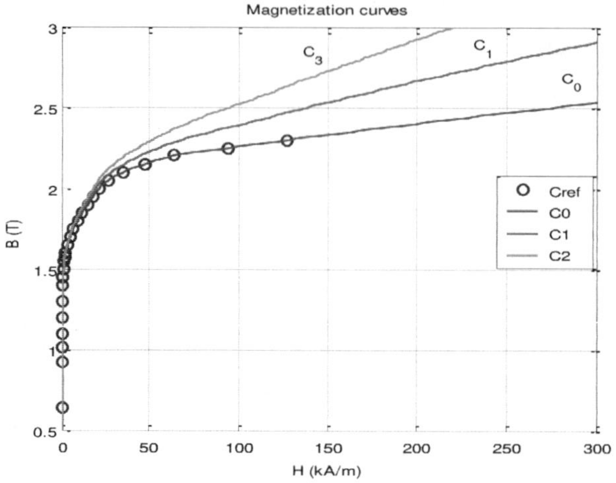

Figure 2. Magnetization curves: Cref –references points, C0 – interpolation and extrapolation on original curve (w2b=0, kFe=1), C1 – interpolation and extrapolation on modified curve with w2b=1, kFe=0.97, C2 – interpolation and extrapolation on modified curve considering w2b=2, kFe=0.97

In the case of yokes saturation at the nominal frequency, losses are very high. Talking about the low frequency generating electrical generators, at which the shaft is coupled directly to the wind turbine, there is a possibility of yoke saturation without any significant increase of iron losses. Flux density in the yoke is calculated with the relation:

$$B_{ref\,1} = B_{ref} + \mu_0 \cdot H_{ref} \cdot \left(\frac{w_{2b} + 1}{k_{Fe}} \right) , \tag{14}$$

where

$$w_{2b} = \frac{h_{st}}{h_y} , \tag{15}$$

the depth of the slot is denoted by h_{st}, and the length of the yoke where the flux density is computed is denoted by h_y. For the case of the teeth:

$$w_{2b} = \frac{w}{b} , \tag{16}$$

w being the length of the slot and b the length of the teeth.

An iterative solving of the circuit is made, starting from the linear inductance around the nominal load, the current is computed and then the curvatures of the inductances are being iteratively computed.

3. FEM Analysis

An approach that includes the influence of magnetic saturation and iron loss using finite-element analysis in the performance prediction of the twin stator winding asynchronous machine was set. The proposed finite-element model provides very good steady-state predictions and can be used for the sizing and design optimization of the machine.

IOP Conf. Series: Materials Science and Engineering **106** (2016) 012029 doi:10.1088/1757-899X/106/1/012029

The asynchronous machine saturation could be increased at low frequency (required in direct driving) without notably iron losses increases in order to have a larger slot area required in two stator winding or for aluminium winding which could replace the copper winding in low cost machines.

The analytical model should be improved for high saturation magnetic core. Increasing the stator slot with 40% will increase the saturation of the stator core (Figure 3) and will change the line field distribution. The leakage inductances computed analytically could be validate through a virtual short circuit using ac finite element analysis where the stator winding currents are given (rated or higher values) at rated frequency and the rotor bar currents are computed as eddy currents (Figure 4). The end coil leakage inductances are not considered in 2D FEM approach.

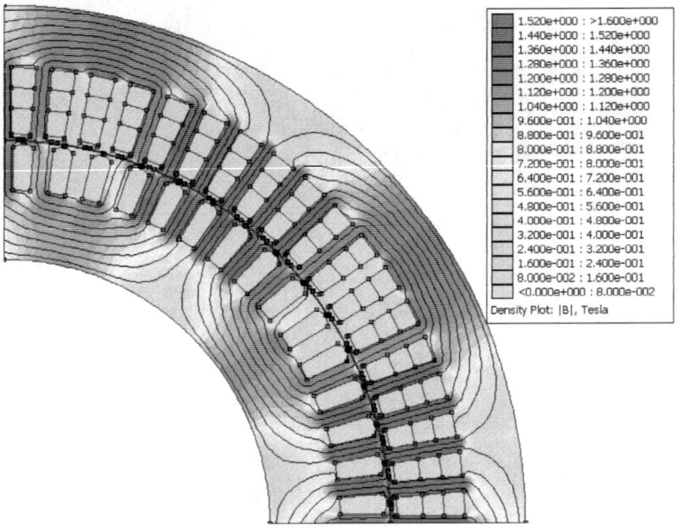

Figure 3. The magnetic field distribution: original 6 kVA, 8 poles, 15 Hz rated frequency induction generator

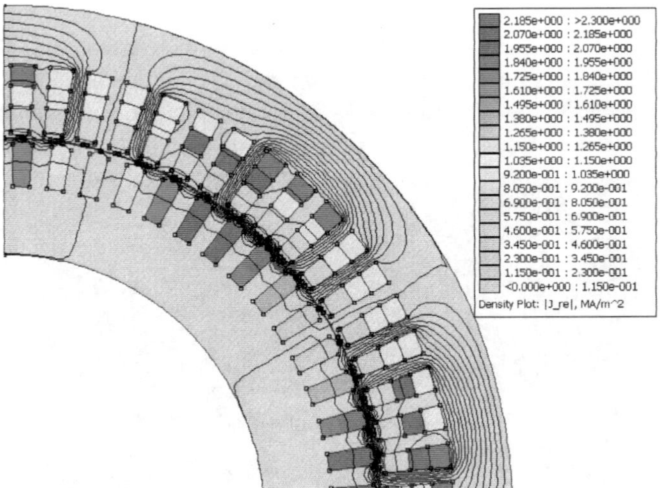

Figure 4. Virtual short circuit test using AC finite element analyses ($f=f_N$=15Hz, I_{mA} = 7A peak current in main winding, $I_{mB}=I_{mC}$=-3.5A)

No load (zero rotor current) FEM investigations at standstill, with dc currents for star connection lead to I_{Ae}= I= -2I_{Be}= -2I_{Ce}. Figure 5 presents the air-gap flux density (produced by main and excitation currents) and Figure 6 shows the normal air-gap magnetic field strength.

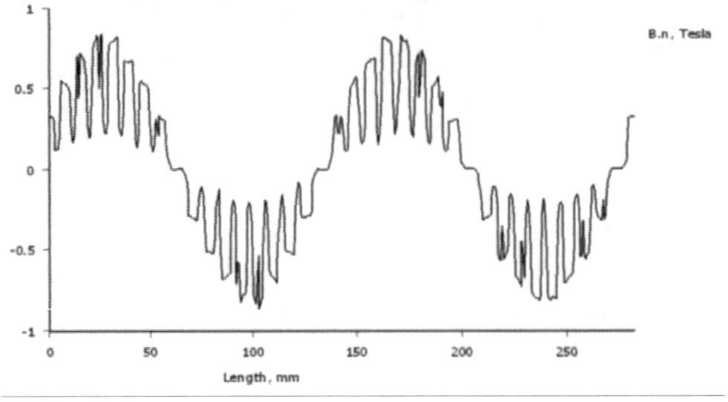

Figure 5. Normal air-gap flux density

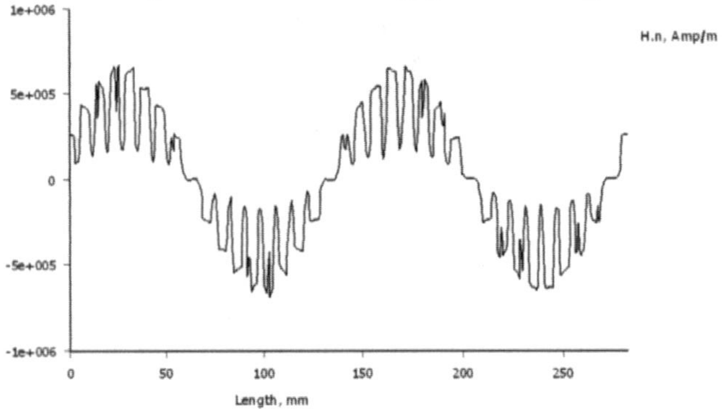

Figure 6. Normal air-gap magnetic field strength

4. Experimental Results

To prove the above design methods, a 6kVA 400V/415 rpm prototype of the DSWA generating system has been developed (Figure 7). The prime mover is simulated by a three-phase cage-type asynchronous machine driven by an inverter of ABB ACS 800-11. In experiments, the value of the auxiliary excitation capacitors is 35 μF, and the value of the filter inductances is 23 mH.

Figure 7. Generator DSWA (left) coupled with prime mover induction motor (right)

Through the experimental tryouts it is desired, in a first step, the computation of the DSWA parameters and characteristics for the stationary regime and their comparison with the values obtained through finite element analysis. In Figure 8 the equivalent phase scheme of the machine is presented. Resistances R_m and R_e are measured in DC current and the inductances will be computed through no load (real and ideal) and short circuit methods.

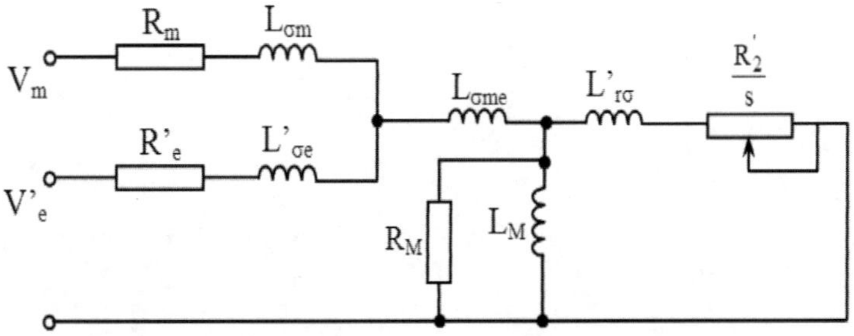

Figure 8. Equivalent phase scheme of the DSWA

3.1. No load probe

The equivalent phase scheme for the no load methods is presented in Figure 9. The equations which characterize the functioning in these conditions are given as follows:

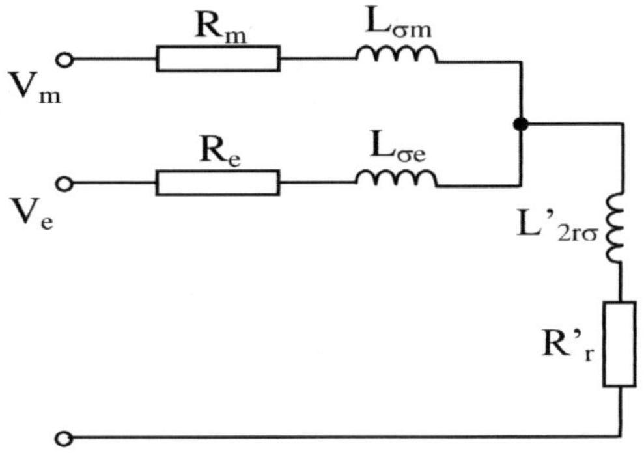

Figure 9. Equivalent phase scheme of DSWA in no load probe

$$\underline{V}_{mo} - (R_m + jX_{\sigma m})\cdot \underline{I}_{1mo} = \underline{V}'_{eo} = k_e \cdot \underline{V}_{eo}, \tag{17}$$

$$\underline{V}'_e - (R'_e + jX'_{\sigma e})\cdot \underline{I}'_{1e} = \underline{V}_{2m}, \tag{18}$$

$$k_e \cdot \underline{V}_e - (R_e + jX_{\sigma e})\cdot k_e^2 \cdot \frac{\underline{I}_{1e}}{k_e} = \underline{V}_{2m}, \tag{19}$$

$$k_e\left[\underline{V}_e - (R_e + jX_{\sigma e})\cdot \underline{I}_{1e}\right] = \underline{V}_{2m}. \tag{20}$$

3.2. Short circuit probe

Equivalent phase scheme in short circuit (Figure 10) and the equations are given as follows:

$$\underline{V}_{msc} - (R_m + jX_{\sigma m})\cdot \underline{I}_{1msc} = \underline{V}'_{esc} = k_e \cdot \underline{V}_{esc}, \tag{21}$$

$$\frac{\underline{V}_{mo} - (R_m + jX_{\sigma m})\cdot \underline{I}_{1mo}}{\underline{V}_{msc} - (R_m + jX_{\sigma m})\cdot \underline{I}_{1msc}} = \frac{\underline{V}_{eo}}{\underline{V}_{esc}}, \tag{22}$$

$$\underline{V}_{mo} - (R_m + jX_{\sigma m})\cdot \underline{I}_{1mo} = \frac{\underline{V}_{eo}}{\underline{V}_{esc}} \cdot \left[\underline{V}_{msc} - (R_m + jX_{\sigma m})\cdot \underline{I}_{1msc}\right], \tag{23}$$

IOP Conf. Series: Materials Science and Engineering **106** (2016) 012029 doi:10.1088/1757-899X/106/1/012029

$$\underline{V}_{mo}-\frac{\underline{V}_{eo}}{\underline{V}_{esc}}\cdot\underline{V}_{msc}=\left(R_m+jX_{\sigma m}\right)\left(\underline{I}_{1mo}-\underline{I}_{1msc}\right), \tag{24}$$

$$R_m+jX_{\sigma m}=\frac{\underline{V}_{mo}-\dfrac{\underline{V}_{eo}}{\underline{V}_{esc}}\cdot\underline{V}_{msc}}{\underline{I}_{1mo}-\underline{I}_{1msc}}, \tag{25}$$

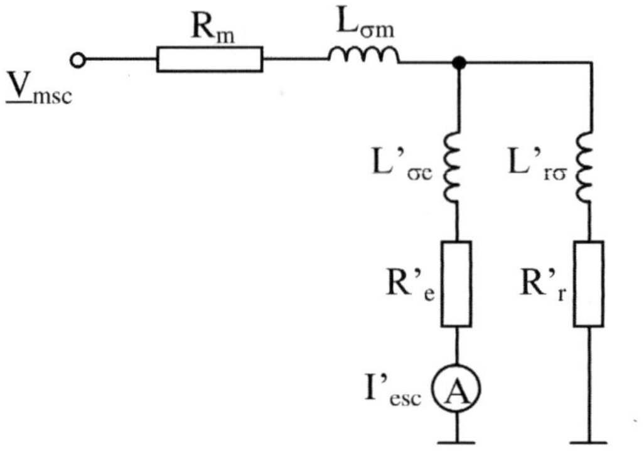

Figure 10. Equivalent phase scheme of the DSWA in short circuit probe

A ratio is made between relations (21) and (25) and the result is:

$$X_e^{'}=\omega L_{\sigma e}^{'}, \tag{26}$$

$$\underline{V}_{msc}-\left(R_m+jX_{\sigma m}\right)\cdot\underline{I}_{1msc}=\left(R_e^{'}+jX_e^{'}\right)\cdot\underline{I}_{esc}^{'}. \tag{27}$$

with the help of the previous measurements and equations the stationary regime parameters are computed for the DSWA.

The magnetizing effect of the main current and that the best arrangement of excitation and main windings are at $\alpha = -\pi/2$ (Figures 11, 12, and 13).

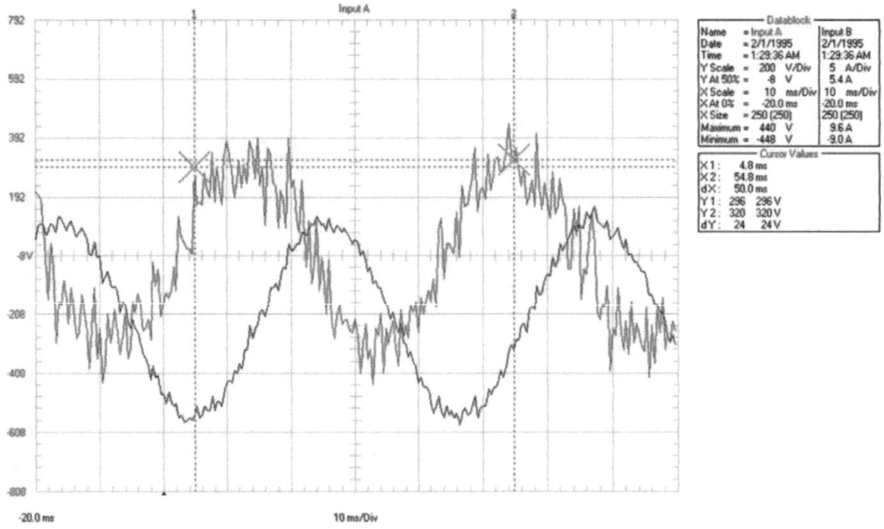

Figure 11. Main voltage and main current: I_{DSWA}=5.82 A, power to network P_{net}=-40%, f=24 Hz, prime mover current I_{PM}=13.3 A, reference torque 32.8 %, speed n=365.5 rpm

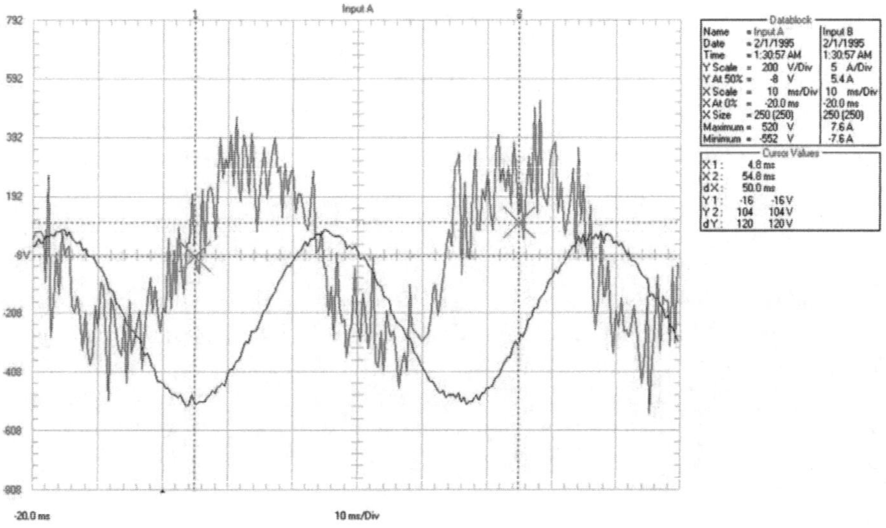

Figure 12. Main voltage and main current: I_{DSWA}=5.12 A, power to network P_{net}=-14.62%, f=24 Hz, prime mover current I_{PM}=12.12 A, reference torque 16.5 %, speed n=362.4 rpm

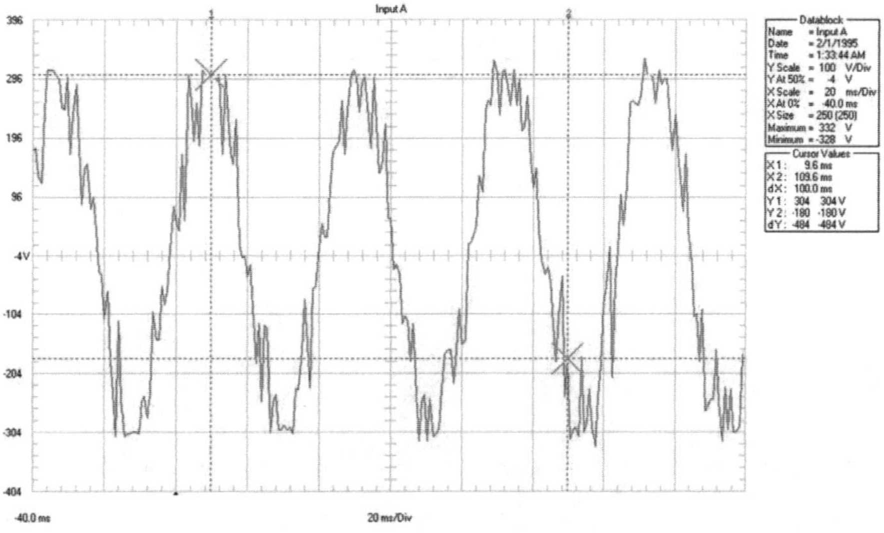

Figure 13. No load excitation voltage

5. Conclusions

In this paper was presented a new type of dual stator windings asynchronous machine operating in generator mode at variable low speed. A mathematical perfectly saturated model should be implemented in flux rotating frame coordinate and then the transients and steady state magnetization inductance could be considered in the model. The FEM validation results are still accurate from despite of large difference between steady state inductance and transient inductances, because the magnetization current magnitude is rather constant during the fundamental currents and voltages. In the experimental tests are prove quantitatively and qualitatively the good correlations with simulations, and the practicality of the proposed system for low power variable speed wind or hydro power plants applications.

References

[1] Bu F, Hu Y, Huang W and Zhuang S 2014 Parameter design and static performance of Dual Stator-Winding Induction Generator Variable Frequency AC Generating System with Capacitive and Inductive Loads, *IEEE Transactions on Industrial Electronics* **61**(8) 3902-3914

[2] Bu F, Hu Y, Huang W , Zhuang S and Shi K 2014 Control Strategy and Dynamic Performance of Dual Stator-Winding Induction Generator Variable Frequency AC Generating System with Capacitive and Inductive Loads, *IEEE Transactions on Power Electronics* **29**(4) 1681-1692

[3] Bu F, Huang W, Hu Y, Shi J and Shi K 2012 A Stand-Alone Dual Stator-Winding Induction Generator Variable Frequency AC Power System, *IEEE Transactions on Power Electronics* **27**(1) 10-13

[4] Bu F, Huang W, Hu Y and Shi K 2012 An Integrated AC and DC Hybrid Generation System Using Dual-Stator-Winding Induction Generator With Static Excitation Controller, *IEEE Transactions on Energy Conversion* **27**(3) 810-812

[5] Bu F, Huang W, Hu Y and Shi K 2011 An Excitation-Capacitor-Optimized Dual Stator-Winding Induction Generator with the Static Excitation Controller for Wind Power Application, *IEEE Transactions on Energy Conversion* **26**(1) 122-131

[6] Bu F, Hu Y, Huang W, Zhuang S and Shi K 2015 Wide-Speed-Range-Operation Dual Stator-Winding Induction Generator DC Generating System for Wind Power Applications, *IEEE Transactions on Power Electronics* **30**(2) 561-573

[7] Nian H and Song Y 2014 Direct power control of doubly fed induction generator under distorted grid voltage, *IEEE Transactions on Power Electronics* **29**(2) 894-905

[8] Zinger D S and Muljadi E 1997 Annualized Wind Energy Improvement Using Variable Speeds, *IEEE Transactions on Industry Applications* **33**(6) 1444-1447

[9] Bansal R C, Bhatti T S and Kothari D P 2003 A bibliographical survey on induction generators for application of non-conventional energy systems, *IEEE Transactions on Energy Conversion* **18**(3) 433-439

[10] Wang D, Ma W, Xiao F, Zhang B, Liu D and Hu A 2005 A novel stand-alone dual stator-winding induction generator with static excitation regulation, *IEEE Transactions on Energy Conversion* **20**(4) 826-835

[11] Li Y, Hu Y, Huang W, Liu L and Zhang Y 2009 The capacity optimization for the static excitation controller of the dual-stator-winding induction generator operating in a wide speed range, *IEEE Transactions on Industrial Electronics* **56**(2) 530 -541

[12] Li Y, Hu Y and Huang W 2005 *Optimal design of dual stator-winding induction generator with variable speed based on improved genetic algorithm*, Proc. 8th Int. Conf. Elect. Mach. Syst., Nanjing, China, September 27-29, pp 2343-2348

[13] Huang W, Hu Y, Bu F and K Shi 2010 *The magneto motive force of a novel dual stator-winding induction generator*, Proc. IEEE Energy Convers. Cong. Expo., Atlanta, USA, September 12-16, pp 3582-3587

[14] Levi E 1995 Impact of iron loss on behaviour of vector controlled induction machines, *IEEE Transactions on Industry Applications* **31**(6) 1287-1296

IOP Conf. Series: Materials Science and Engineering **106** (2016) 012030 doi:10.1088/1757-899X/106/1/012030

Compensator configurations for load currents' symmetrization

D Rusinaru[1], L G Manescu[1] and R C Dinu[1]

[1]University of Craiova, Faculty of Electrical Engineering, Electrical Engineering, Energetics and Aerospace Department, 107 Decebal Street, 200440 Craiova, Romania

E-mail: rcdinu@elth.ucv.ro

Abstract. This paper approaches aspects regarding the mitigation effects of asymmetries in 3-phase 3-wire networks. The measure consisting in connecting of load current symmetrization devices at the load coupling point is presented. A time-variation of compensators parameters is determined as a function of the time-recorded electrical values. The general sizing principle of the load current symmetrization reactive components is based on a simple equivalent model of the unbalanced 3-phase loads. By using these compensators a certain control of the power components transits is ensured in the network. The control is based on the variations laws of the compensators parameters as functions of the recorded electrical values: $[B]=[T]\cdot[M]$. The link between compensator parameters and measured values is ensured by a transformation matrix [T] for each operation conditions of the supply network. Additional conditions for improving of energy and efficiency performance of the compensator are considered: i.e. reactive power compensation. The compensator sizing algorithm was implemented into a MATLAB environment software, which generate the time-evolution of the parameters of load current symmetrization device. The input data of application takes into account time-recording of the electrical values. By using the compensator sizing software, some results were achieved for the case of a consumer connected at 20 kV busbar of a distribution substation, during 24 hours measurement session. Even the sizing of the compensators aimed some additional network operation aspects (power factor correction) correlated with the total or major load symmetrizations, the harmonics aspects of the network values were neglected.

1. Introduction

Different types of perturbations distort the power supplying of the actual loads. The loads are also responsible for unfavorable effects on the power system operation and consequently required a great number of study initiatives. Their purpose is to clarify the power related phenomenon or finding efficient methods for load symmetrization.

In case of an unbalanced operation of power network the symmetrization can be achieved either by a mono-phase loads repartition on the three phases of the network, or by using special devices, e.g. compensators for load currents symmetrization [1-3].

A wide-spread solution for symmetrization of the electrical values' in the unbalanced power systems is offered by SVC's, having an independent control of the each phase reactive powers.

The asymmetry mitigation is one of the measures applied for optimization of processes in power systems. The design of these devices aims also some other aspects regarding the quality and efficiency of delivered electricity (i.e. harmonics, power factor).

2. Load currents' symmetrization

Designing of the most efficient devices for power quality disturbances' mitigation registered already multiple alternatives, even in fact the principle is almost a common one: the power flow control on the network phases [4], [5].

This paper presents a solution for a symmetrization device used in 3-phase 3-wire distribution networks. This solution was developed based on the symmetrical components theory.

The general strategy for load currents symmetrization can be applied (Figure1) by taken into account the consumers' operational particularities.

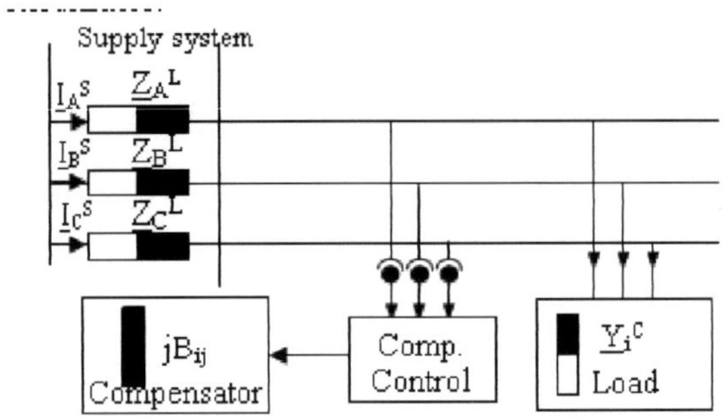

Figure 1. Supply configuration for an unbalanced load assisted by a load currents symmetrization compensator

For the load current symmetrization a SVC configuration (Δ or open-Δ□□) can be connected in the load common coupling point (CCP). The compensator is designed to ensure different reactive power injections on each phase, depending on the measured values.

In the case of unbalanced loads in 3-phases 3-wires networks the compensator is designed so that the negative sequence components of the load currents to be annulled. The compensator-load ensemble will appear as symmetrical in CCP.

Mathematically speaking, the symmetrization principle can be described by the following equations:

$$\Re e\left(\underline{I}_-^C + \underline{I}_-^\Delta\right) = 0; \quad \Im m\left(\underline{I}_-^C + \underline{I}_-^\Delta\right) = 0, \tag{1}$$

where, $\underline{I}_-^C$ is the negative component of the load current and $\underline{I}_-^\Delta$ - negative component current trough the compensator's branches.

Establishing of the most efficient configuration of the symmetrization equipments totally depends on the load representation model, adjusted to the network operation characteristics.

In this paper, the configuration of the symmetrization equipment is based on a simplified model of the unbalanced three-phase load.

For simplifying reasons, only the hypothesis of symmetrical voltages is considered:

$$\underline{U}_A = \underline{a}\underline{U}_B = \underline{a}^2\underline{U}_C = \underline{U}_+, \tag{2}$$

where, $\underline{U}_i$ are the supply phase voltages (i = A, B, C) and U_+ - positive component voltage.

An unbalanced load model is developed as in Figure 2 [3] and it contains an equivalent load two phases connected through and admittance $\underline{Y}_m$ given by eq. (3) and a symmetrical equivalent 3-phase load represented by three phase admittances $\underline{Y}_e$.

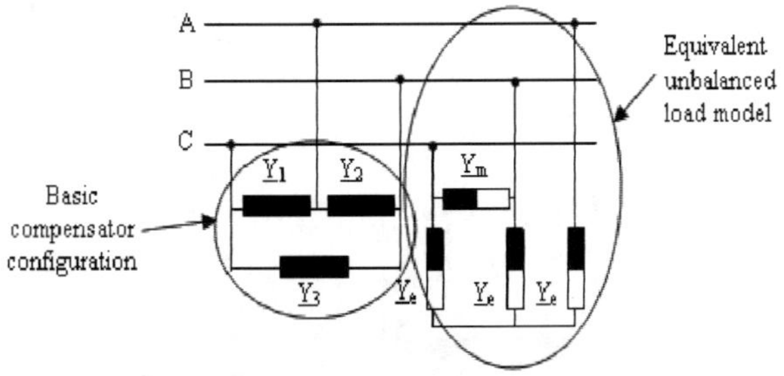

Figure 2. The equivalent electrical configuration of the unbalanced consumer – compensator ensemble

$$\underline{Y}_m = -\left(a\underline{S}_A^* + \underline{a}^2 \underline{S}_B^* + \underline{S}_c^*\right)/U_+^2 = \\ = \frac{\left[\left(P_A - \sqrt{3}Q_A + P_B + \sqrt{3}Q_B - 2P_C\right) + j\left(-\sqrt{3}P_A - Q_A + \sqrt{3}P_B - Q_B + 2Q_C\right)\right]}{2U_+^2}, \quad (3)$$

with, $\underline{S}_i$ – complex phase powers at CCP and P_i, Q_i – active, reactive phase powers at CCP.

The symmetrization configuration can be of Δ-type, or it can be endorsed simply by connecting two reactive components. So that the displacements from a balanced load behavior can be mitigated, and an equivalent symmetrical structure can be achieved. The solution of system (1) gives the susceptances of the on-line symmetrization compensator. A representation of compensator parameters depending on time-recording of electrical values is given by eq. (4).

$$[B] = [T]\cdot[M], \quad (4)$$

where, $[B] \overset{not}{=} [B_1 \quad B_2 \quad B_3]_t$ is the column vector of the SVC's susceptances; $[M] \overset{not}{=} \left[P_A^C/\left(U_A^C\right)^2 \quad P_B^C/\left(U_B^C\right)^2 \quad P_C^C/\left(U_C^C\right)^2 \quad \left|Q_A^C/\left(U_A^C\right)^2 \quad Q_B^C/\left(U_B^C\right)^2 \quad Q_C^C/\left(U_C^C\right)^2\right]_t$ - the column vector of the measured electrical values in the load connection bus and *[T]* – susceptances / electrical values transformation matrix.

An adjustment of this measure for the power factor correction at load's CCP is an additional criterion supplementing the undetermined equations system (1) from mathematically point of view.

3. Load currents symmetrization without additional objective

If only a simple load currents symmetrization is aimed, the compensator configuration can be of Steinmetz type [7], [8] or contains only two reactive elements – see Figure 3. The compensator's elements, $\underline{Y}_1$ and $\underline{Y}_2$, can be placed on the branches complementary to that corresponding to the equivalent 1-phase load. Therefore, the chosen configuration will cancel the negative sequence current ($\underline{I}_- = 0$).

The phase currents at the load coupling point in the compensator presence can be written as:

$$\left[I_f\right] = \left[Y_{l-c}\right]\cdot\left[U_f\right], \quad (5)$$

where, $\left[I_f\right] \overset{not}{=} \left[\underline{I}_A \quad \underline{I}_B \quad \underline{I}_C\right]_t$ is the phase currents column vector and admittance matrix of the load – compensator ensemble is $\left[Y_{l-c}\right] \overset{def}{=} \begin{bmatrix} \underline{Y}_e + \underline{Y}_1 + \underline{Y}_2 & -\underline{Y}_2 & -\underline{Y}_1 \\ -\underline{Y}_2 & \underline{Y}_e + \underline{Y}_2 + \underline{Y}_m & -\underline{Y}_m \\ -\underline{Y}_1 & -\underline{Y}_m & \underline{Y}_e + \underline{Y}_1 + \underline{Y}_m \end{bmatrix}$.

The condition (1) for compensator sizing can be written as in eq. (6):

$$\underline{I}_{-} = -\left(a\underline{Y}_{1} + \underline{a}^{2}\underline{Y}_{2} + \underline{Y}_{m}\right)U_{+} = 0. \tag{6}$$

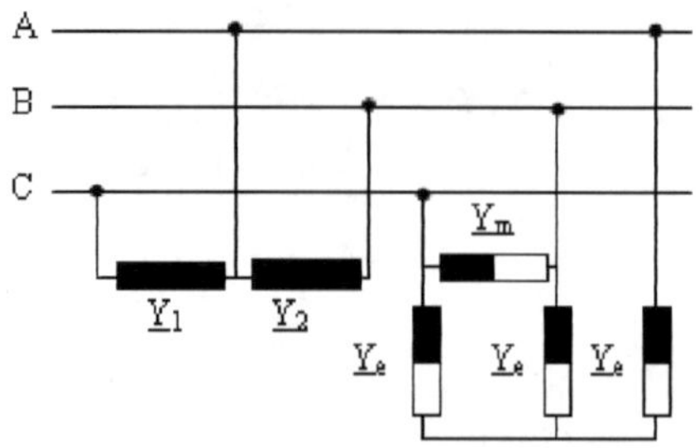

Figure 3. Compensator for load current symmetrization (no additional measures)

So that, the compensator's parameters are given by the following equations:

$$B_1 = G_m / \sqrt{3} + B_m;\ B_2 = -G_m / \sqrt{3} + B_m, \tag{7}$$

with, $G_m = \Re e(\underline{Y}_m)$; $B_m = \Im m(\underline{Y}_m)$ as parameters of equivalent two-phases admitance given by (3).

The transformation susceptances/measured electrical values matrix [T] is given by:

$$[T] = \left[\begin{array}{ccc|ccc} -1/\sqrt{3} & 2/\sqrt{3} & -1/\sqrt{3} & -1 & 0 & 1 \\ -2/\sqrt{3} & 1/\sqrt{3} & 1/\sqrt{3} & 0 & -1 & 1 \end{array}\right], \tag{8}$$

The active power on the each phase of the load-compensator group is given by eq. (9) [6]:

$$\begin{aligned} P_A^{ce} &= U_+\left[\sqrt{3}(I_{1,a} + I_{2,a}) - I_{1,r} + I_{2,r}\right]/2 \\ P_B^{ce} &= U_+\left[-\sqrt{3}(I_{2,a} - 2I_{m,r}) - I_{2,r}\right]/2 \\ P_C^{ce} &= \frac{1}{2}U_+\left[-\sqrt{3}I_{1,a} + I_{1,r} + 2I_{m,r}\right]/2 \end{aligned} , \tag{9}$$

where, P_i^{ce} are the phase active powers of the set compensator/equivalent mono-phase load and I_{1a}, I_{1r}, I_{2a}, I_{2r}, I_{ma}, I_{mr} is active and reactive currents on the branches of the ensemble compensator/equivalent mono-phase load.

In the compensator presence the consumer absorbs from the network an active power P_{tot} and behaves as a symmetrical load, since $P_A^{ce} + P_B^{ce} + P_C^{ce} = 0$. The three phase active powers are redistributed in the system, without changing their flow sense (from load to other balanced components of the network).

$$P_{tot} = P_A + P_B + P_C = 3G_e U_+^2. \tag{10}$$

4. Mathematical modeling through Least Squares Method

In this case, a compensator configuration of Δ-type is recommended, as in Figure 2.

The bus admittance matrix of ensemble load-compensator can be written as:

$$[Y_{l-c}] = \begin{bmatrix} \underline{Y}_{AA} & -\underline{Y}_2 & -\underline{Y}_1 \\ -\underline{Y}_2 & \underline{Y}_{BB} & -\underline{Y}_3 - \underline{Y}_m \\ -\underline{Y}_1 & -\underline{Y}_3 - \underline{Y}_m & \underline{Y}_{CC} \end{bmatrix}. \tag{11}$$

with $\underline{Y}_{AA} \overset{not}{=} \underline{Y}_e + \underline{Y}_1 + \underline{Y}_2; \underline{Y}_{BB} \overset{not}{=} \underline{Y}_e + \underline{Y}_2 + \underline{Y}_3 + \underline{Y}_m; \underline{Y}_{CC} \overset{not}{=} \underline{Y}_e + \underline{Y}_1 + \underline{Y}_3 + \underline{Y}_m$

The correspondent symmetrical components are given by:

$$\underline{I}_- = -\left(a\underline{Y}_1 + a^2\underline{Y}_2 + \underline{Y}_3 + \underline{Y}_m\right)\cdot U_+; \ \underline{I}_+ = \left(\underline{Y}_1 + \underline{Y}_2 + \underline{Y}_3 + \underline{Y}_m\right)\cdot U_+ . \qquad (12)$$

The compensators' parameters result from by annulling the negative sequence current, in assisted by the reactive power compensation, as in eq. (13):

$$\{\Re e(\underline{I}_-) = 0; \quad \Im m(\underline{I}_-) = 0; \quad \Im m(\underline{I}_+) = min \qquad (13)$$

In this case, the compensation of reactive power at the load CCP is considered, aiming a minimum power factor with a value imposed by the network operator, λ_p^{impus}. This value imposes consequently the argument of positive component current [5], resulting from a cos-function given by (14):

$$\cos\varphi^+_{U1,comp} = \frac{G_m}{\sqrt{G_m^2 + (B_1 + B_2 + B_3 + B_m)^2}} = \lambda_p^{impus} . \qquad (14)$$

The three compensator's susceptances result as in eq. (15):

$$B_1 = \frac{2G_m}{\sqrt{3}}\cdot K_2; \ B_2 = \frac{2G_m}{3} K_1; \ B_3 = \frac{G_m}{3}\cdot K_1 - B_m , \qquad (15)$$

with $K_1 \overset{not}{=} \sqrt{1/\left(\lambda_p^{impus}\right)^2 - 1}; \quad K_2 \overset{not}{=} 1 + K_1/\sqrt{3}$.

The transformation susceptances/measured electrical values matrix [T] is:

$$[T] = \left[\begin{array}{ccc|ccc} K_2/\sqrt{3} & K_2/\sqrt{3} & -2K_2/\sqrt{3} & -K_2 & K_2 & 0 \\ K_1/3 & K_1/3 & -2K_1/3 & -K_1/\sqrt{3} & K_1/\sqrt{3} & 0 \\ \sqrt{3}+K_1/6 & -\sqrt{3}+K_1/6 & -K_1/3 & 1-K_1/2\sqrt{3} & 1+K_1/2\sqrt{3} & -2 \end{array}\right] \qquad (15)$$

5. The model simulation results

For each of the previous cases a set of MATLAB developed algorithms was achieved. These algorithms allow to draw the the time-evolution of the parameters of load current symmetrization devices. The application takes into account time-recording of the electrical values.

The algorithms were applied for the case of a consumer connected at 20 kV busbar of a distribution substation, for a 24 hours measurement session (Figure 4. a...c).

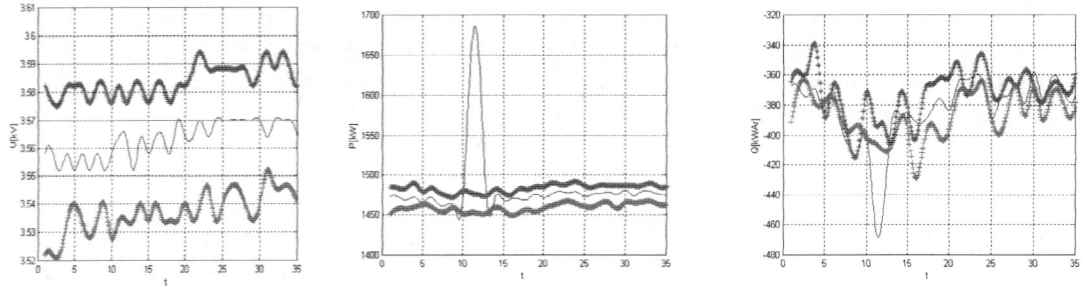

a. b. c.

Figure 4. Time variation of the state values and compensator elements (U_A – reference):
a. voltages variation; b, c – active/reactive powers variation

The curves represented by these algorithms can supply the on-line control strategy of the compensators' parameters for any dynamic load with an estimated evolution. The results of proposed cases simulations were concluded in Table 1.

Table 1. The results of symmetrization compensators behavior

Interest values	Case &3 (Figure 5)	Case &4 (Figure 6)
I_+^C (A)	0.074...133.328 (initial values)	
I_-^C (A)	0.058...4.452 (initial values)	
Suscept. Δ□[mS]:		
B_{AB}^{Δ}	-0.420...0.270	-0.790... 0.015
B_{BC}^{Δ}	-1.850..0..240	-1.170...0.250
B_{CA}^{Δ}	-	-0.080...0.800

As it can be observed, the variation of the state values in the load coupling point results in a variation of the compensator's parameters in the inductive field, as well as in the capacitive one (Figure 5).

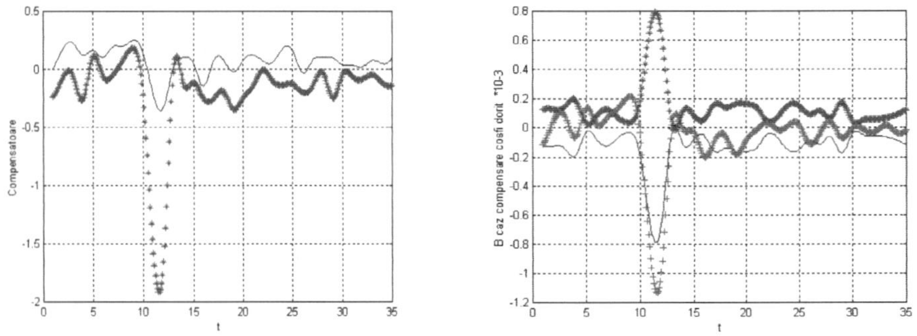

a. b.

Figure 5. Time variation of the symmetrization compensator elements : a. case &3; b – reactive power compensation within an imposed PF (case &4)

Each additional condition refering to the performances of the consumer-compensator set can lead to a majoring of the variation limits for the compensator parameters.

6. Conclusions

The paper proposes some configurations of devices used for load currents symmetrization in 3-phases 3-wires networks, and additional for improving of some energy and efficiency aspects.

The general sizing principle is based on a simple equivalent model of the unbalanced 3-phase loads. By using these compensators a certain control of the power flow is ensured in the load CCP. The control is based on the variations laws of the compensators parameters as functions of the measured electrical values. The link is ensured by a transformation matrix *[T]*, with different representation for each considered case.

The sizing of the compensators aimed also some additional energetical aspects (power factor correction) correlated with the total or major load symmetrization, the harmonics aspects of the state values being neglected. The structure of these compensators determines in most of the cases important current or voltage distortions. This disadvantage can be overcome by choosing the compensator configuration based on a correlated sizing, more precisely by considering simultaneous the influence of asymmetries and harmonics [8], [9].

References

[1] Ghosh A and Joshi A 2000 A New Approach to Load Balancing and Power Factor Correction in Power Distribution System, *IEEE Trans. on Power Delivery* **15**(1) 417-422

[2] Chen C C and Hsu Y Y 2000 A Novel Approach to the Design of a Shunt Active Filter for an Unbalanced Three-phase Four-wire System under Nonsinusoidal Conditions, *IEEE Trans. in Power Delivery* **15**(4) 1258-1264

[3] Lee S Y and Wu C J 1993 On-line reactive power compensation schemes for unbalanced three/phase four wire distribution feeders, *IEEE Trans. Power Delivery* **8**(4) 1958-1965

[4] Rusinaru D, Manescu L G and Merfu M 2013 The Load Unbalance Influence on the Power Factor Value in Three-Phase Distribution Networks, *Recent Advances in Electrical Engineering Series* **22** 314-319

[5] Rusinaru D 2005 *Unbalaced permanent operation of the power networks (in Romanian)*, Universitaria Publ., Craiova, Romania

[6] Buta A and Pana A 2000 *Symmetrization of the loads in distribution power networks (In Romanian)*, Orizonturi Universitare Publ., Timisoara, Romania

[7] Chindris M, Cziker A and Stefanescu S 2001 The study of the performances of some balancing systems for biphasic loads (in Romanian), *Energetica Rev.* **49**(6) 273-278

[8] Czarnecki L S 1994 Supply and Loading Quality Improvement in Sinusoidal Power Systems with Unbalanced Loads Supplied with Asymmetrical Voltage, *Electrical Engineering - Archiv fűr Electrotechnik* **77**(3) 169-177

[9] Häfner J 1999 *On the Reduction of Harmonics in Power Systems – A New Hybrid Power Filter Approach*, Technische Universität Berlin, Germany, Doctoral Thesis

IOP Conf. Series: Materials Science and Engineering **106** (2016) 012031 doi:10.1088/1757-899X/106/1/012031

Computer-based mechanical design of overhead lines

D Rusinaru[1], C Bratu[1], R C Dinu[1] and L G Manescu[1]

[1]University of Craiova, Faculty of Electrical Engineering, Electrical Engineering, Energetics and Aerospace Department, 107 Decebal Street, 200440 Craiova, Romania

E-mail: rcdinu@elth.ucv.ro

Abstract. Beside the performance, the safety level according to the actual standards is a compulsory condition for distribution grids' operation. Some of the measures leading to improvement of the overhead lines reliability ask for installations' modernization. The constraints imposed to the new lines components refer to the technical aspects as thermal stress or voltage drop, and look for economic efficiency, too. The mechanical sizing of the overhead lines is after all an optimization problem. More precisely, the task in designing of the overhead line profile is to size poles, cross-arms and stays and locate poles along a line route so that the total costs of the line's structure to be minimized and the technical and safety constraints to be fulfilled.The authors present in this paper an application for the Computer-Based Mechanical Design of the Overhead Lines and the features of the corresponding Visual Basic program, adjusted to the distribution lines. The constraints of the optimization problem are adjusted to the existing weather and loading conditions of Romania. The outputs of the software application for mechanical design of overhead lines are: the list of components chosen for the line: poles, cross-arms, stays; the list of conductor tension and forces for each pole, cross-arm and stay for different weather conditions; the line profile drawings.The main features of the mechanical overhead lines design software are interactivity, local optimization function and high-level user-interface.

1. Introduction

In the case of the overhead lines, the absence of further insulation beside the small sections connected on the line supports, as well as the relatively facile construction lead to a low-expensive electric circuit compared to an underground cable, mostly at medium and high voltage levels. While in normal environmental conditions the ratio of underground cable to overhead line costs is within 25:1 to 5:1, in areas of difficult terrain, there is an even bigger cost advantage with overhead lines. In the more sparsely populated areas, the overhead lines routes are more opportune. On the other hand, the majority of the very/high voltage routes are predominantly overhead, also because of the large cost difference to the cables.

Overhead conductors must be installed at a height providing an adequate electrical clearance from the ground, from nearby buildings or vegetation, and over roads and railways. The domain regulations set out the minimum clearance allowing for factors as the swing of conductors in prescribed windy conditions. The line height is determined by the maximum conductor sag between towers. This one is influenced by the span length, the conductor type, the maximum permissible conductor stress, the maximum conductor temperature, and the air ambient temperature. Calculation of overhead lines mechanics should take into account all the above criteria [1-3].

In general, the overhead conductor elongates by the permanent mechanical forces during its lifetime. Additional to this irreversible elongation, the sag of overhead conductors will increase with temperature caused by the electrical load and specific environmental conditions or other rope forces, e.g. snow and ice load. During the planning and construction, a lower sag than necessary gives a safety distance to the regulation clearance. With sag design or verification of planned or aged overhead lines by measurement or calculation, the reliability of the overhead lines can be improved. Depending on the system voltage and object type, the minimum phase to earth clearance can be ensured, too.

2. The design objective

In general, the design of the overhead lines has the following steps [1], [2], [4]:

1. input data collecting
2. route selection
3. conductor type selection
4. poles type selection
5. setting of the basic span length
6. pole positions nomination
7. choosing of tower heights and circuit heights at supports
8. drawing of ground line profile
9. drawing of circuit profile
10. checking of the vertical and horizontal clearance
11. checking of the structure stability to the mechanical forces
12. other requirements.

Some other initial conditions can be initially assumed as for instance, the poles height and size. These ones may later need to be amended as the design is checked and gradually refined. More options will be tried iteratively until an optimum arrangement is obtained.

The input data required by the profiling overhead algorithm should be organized in an appropriate shape file including all relevant requirements, constraints and background information as the following: existing and proposed schematics; planning requirements; future development; authorities requirements; coordination with lighting system; maps.

The design purpose is optimizing the profile of an overhead line by minimizing the line construction costs. The algorithm is adjusted for a software implementation allowing to the designer to find a first constructive solution.

The procedure for profiling the overhead lines firstly considers information about ground level coordinates of the line routes and superimposes the pole positions on the ground profile and their dimensions. Based on this information, the computer-based algorithm calculates the sag curves and draws the line profile under different weather conditions, e.g. minimum temperature of $-30\ ^0$C (without wind and ice), maximum temperature of $+40\ ^0$C, medium temperatures ($+15\ ^0$C, $+20\ ^0$C, $+30\ ^0$C) without/with wind, ice load conditions at $-5\ ^0$C without wind or with windy condition.

The sag template for drawing the circuit profile is established taking into account the permanent and the variable loading conditions. The profile is checked to ensure that:

- the circuit profile does not cross beyond the ground line offset line;
- adequate clearance is kept between neighborhood circuits or relative to other structures.

In case the ground clearance is not fulfilled, the program considers to increase the pole height or to reduce the span length by introducing supplementary poles so that the cost of line construction is minimized.

The constraints of the minimization problem are:

- the horizontal tension on the conductor should be less than a maximum allowable value, fulfilled for the distance between two poles less than the critical span of the conductor;
- there must be fulfilled the national safety regulations (clearance);
- the line route often has areas where the pole location is forbidden.

3. The design program structure

The software program entitled *Mechanical design of overhead lines* (in Romanian), is a Visual Basic application, with a friendly interface and Excel and Mathcad export/import facilities. The software is designated to profile the distribution overhead lines fulfilling the clearance restrictions and to obtain a minimization of the line construction by a proper selection of the poles position and heights [5]. There should be mentioned this version of the design program does not have global optimization algorithm for pole location. The main structure of the overhead line profiling software is given in Figure 1, and its modules are described as following:

I. Firstly, the ground line profile is plotted at a convenient scale (e.g. 1:1000) using the survey data for the line route profile. The survey data consists of distance and slope information for constitutive ground segments, between which the ground is considered with constant slope. A pair of trigonometric formulae is used in this stage:

$$x_{i+1} = x_i + d_i \cdot cos\,\theta_i; \quad y_{i+1} = y_i + d_i \cdot sin\,\theta_i, \tag{1}$$

Where, x_{i+1} , x_i is the horizontal coordinate at the *i-th* ground segment end, respectively ground segment start, [m], y_{i+1} , y_i is the vertical height at the *i-th* ground segment end, respectively ground segment start, [m], d_i is the measured length of the *i-th* ground segment, [m] and θ_i is the slope of the *i-th* ground segment, [deg].

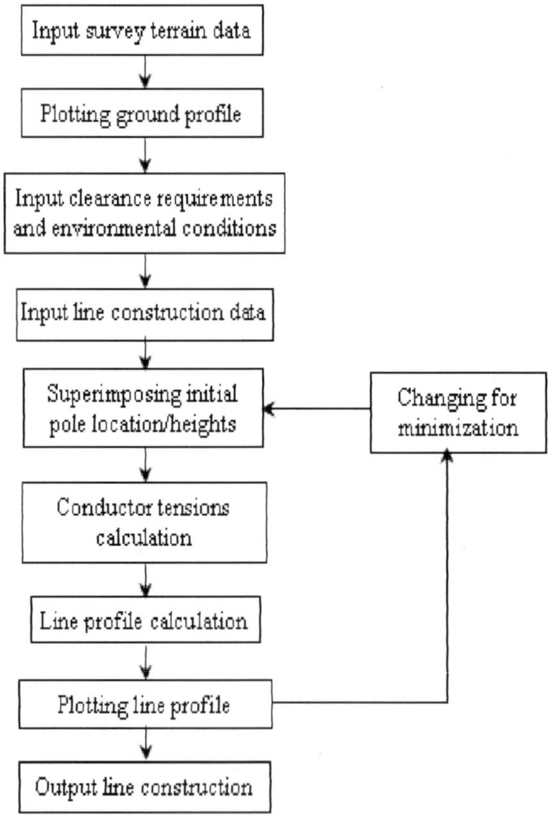

Figure 1. Algorithm structure for profiling of overhead lines

II. The coordinates are set for superimposed initial locations of the poles on the line route, for certain type of poles. In this stage, the same height is considered for all poles (e.g. 16 m).

III. There are also introduced the clearance requirements and weather conditions, as well as pole costs expressed in monetary values on height unit.

IV. The line route is considered as divided in segments for which the equivalent span is calculated, according to the following steps:

IV.1. The conductor loadings for different weather conditions, g_k *[daN/m·mm²]*, are determined;

IV.2. The tensions of the conductor under corresponding weather conditions are determined iteratively, using a Newton-Raphson algorithm applied to the state equation (2), given as in [3], [4], [6]:

$$\sigma_n - \frac{B_n}{\sigma_n^2} = A \tag{2}$$

where σ_n is the conductor tension for the considered weather condition (n-named), [daN/mm^2], B_n is coefficient of state equation variable for weather condition n, given by the equation (3) and A is free term of the state equation given by the equation (4):

$$B_k = a^2 g_k^2 \cdot E / 24, \tag{3}$$

where a is the span of the tension line segment, [m], g_k is the conductor loadings for k weather condition, [daN/m·mm^2], according to [3] and E is elastic coefficient of the conductor, [daN/mm^2],

$$A = \sigma_{ref} - \frac{B_{ref}}{\sigma_{ref}^2} - E \cdot \alpha \cdot (\theta_n - \theta_{ref}) \tag{4}$$

where σ_{ref} is the conductor tension for the reference weather conditions, [daN/mm^2], B_{ref} is coefficient of state equation variable for reference weather condition, given by the equation (3), α is temperature coefficient of the conductor, [^{0}C^{-1}] and θ_n, θ_{ref} is temperatures of conditions n, respectively reference, [^{0}C].

IV.3. Using a parabolic function, the sag curve (5) for support at different level is determined:

$$f(x) = 4 \cdot f \cdot x / a \cdot (1 - x / a). \tag{5}$$

Where, f is the curve sag, [m].

$$f = \frac{a^2 \cdot g}{8 \cdot \sigma} \tag{6}$$

IV.4. The maximum span length is determined:

$$a_{cr} = \sigma_{maxadm} \cdot \sqrt{\left(24\left(\theta_{-5^0} - \theta_{min}\right)\right) / \left(\left(\frac{g_{cond+ice}}{\sigma_{Al,-5^0C+ice}}\right)^2 - \left(\frac{g_{cond}}{\sigma_{Al,min}}\right)^2\right)}, \text{ [m]}. \tag{7}$$

where, θ_{min} is the minimum temperature (-30^{0}C), g_{cond}, $g_{cond+ice}$ is loadings of the conductor under its own weight, respectively with ice weight and $\sigma_{Al,-5C+ice}$, $\sigma_{Al,min}$ is aluminium conductor tensions for –5^{0}C + ice condition, respectively minimum temperature, according [3].

V. The poles position is changed on those segments where the critical span length is exceeded.

VI. The line route is plotted and the ground clearance is checked.

VII. The line construction costs are determined.

VIII. The pole height is reduced at those segments where the ground clearance is allowing.

IX. The new line construction costs are determined.

4. The model simulation results

The software application outputs consist in the following data:

- list of poles chosen for line construction and their costs;
- list of conductor tensions for different weather conditions;
- list of span lengths for the ground profile segments;
- list of sags for ground profile segments for different weather conditions;
- drawing of the ground profile (by constitutive segments);
- drawing of the line profile for different weather conditions.

The program *Mechanical design of overhead lines* has a high level of interactivity given by an intuitive graphic interface. Figure 2 shows the screen shot indicating the resulting equivalent tension of a sample conductor for different weather conditions, while in Figure 3 one of the final screen shots of the application is shown.

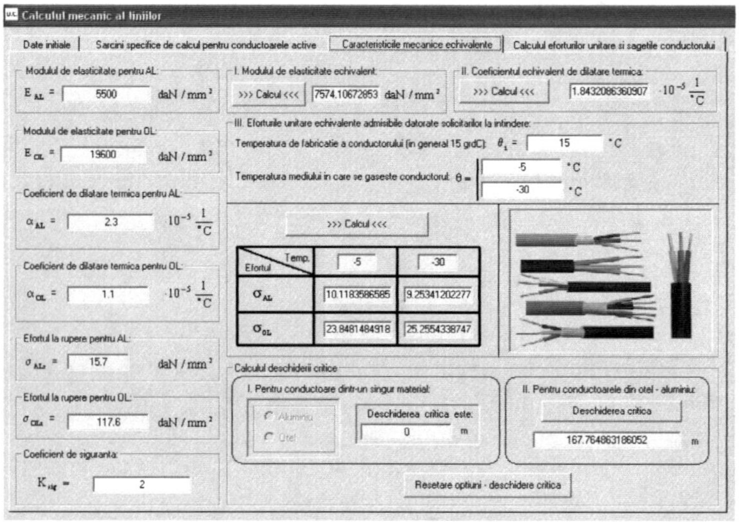

Figure 2. Screen shot with equivalent tensions on the conductor

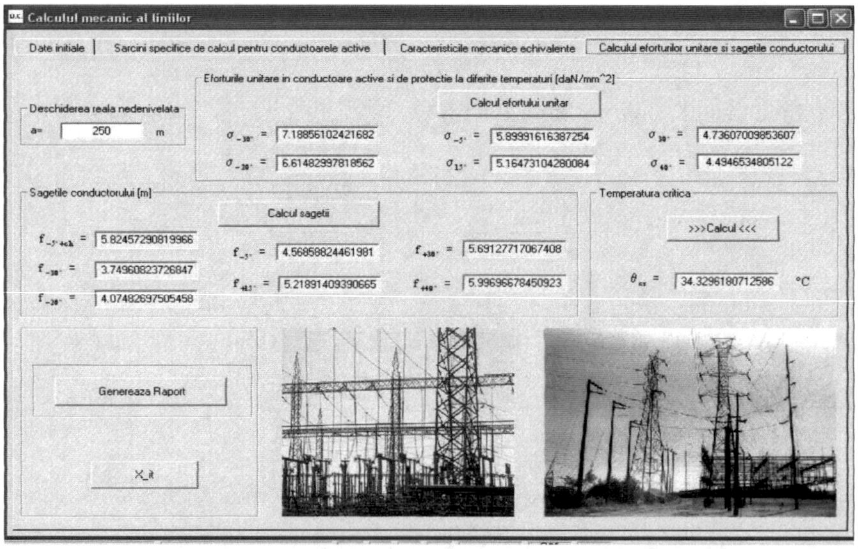

Figure 3. Screen shot of *Mechanical design of overhead lines* with segment line sag values

The output data lists are also organized in Excel reports containing the values generated in different stages of line mechanical calculus.

Figure 4 shows a sample of the sag curve generated by software for two constitutive segments of the conductor profile and the ground clearance curve. If there is no intersection between these two curves, the ground clearance requirements are fulfilled.

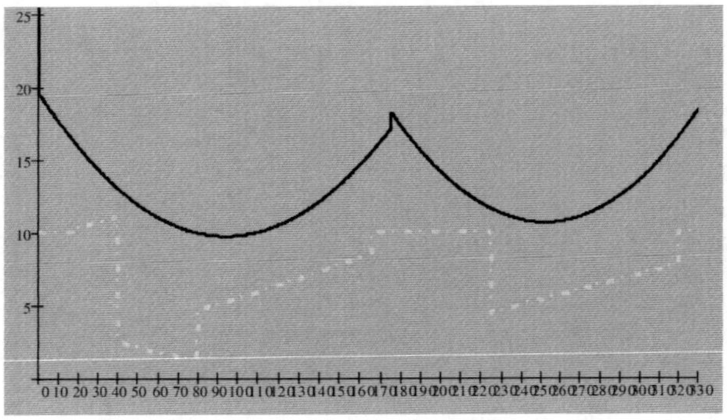

Figure 4. Software line profile output

5. Conclusions

The software *Mechanical design of overhead lines* is an efficient tool for profiling of overhead distributions lines.

The design software application includes modules for calculus of line tensions, line sag curves, graphical outputs of ground and line profiles, as well as an iterative algorithm for minimization of poles heights with fulfilling of clearance restrictions.

By using such of designing tool, the overhead lines sizing can become more accurate, and consequently, more costly efficient. The designing time can be sensible reduced and the interactivity features can by more supportive for designers.

References

[1] Partanen J, Pajala H and Kumpulainen L 1988 *A Personal Computer Based Design System for Profiling Overhead Lines*, Proc. of the International Conference on Overhead Line Design and Construction: Theory and practice, London, UK, November 28-30, pp 146-150

[2] Lakervi E and Holmes E J 1989 *Electricity distribution network design,* Peter Peregrinus Ltd., London, UK

[3] Iacobescu Gh, Iordănescu I, Eremia M, Dumitriu C and others 1977 *Electrical networks. Problems (in Romanian),* EDP Publ., Bucharest, Romania

[4] Kiessling F, Nefzger P, Nolasco J F and Kaintzyk U 2003 *Overhead power lines: planning, design, construction*, Springer, Berlin, Germany

[5] *** 2007 www.energex.com.au: *Overhead design manual,* ENERGEX

[6] *** 2014 *EDSA Design Base. User Guide.* EDSA Microcorporation, USA

Researches upon cavitation erosion behavior of some duplex steels

I Bordeasu[1], M O Popoviciu[2], I Mitelea[1], L M Micu[1], C Bordeasu[1], C Ghera[1] and A Iosif[3]

[1]Politehnica University of Timisoara, Faculty of Mechanical Engineering, Mihai Viteazul Street, Nr. 1, 300222 Timisoara, Romania
[2]Academy of Romanian Scientists, Mihai Viteazul Street, Nr. 1, 300222 Timisoara, Romania
[3]Politehnica University of Timisoara, Faculty of Civil Engineering, Traian Lalescu Street, Nr. 2, 300223 Timisoara, Romania

E-mail: ilarica59@gmail.com

Abstract. This paper presents the cavitation erosion behavior of two stainless steels having a duplex structure formed by austenite and ferrite. The conclusions were obtained by using both the cavitation erosion characteristic curves and the pictures of the eroded surfaces obtained with performing optic microscopes. The researches were focused upon the optimal correlation between the cavitation erosion resistance and the rate of the two structural constituents. The tests were done with T2 facility, with ceramic crystals, which integrally respects the ASTM G32-2010 Standard. The obtained results present the cumulative effect upon cavitation erosion of the chemical composition, mechanical properties and the structural constituents. The results of the researches are of importance for the specialists which establishes the composition of the stainless steels used for manufacturing hydraulic machineries or other devices subjected to cavitation erosion.

1. Introduction

Duplex stainless steel is a family of materials with austenite-ferrite structure, in which the chemical composition, severally controlled, contains between 21-25% Cr and 5-7% Ni. In smaller quantities they may contain also molybdenum and nitrogen. The common heat treatment is quenching in water.

Those steels are used in fields where: corrosion resistance is important, inclusively by pitting, the welding is easy to be performed and the mechanical characteristics are maintained approximately unchanged at temperatures between (-50÷+280) °C [1]. Taking into account the enounced characteristics we put the question of employing those stainless steels also for manufacturing and repair works of the hydraulic machineries elements subjected to cavitation erosion.

2. Researched materials

The researched steels **X2CrNiN23-4** and **X2CrNiMoN22-5-3**, in conformity with 100088-1, 2, 3-2005 Standard [2], have the following chemical compositions and mechanical properties:

- steel **X2CrNiN23-4**: <0.03%C, (21.5-24.5)%Cr, (3.0-5.0) %Ni, (0.05-0.6)%Mo, (0.05-0.2)%N, <1.0% Si, <2.5%Mn, (0.05-0.06)%Cu, <0.04%P, <0.03%S, rest iron; ultimate strength R_m = 670 MPa, yield limit $R_{p0.2}$ = 450 MPa, Brinell hardness = 210 daN/mm^2

- steel **X2CrNiMoN22-5-3**: <0.03%C, (22-23)%Cr, (4.5-6.5)%Ni, (3-3.5)%Mo, (0.14-0.20)%N, <1.0% Si, <2.0%Mn, <0.03%P, <0.02%S, rest iron, ultimate strength R_m = 750 MPa, yield limit $R_{p0\cdot2}$ = 510 MPa, Brinell hardness = 230 daN/mm^2.

For the employed samples, the actual chemical compositions and mechanical properties were determined in the Timisoara "Politehnica" Laboratories. These values are:

- steel **X2CrNiN23-4**: 0.02%C, 21.17%Cr, 4.27%Ni, 0.49%Mo, 0.06%N, 0.95% Si, 2.43%Mn, 0.057%Cu, 0.035%P, 0.013%S, rest iron, ultimate strength R_m = 667 MPa, yield limit $R_{p0\cdot2}$ = 454MPa, Brinell hardness =203 daN/mm^2
- steel **X2CrNiMoN22-5-3**: 0.02%C, 22.72%Cr, 5.23%Ni, 3.14%Mo, 0.19%N, 0.22% Si, 0.47%Mn, 0.03%P, 0.013%S, rest iron, ultimate strength R_m =743 MPa, yield limit $R_{p0.2}$ =504 MPa, Brinell hardness =217 daN/mm^2

The pitting corrosion index PRE, computed with the relation (1) [1]:

$$PRE = \%Cr + 3.3 \cdot \%Mo + 16 \cdot \%N \tag{1}$$

is 23,747 for the steel **X2CrNiN23-4** and 36,122 for the steel **X2CrNiMoN22-5-3**, both presenting a high resistance to pitting corrosion.

Using the Schäffler diagram (Figure 1) and the equivalent values for chromium (CrE) and Nickel (NiE), determined with the specific relations [3], [4], the approximate structure of the researched steels is: 18 % austenite and 82 % ferrite for **X2CrNiN23-4** (CrE ≅ 23.85 %, NiE ≅ 6.085 %) and 10 % austenite and 90 % ferrite for **X2CrNiMoN22-5-3** (CrE ≅ 26.19 %, NiE ≅ 6.065 %). The microscopic aspects of the basic structure of those steels are presented in Figure 2.

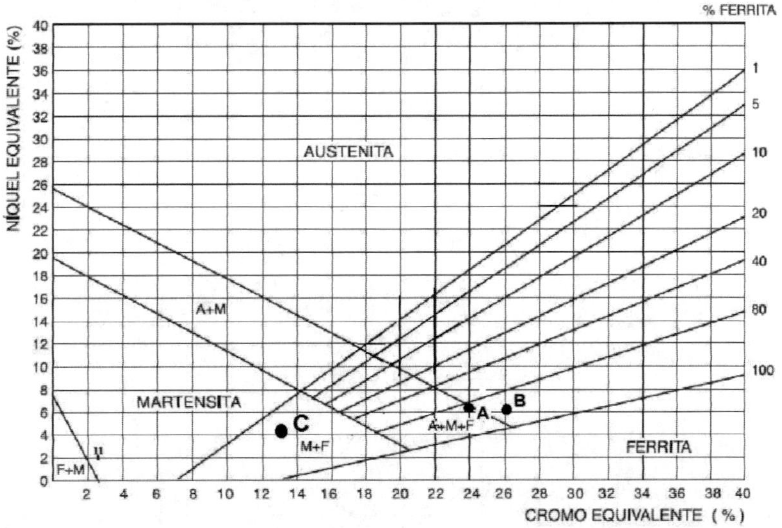

Figure 1. Positions of the researched steels in the Schäffler diagram
A-stainless steel **X2CrNiN23-4**; **B**-stainless steel **X2CrNiMoN22-5-3**; **C**- comparison stainless steel **OH12NDL**

As can be seen in the Schäffler diagram, the chemical components of the studied steels conducted to important structural differences. The steel **X2CrNiN23-4** (symbolized with **A**) being in the zone with austenite and martensite. On the contrary the steel **X2CrNiMoN22-5-3** (symbolized with **B**) is in the zone with austenite and ferrite. In the same diagram with the symbol C is marked the martensitic stainless steel **OH12NDL**, used on a large scale in the former Soviet Union but also in Romania for manufacturing hydraulic turbines runners. This steel presented a good behavior to cavitation erosion.

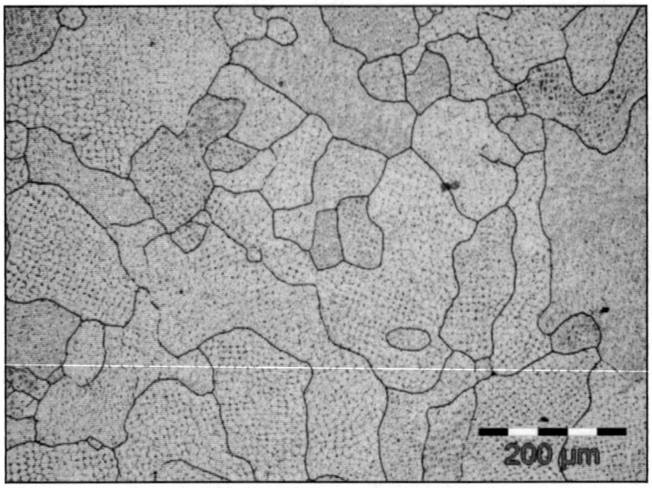

Steel **A** (**X2CrNiN23-4**)

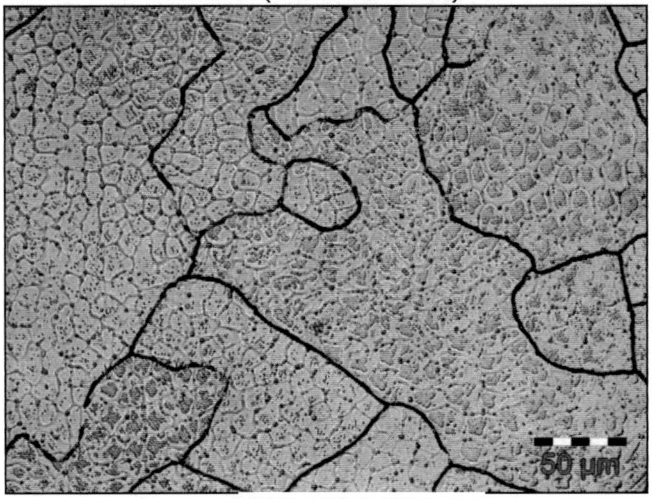

Steel **B** (**X2CrNiMoN22-5-3**)

Figure 2. Basic structure

3. Researched procedure. Experimental results

The cavitation erosion behavior was realized in the Timisoara Poytechnic University Cavitation Laboratory in the test facility T2 with piezoelectric crystals [5], [6], [7], [8]. The device was realized in conformity wit the ASTM G32-2010 Standard [9]. The device is provided with a lot of instrument to maintain the running parameters at constant values. Both test maximum test exposure, the intermediary measuring intervals, the weighing procedure the number of tested specimens, were chosen in conformity with the procedure adopted in our [5], [6], [10]. Because the results may be applied for the use of the steels for manufacturing or repair works, especially by welding, of hydraulic machineries details the liquid used for tests was drinking water from urban water supply system at a temperature of 22 ± 1^0C.

The previous researches [11] put into evidence that for 165 minutes total exposure time, accepted by our laboratory as sufficient for attaining a constant erosion rate, the corrosive effect of the drinking water is negligible and deionization of water is not necessary. Some tests recorded previously show that the results are not influenced by the quality of water, especially for corrosion resistant materials such as the steels **X2CrNiN23-4** and **X2CrNiMoN22-5-3**.

With the purpose to appreciate the time evolution of cavitation erosion behavior in Figures 3 and 4 is presented the evolution in time of the mean depth erosion (MDE) and the mean depth erosion rate MDER.

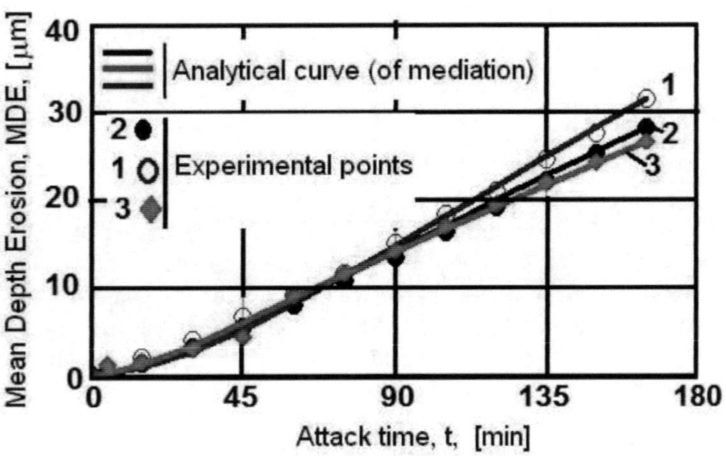

Figure 3. Evolution of mean depth erosion with time
1- Stainless Steel **X2CrNiMoN22-5-3**; 2 - Stainless Steel **X2CrNiN23-4**;
3 - Comparison Steel **OH12NDL**

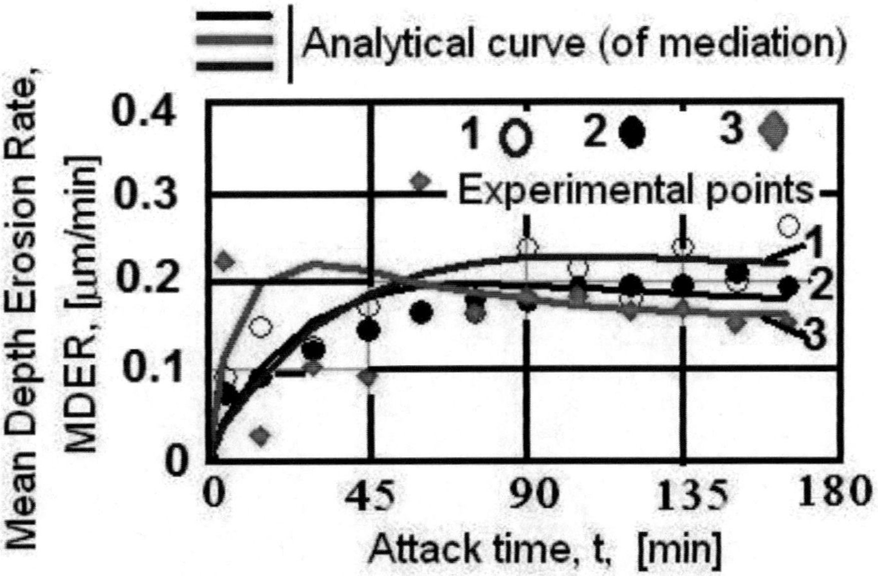

Figure 4. Evolution of MDER with exposure time
1- Stainless Steel **X2CrNiMoN22-5-3**; 2 - Stainless Steel **X2CrNiN23-4**;
3 - Etalon Stainless Steel **OH12NDL**

The curves of the time evolution in Figures 3 and 4 were obtained analytically, using the Bordeasu method [12]. The mean points (mean of three measurements) are also presented in Figures 3 and 4. The shape of the curves is specific for materials with good cavitation erosion behaviors [3], [4], [13].

It can be seen in Figures 3 and 4 that in comparison with the reference standard OH12NDL, both researched steels have smaller resistance. From Figure 1 result that after 165 minutes of exposure the mean depth erosion for the steel **X2CrNiMoN22-5-3** is with 15.74 % greater and for the steel **X2CrNiN23-4** is with 5.88 % greater than that of the standard steel.

Taking into consideration the parameter $1/MDER_{165}$ ($MDER_{165}$ being the level value of the curve MDER(t), where the curve have the tendency to remain horizontal), it results that: the steel **X2CrNiN23-4** in comparison with the standard steel OH12NDL has a cavitation resistance diminished with approximate 25% and the steel **X2CrNiMoN22-5-3** a cavitation resistance diminished with approximate 10%. The principal motive of those reduced cavitation erosion resistance is the microstructure constitution, where ferrite is the wickets component but also the stable austenite is eroded in a greater measure than martensite. The standard steel has approximate 74% martensite and only 26% ferrite [3], [6], in comparison with the researched steels (82 % for the steel **X2CrNiN23-4 and 90 % for X2CrNiMoN22-5-3)**.

As can be seen in Figures 3 and 4 from the researched stainless steels the better behavior present the steel **X2CrNiN23-4** (symbolized by **A** in Schäffler diagram) has a cavitation resistance approximate with 17 % greater than the steel **X2CrNiMoN22-5-3**. The explanation is the greatest proportion of austenite **(18 %)** and the posible transformation of the austenite in martensite, as a consequence of here position at the boundary between austenite and martensite. This structure gives increase of the cavitation resistance of **X2CrNiMoN22-5-3,** even if the mechanical characteristics are worse than those for **X2CrNiN23-4**. This conclusion is sustained also by the good behavior of the steel **OH12NDL** which has worse mechanical characteristics with exception of the Brinell Hardness (ultimate strength Rm=650MPa, yield limit $R_{p0.2}$ = 400 MPa, Brinell Hardness HB= 225 daN/mm^2, [3], [6]). As a consequence, for close values of mechanical characteristics the microstructure is extremely important for cavitation erosion resistance.

In Figures 5-8 there are presented images of the roughness before and after the total cavitation exposure. Figures 5 and 7 present the roughness obtained before cavitation, with a MITUTOYO device at the Timisoara National Institute for Research-Development in Welding and Material Tests. The images of Figures 6 and 8 show the aggressively cavitation erosion expressed in the level of the roughness Ra, Rz si Rt, Figures 6a si 8a, measured after three directions Figures 6c and 8c.

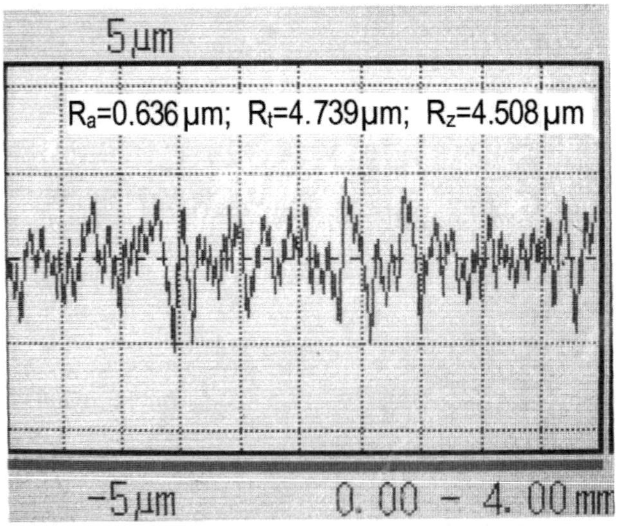

Figure 5. Surface roughness before cavitation exposure **X2CrNiN23-4**

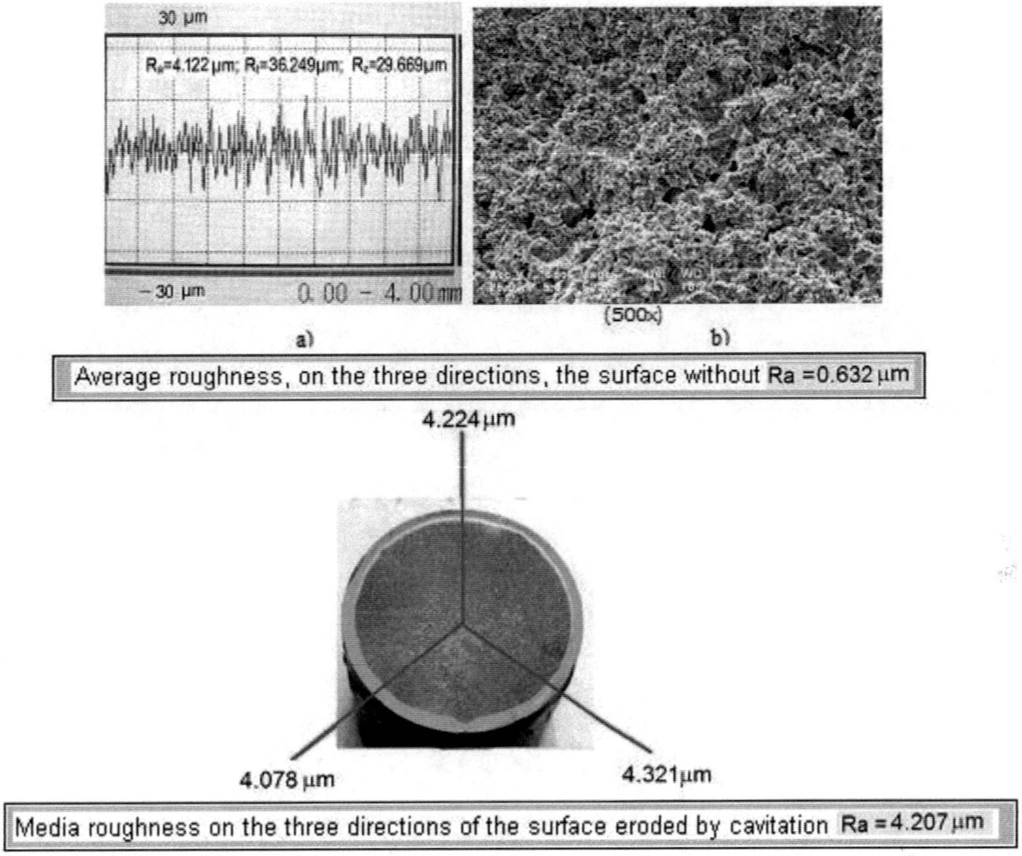

Figure 6. Roughness and microstructure for **X2CrNiN23-4** after cavitation exposure

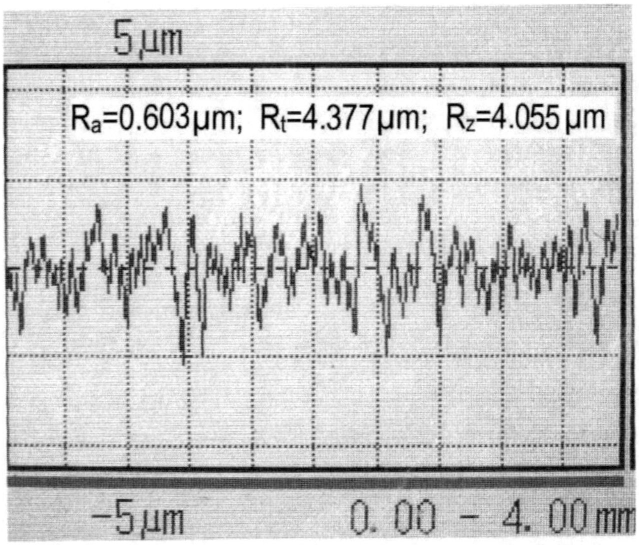

Figure 7. Surface roughness before cavitation exposure **X2CrNiN23-4**

IOP Conf. Series: Materials Science and Engineering **106** (2016) 012032 doi:10.1088/1757-899X/106/1/012032

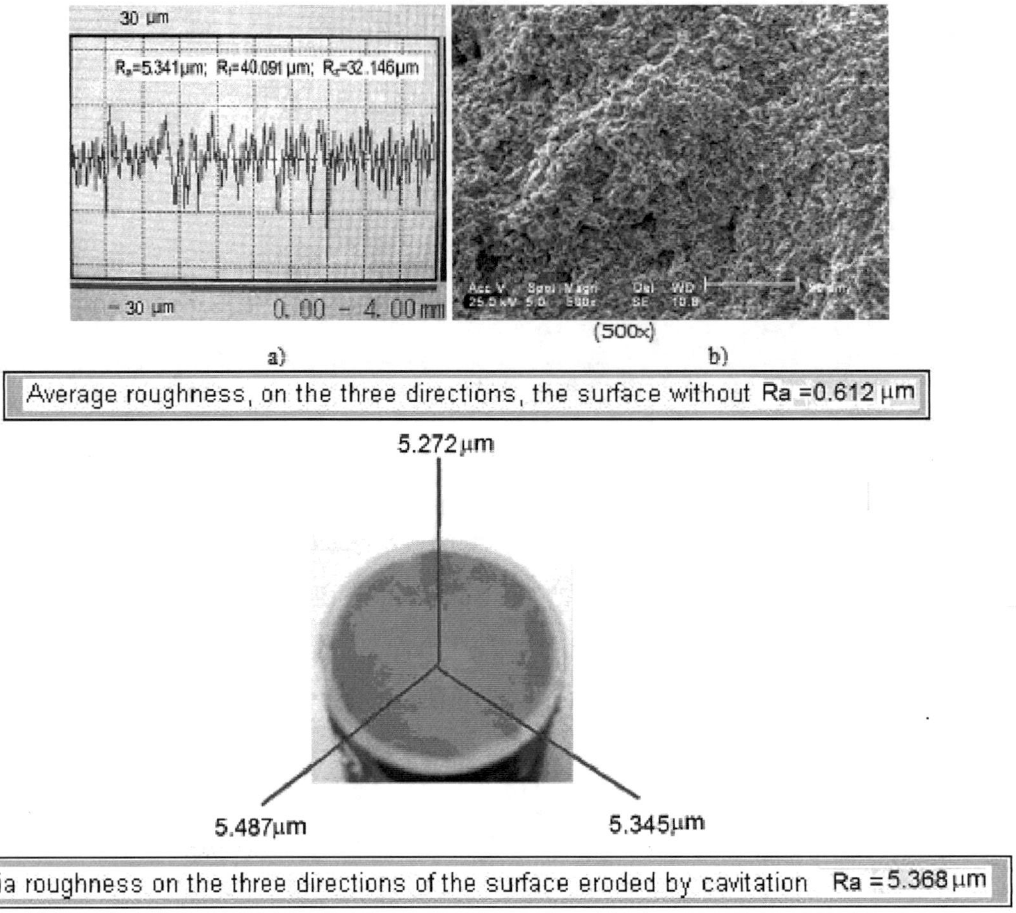

Figure 8. Roughness and microstructure for **X2CrNiMoN22-5-3** after cavitation exposure

As can be seen from comparisons of the roughness parameters (Figures 6a and 8a), but also from the microstructure photos (Figures 6b and 8b), the level of deterioration of the steel **X2CrNiMoN22-5-3** is greater that that of the steel **X2CrNiN23-4**. These observations correspond to those seen in the evolutions of the MDE and MDER. The better behavior of the steel **X2CrNiN23-4** is given by the transformation of austenite in martensite under the influence of the stresses determinate by the cavitation bubbles implosions.

4. Conclusions

a. With different chemical compositions and the structure formed from austenite and ferrite, the stainless steels **X2CrNiMoN22-5-3** and **X2CrNiN23-4** present visible differences of the cavitation erosion behavior. The difference is given by a greater proportion of austenite. The steel **X2CrNiN23-4** has a greater proportion of austenite (approximate 18 %) and has also a greater cavitation erosion resistance in comparison with the steel **X2CrNiMoN22-5-3** (approximate 10 % austenite).
b. Even if the cavitation erosion resistance is smaller than that of the standard steel **OH12NDL**, the cavitation erosion resistance is acceptable and those steels can be used for repair works, through welding, of the eroded zone of the runners or runner of hydraulic machineries.
c. In the case of duplex steels (with the structure formed from austenite and ferrite) a good structure proportion confers a greater cavitation erosion resistance regardless of the principal mechanical (R_m, $R_{p0.2}$ si HB).

References

[1] ***www.bssa.org.uk/topics.php?article=111

[2] ***http://www.outokumpu.com/en/stainless-steel/grades/austenitic/Pages/default.aspx

[3] Bordeasu I 2006 *Cavitation Erosion of Materials,* Timisoara Polytechnic Publishing House, Timisoara, Romania (in Romanian)

[4] Karabenciov A 2013 *Cercetări asupra eroziunii produse prin cavitaţie vibratorie la oţelurile inoxidabile cu conţinut constant în nichel şi variabil de crom*, Politehnica University of Timisoara, Romania, Doctoral Thesis

[5] Bordeasu I, Popoviciu M O, Mitelea I, Ghiban B, Ghiban N, Sava M, Duma S T and Badarau R 2014 Correlations between mechanical properties and cavitation erosion resistance for stainless steels with 12% Chromium and variable contents of Nickel, *IOP Conf. Ser.: Mater. Sci. Eng.* **57** 012006

[6] Bordeasu I, Popoviciu M O, Mitelea I, Ghiban B and Ghiban N 2014 Cavitation erosion resistance of two steels with the same percentage of Chromium and Nickel but different Carbon content, *IOP Conf. Ser.: Mater. Sci. Eng.* **57** 012007

[7] Oanca O, Bordeasu I, Mitelea I and Crăciunescu C 2013 *Phenomenology of Degradation by Cavitation for Heat Treated CuNiAlFe Bronzes*, 22-th International Conference on Metallurgy and Materials, Brno, Czech Republic, May 15-17, pp 1561-1566

[8] Oanca O, 2014 *Tehnici de optimizare a rezistenţei la eroziunea prin cavitaţie a unor aliaje CuAlNiFeMn destinate execuţiei elicelor navale*, Politehnica University of Timisoara, Romania, Doctoral Thesis

[9] ***2010 *Standard method of vibratory cavitation erosion test*, ASTM, Standard G32-10

[10] I Bordeasu, M O Popoviciu, L M Micu, O V Oanca, D Bordeasu, A Pugna and C Bordeasu 2015 Laser beam treatment effect on AMPCO M4 bronze cavitation erosion resistance, *IOP Conf. Ser.: Mater. Sci. Eng.* **85** 012005

[11] Bordeasu I, Popoviciu M, Mitelea I, Ghiban B, Balasoiu V and Tucu D 2007 Chemical and mechanical aspects of the cavitation phenomena, *Revista de chimie* **58**(12) 1300-1304

[12] Bordeasu I, Popoviciu M O, Balasoiu V and Patrascoiu C 2004 An Analytical Model for the Cavitation Erosion Characteristic Curves, *Scientific Bulletin "Politehnica" University of Timişoara, Transaction of Mechanics* **49**(63) 253-258

[13] Franc J P 1995 *La Cavitation. Mecanismes phisiques et aspects industriels,* Press Universitaires de Grenoble, Grenoble, France (in French)

IOP Conf. Series: Materials Science and Engineering **106** (2016) 012033 doi:10.1088/1757-899X/106/1/012033

Researches concerning influence of magnesium, aluminum and titanium lime on steel desulfurization

V Putan[1], A Putan[1], A Josan[1] and L Vilceanu[1]

[1]Politehnica University of Timisoara, Department of Engineering and Management, 5 Revolution Street, 331128 Hunedoara, Romania

E-mail: vasile.putan.fih.upt.ro

Abstract. The paper presents the results of laboratory experiments on steel desulphurisation with slag from the system MgO-Al_2O_3- TiO_2. To determine the influence, on the desulphurisation process, of the titanium oxide added in calcium aluminate slag, we experimented, in the laboratory phase, the steel treatment with a mechanical mixture consisting of lime, aluminous slag and slag obtained from the titanium making process through the aluminothermic technology. The steel melting was carried out in an induction furnace of 10 kg capacity, existent in the "Metallic Melts" laboratory of the Engineering Faculty of Hunedoara. During the research, we aimed to establish correlation equations between the sulphur distribution coefficient and the slag components (MgO, Al_2O_3, TiO_2). The data obtained in the experiments were processed in MATLAB programs, resulting multiple correlation equations, which allowed the elucidation of some physical-chemical phenomena specific to the desulphurisation processes.

1. General considerations

The steel refining with liquid slag or various powder mixtures of synthetic slag is based on the intensification of the unwanted impurities (sulphur, non-metallic suspensions & oxygen) passage from the liquid steel in the slag, mainly by diffusion, or partly through the entrainment of some suspensions by settling the synthetic slag particles found in the treated steel bath. The synthetic slag can be also obtained by adding mechanical mixture directly in the casting ladle; in this case, for compensating the cooling of the steel in the casting ladle due to the addition of materials (melting and superheating), the steel temperature should be at least 20-40°C higher than the normal one [1]. In the practice of deoxidation with synthetic slag, we usually use slag that correspond to the binary systems CaO-Al_2O_3, CaO-TiO_2 and CaO-CaF_2, or to the ternary systems CaO-SiO_2-Al_2O_3, CaO-CaF_2-Al_2O_3 and MgO-Al_2O_3-TiO_2, [2].

The viscosity of the synthetic slag has significant influence on the development of physical and chemical processes during the treatment of the liquid steel, interfering with significant weight on the emulsifying capacity of slag. The increase of the slag viscosity from 0,15 to 0,45 Ns/m^2 (from 1,5 to 4,5 Poise) determines the decrease with approx. 30% of the steel-slag interaction surface. Such increasing of the calcium aluminate slag viscosity can be seen when its temperature is decreasing (for example, from 1600°C to 1470°C). Therefore, it is very important to ensure, during processing the steel with liquid slag, the optimum thermal regime specific to the chosen slag type and to realize its convenient fluidity (viscosity).

The viscosity of the synthetic slag is also influenced by other components; it increases significantly with the increasing of the SiO_2 content, while MgO contents up to 8% are favorable. At temperatures higher than 1500°C, the viscosity is slightly decreasing when adding TiO_2 in the calcium aluminate slag [3].

Similarly, the bigger is the contact surface between the synthetic slag and the metallic bath, the faster is the passage of the significant elements to the slag, the contact surface being, along with the viscosity, another determinant element in treating the steel with synthetic slag [4].

2. Laboratory experiments

To determine the influence, on the desulphurisation process, with slag from the system MgO-Al_2O_3-TiO_2, we performed laboratory experiments, i.e. we treated the steel with liquid synthetic slag obtained by melting the mixture consisting of magnesia, aluminate slag and slag obtained from the titanium making process through the aluminothermic technology.

The steel melting was carried out in an induction furnace of 10 kg capacity and the slag melting was carried out in a crucible furnace, both existent in the "METALLIC MELTS" laboratory of the Engineering Faculty of Hunedoara (Figure 1).

The charge to be melted consisted of steel samples (samples of steel for tubes, taken from the casting ladle before the LF treatment, i.e. before introducing the steel in the LF). The steel quantity obtained was 10 kg/heat, and the addition of liquid slag was 3% (300 g/heat). The synthetic slag was added directly in the casting ladle; so, the slag reached the ladle before the steel, ensuring a good mix between the two melts. To determine the sulphur distribution coefficient, we took steel and slag samples before and after the treatment, to find the sulphur content and the chemical composition. We also measured the steel and slag temperature before and after the treatment.

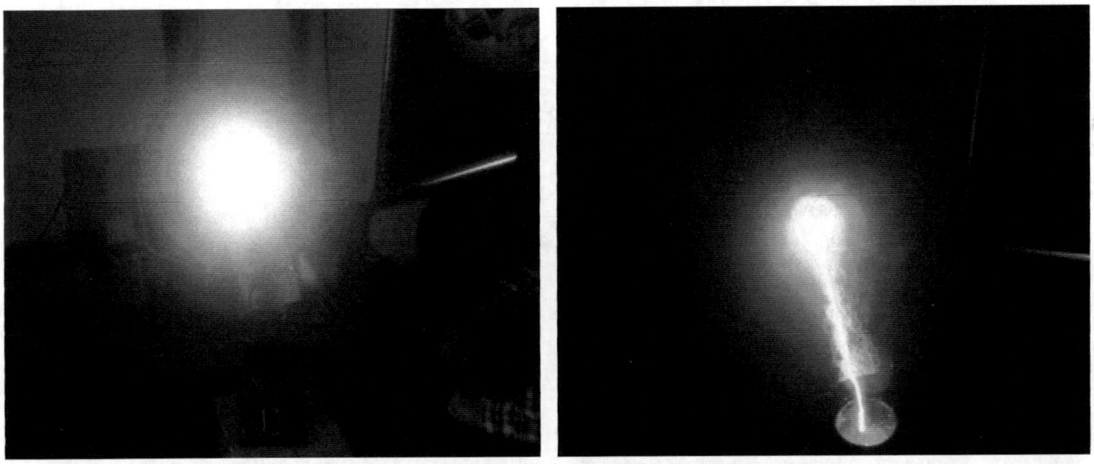

Figure 1. Issues during laboratory experimentation phase

The charge to be melted consisted **of steel samples** (samples of steel for tubes, taken from the casting ladle before the LF treatment, i.e. before introducing the steel in the LF). The steel quantity obtained was 10 kg/heat, and the addition of liquid slag was 3% (300 g/heat). The synthetic slag was added directly in the casting ladle; so, the slag reached the ladle before the steel, ensuring a good mix between the two melts. To determine the sulphur distribution coefficient, we took steel and slag samples before and after the treatment, to find the sulphur content and the chemical composition. We also measured the steel and slag temperature before and after the treatment.

3. Results obtained from processing the experimental data

By processing the data in the MATLAB program, we obtained multiple correlation equations L_S=f((%Al_2O_3), (%MgO), (%TiO_2)), and, by graphically represented them, we obtained the correlation surfaces.

In these multiple correlation equations L_S is sulfur partition coefficient defined as

$$L_S = (S)/[S] \tag{1}$$

where: (S) - the sulfur content in slag;
[S] - the sulfur content in steel.

Since this hyper surface cannot be represented in 4-dimensional space, recourse to replacement, in succession, by an independent variable with its mean value. These areas, which belong to the 3-dimensional space can be represented and interpreted by technologists (Figure 2 to 7).

To establish the optimum chemical composition range, we analysed the regression surfaces for finding the value of the L_S, desirable above the average value obtained from the data afferent to the analysed heats. The average value for sulfur partition coefficient is L_{Smed} = 204,01. Independent parameters are likely to vary within certain limits of technological fields, such as values for L_S to be higher that 220 (shaded areas).

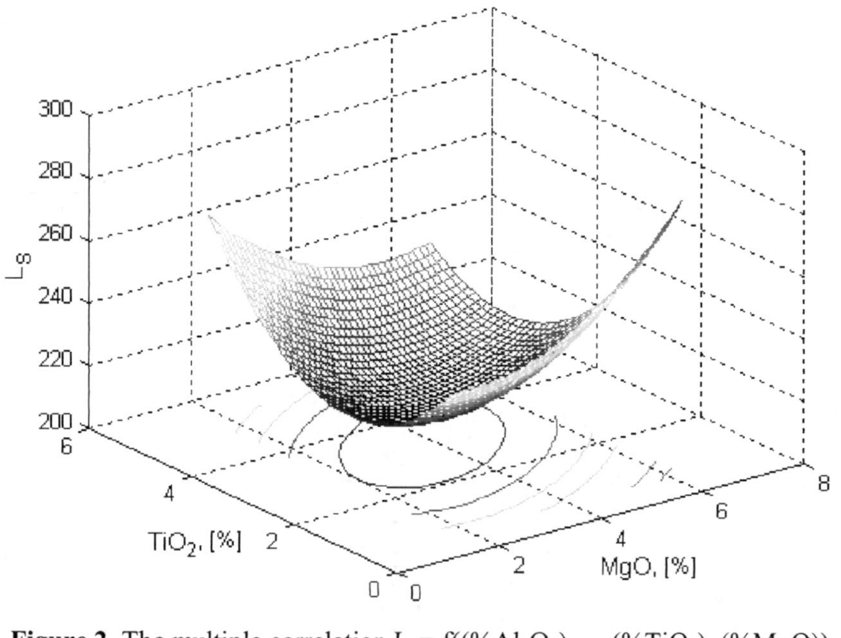

Figure 2. The multiple correlation L_S= f((%Al_2O_3)$_{med}$, (%TiO_2), (%MgO)) - regression surface

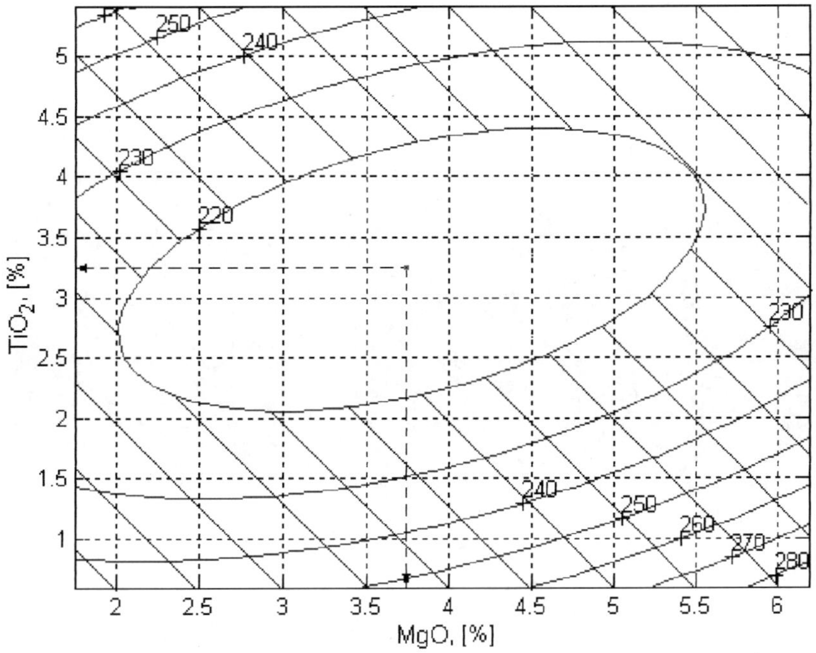

Figure 3. The multiple correlation L_S= f((%Al_2O_3)$_{med}$, (%TiO_2), (%MgO)) - level curves

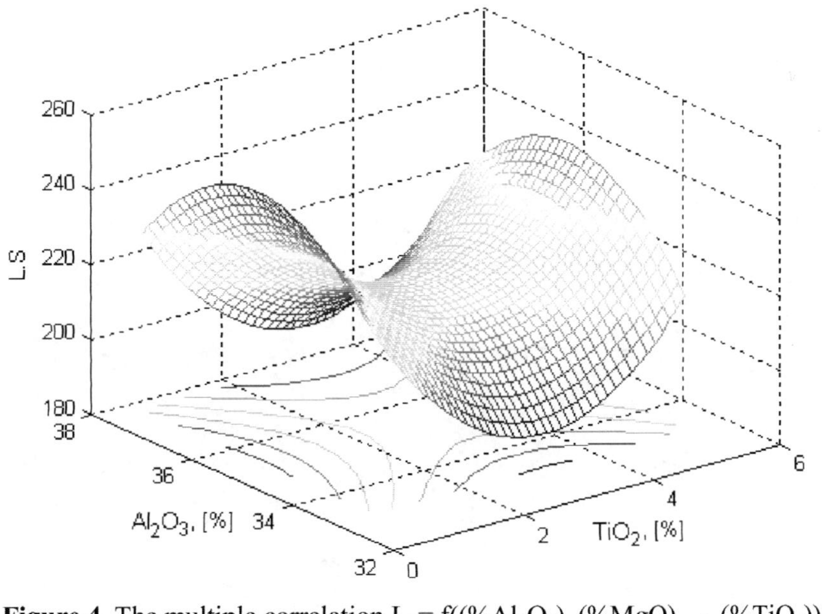

Figure 4. The multiple correlation L_S= f((%Al_2O_3), (%MgO)$_{med}$, (%TiO_2)) - regression surface

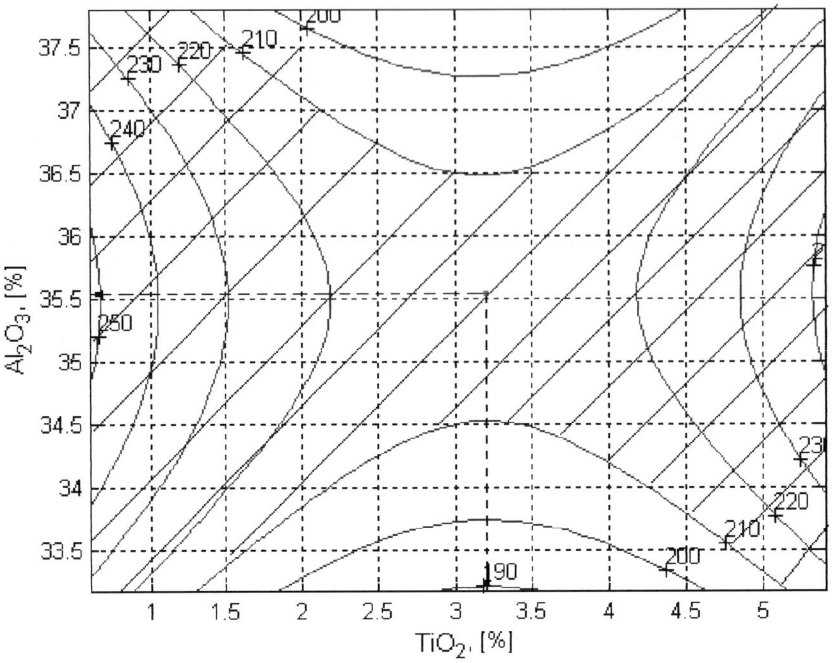

Figure 5. The multiple correlation L_S= f((%Al_2O_3), (%MgO)$_{med}$, (%TiO_2)) - level curves

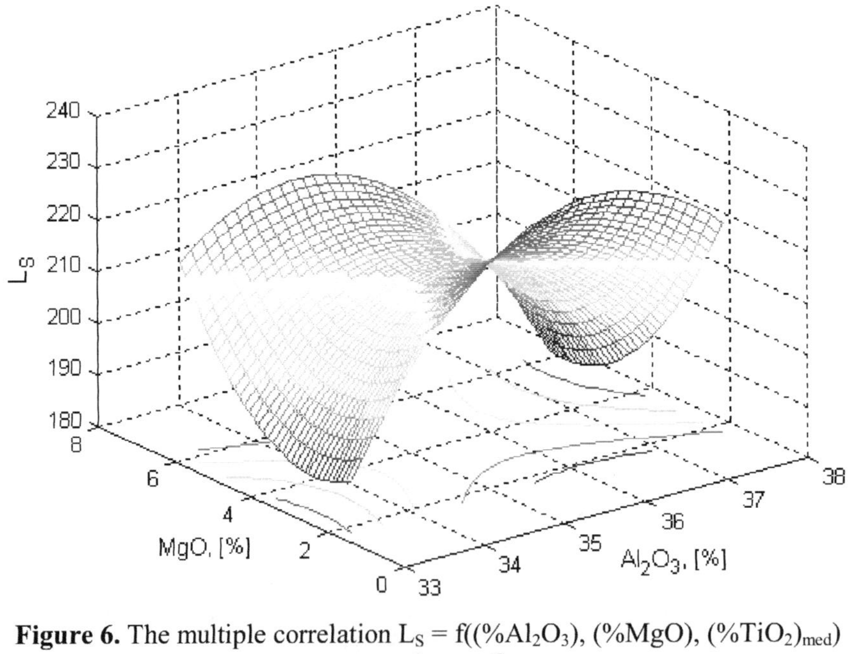

Figure 6. The multiple correlation L_S = f((%Al_2O_3), (%MgO), (%TiO_2)$_{med}$) - regression surface

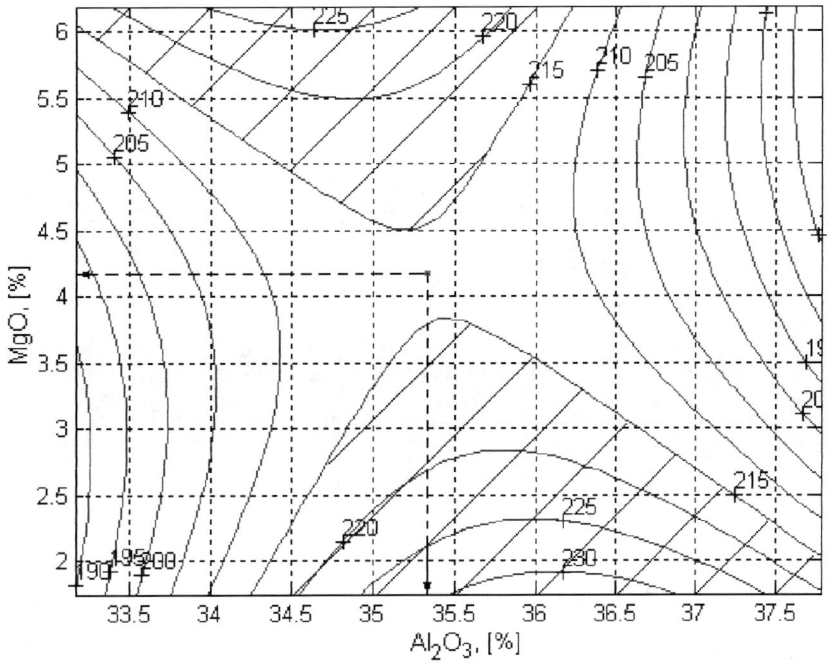

Figure 7. The multiple correlation $L_S = f((\%Al_2O_3), (\%MgO), (\%TiO_2)_{med})$ - level curves

4. Conclusion:

Based on the experiments, on the results obtained from data processing and on the technical analysis of these data, we concluded the followings:

- the chemical composition of the slag has a significant influence on the L.S., either indirectly, due to the viscosity, or directly, due to the affinity of the oxide cautions to the sulphur anions;
- the distribution report of sulfur it is strongly influenced by major oxides slag used in the experiments, namely: MgO, Al_2O_3, TiO_2 ;
- graphic representations for determining values for independent parameters (for oxides of slag) so as to result the sulfur distribution ratio desired value;
- having regard to the joint influence of oxides on the distribution ratio for the content of oxides whose values lie within technology, it can get to L_S values better than average;
- correlations obtained are useful both for research and for practice of steel or its treatment with synthetic slag in LF type installations.

References

[1] Tripşa I and Pumnea C 1981 *Steel deoxidation*, Ed. Tehnică, Bucureşti, Romania (in Romanian)
[2] Vacu S and all 1980 *Elaboration of alloy steel,* Vol. I, Ed. Tehnică, Bucureşti, Romania (in Romanian)
[3] Heput T, Ardelean E and Kiss I 2005 Some influence of the viscosity of synthetic slags used in continuous steel casting, *Revista de Metalurgia* **41**(3) 220-226
[4] Heput T, Ardelean E, Socalici A, Maksay St. and Găvănescu A 2007 Steel desulphurization with synthetic slag, *Revista de Metalurgia* **43**(3) 181-186

IOP Conf. Series: Materials Science and Engineering **106** (2016) 012034 doi:10.1088/1757-899X/106/1/012034

Simulation of solidification process for billet with ϕ350mm section, continuous casted

E Ardelean[1], A Lăscuţoni[2], M Ardelean[1], A Socalici[1] and T Heput[1]

[1]Politehnica University of Timisoara, Engineering and Management Department, 5 Revolution Street, 331128 Hunedoara, Romania
[2]Technical College of Railway Transportation „Anghel Saligny", 136 National Street, 335900 Simeria, Romania

E-mail: erika.ardelean@fih.upt.ro

Abstract. The quality of continuous casting product depends both on the quality of the steel, and on technological parameters adopted during the casting and how solidification of billets is conducted. A simulation of the solidifying process is very useful in the industrial casting practice, providing specialists with information about the phenomena during the process and the manner in which certain parameters may vary in order to obtain the desired effects. This paper presents a two-dimensional simulation model that can be used in the continuous casting process, when micro-coolers are used in order to control thermal regime during solidification.

1. Introduction

Mathematical modeling and solidification simulation of continuously cast semi-finished products is presented in the literature [1-16], each model having specific particularities; can be studied general models or models designed specifically for certain types and size [1], [13].

Mathematical modeling of cooling and solidification process of continuous cast round blanks, shown below, is based on the mathematical description of this phenomenon. The solution to this problem is actually the solution of non-steady state heat transfer.

The method is based on the differential equation of heat transfer in finite difference equations. The differential equation of heat transfer by three axes has the form:

$$\frac{\partial t}{\partial \tau} = a \cdot \nabla^2 t \qquad (1)$$

where: t – temperature, (°C) (or (K));
- τ - time, (s);
- a - thermal diffusivity, (m^2/s).

Scalar Laplacian in cylindrical coordinates is:

$$\nabla^2 t = \frac{1}{r}\frac{\partial}{\partial r}\left(\frac{\partial t}{\partial r}\right) + \frac{1}{r^2}\frac{\partial^2 t}{\partial \varphi^2} + \frac{\partial^2 t}{\partial z^2} \qquad (2)$$

or :

$$\nabla^2 t = \frac{\partial^2 t}{\partial r^2} + \frac{1}{r}\frac{\partial t}{\partial r} + \frac{1}{r^2}\frac{\partial^2 t}{\partial \varphi^2} + \frac{\partial^2 t}{\partial z^2} \qquad (3)$$

Neglecting the heat transfer in the longitudinal direction is obtained:

$$\frac{\partial t}{\partial \tau} = a \cdot \left(\frac{\partial^2 t}{\partial r^2} + \frac{1}{r}\frac{\partial t}{\partial r} + \frac{1}{r^2}\frac{\partial^2 t}{\partial \varphi^2} \right) \tag{4}$$

Taking into account the temperature dependence of thermal conductivityλ, can be insert a lower temperature that include this variation:

$$\Phi = \int_{t_0}^{t} \frac{\lambda}{\lambda_0} dt \tag{5}$$

where λ and λ_0 are thermal conductivity at arbitrary t respectively t_0 temperature. It follows:

$$\frac{\partial \Phi}{\partial \tau} = a(\Phi) \cdot \left(\frac{\partial^2 \Phi}{\partial r^2} + \frac{1}{r}\frac{\partial \Phi}{\partial r} + \frac{1}{r^2}\frac{\partial^2 \Phi}{\partial \varphi^2} \right) \tag{6}$$

$$\frac{\partial \Phi}{\partial t} \cdot \frac{\partial t}{\partial \tau} \cdot \frac{\rho \cdot c}{\lambda} = \frac{\partial^2 \Phi}{\partial r^2} + \frac{1}{r}\frac{\partial \Phi}{\partial r} + \frac{1}{r^2}\frac{\partial^2 \Phi}{\partial \varphi^2} \tag{7}$$

$$\frac{\lambda}{\lambda_0} \cdot \frac{\partial t}{\partial \tau} \cdot \frac{\partial H}{\partial t} \cdot \frac{\rho}{\lambda} = \frac{\partial^2 \Phi}{\partial r^2} + \frac{1}{r}\frac{\partial \Phi}{\partial r} + \frac{1}{r^2}\frac{\partial^2 \Phi}{\partial \varphi^2} \tag{8}$$

$$\frac{\partial H}{\partial \tau} = \frac{\lambda_0}{\rho} \cdot \left(\frac{\partial^2 \Phi}{\partial r^2} + \frac{1}{r}\frac{\partial \Phi}{\partial r} + \frac{1}{r^2}\frac{\partial^2 \Phi}{\partial \varphi^2} \right) \tag{9}$$

2. The two-dimensional solidification and cooling of the semi-finished product

To achieve two dimensional mathematical modelling of solidification for a 350mm diameter round billet continuous casted is considered a section of crystallizer - billet which is divided by a mesh network. The simulation it also allows introducing a varying percentage of micro-coolers, for a more rigorous temperature control and achieving an adequate casting structure.

The mathematical model is based on the following simplifying assumptions:

- crystallizer section shall be considered equivalent to a circular section;
- it is considered that heat loss is uniform on all surfaces crystallizer;
- crystallizer thermal conductivity of the material is considered constant;
- is neglected heat transmission to the longitudinal axis, considering that it occurs only in the horizontal section of the wire;
- density variation is neglected;
- is considered zero moment when the liquid steel temperature is uniform and micro-coolers temperature is equal to ambient temperature;
- the thermal conductivity of the slag layer on the surface of the wire is considered to be constant;
- the release of melting latent heat occurs between liquidus - solidus temperature directly proportional.

In order to achieve solidification simulation of continuously semi-finished casted parts, it compiled a computer program in C ++, using the characteristics of a steel grade for steel pipes (S 235, according to EN 10297).

To achieve the simulation, first opens the program's main interface, shown in Figure 1. The simulation was performed for 350mm diameter round semi continuous casting. The program allows cooling without micro-coolers blank, uniform distribution of micro-coolers a random distribution respectively a repeatable random distribution; in this simulation is used repeated random distribution of micro-coolers - Figure 2. Micro-coolers repeatable random addition into the crystallizer was done in three variants: the addition of 1%, 2% and 3% micro-coolers and with different thicknesses of their diameter. The simulation is performed only for the primary and secondary cooling, not for the whole path of the wire in the continuous casting installation. This explains the high values of the steel temperature inside the wire, which decrease below the solidus temperature until debiting semi-finished product.

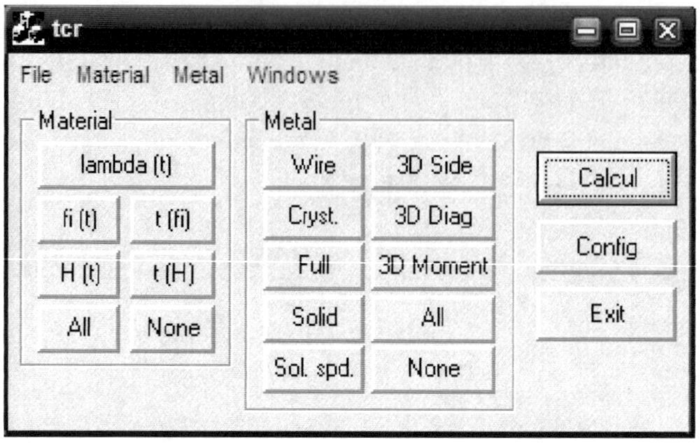

Figure 1. The program main interface

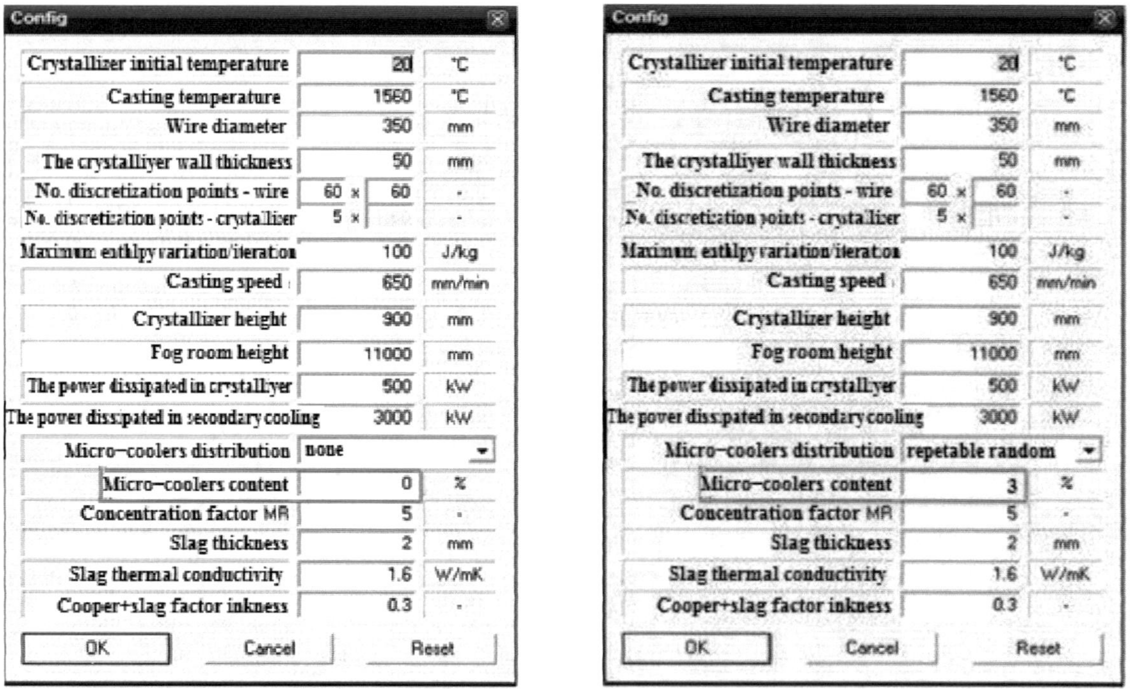

Figure 2. Data`s configuration

For a better illustration of how operate the program, we have been made a screen captions at different time points (beginning, middle and end of simulation), which can give information concerning the temperature of the wire and the crystallizer, according to Figure 3 or Figure 4.

From Figure 5 it is noted that about at 4 minutes after the liquid steel contact with the crystallizer, the variation speed of radius relative is stabilized, the solidification process respectively the heat disposal resulting normal.

IOP Conf. Series: Materials Science and Engineering **106** (2016) 012034 doi:10.1088/1757-899X/106/1/012034

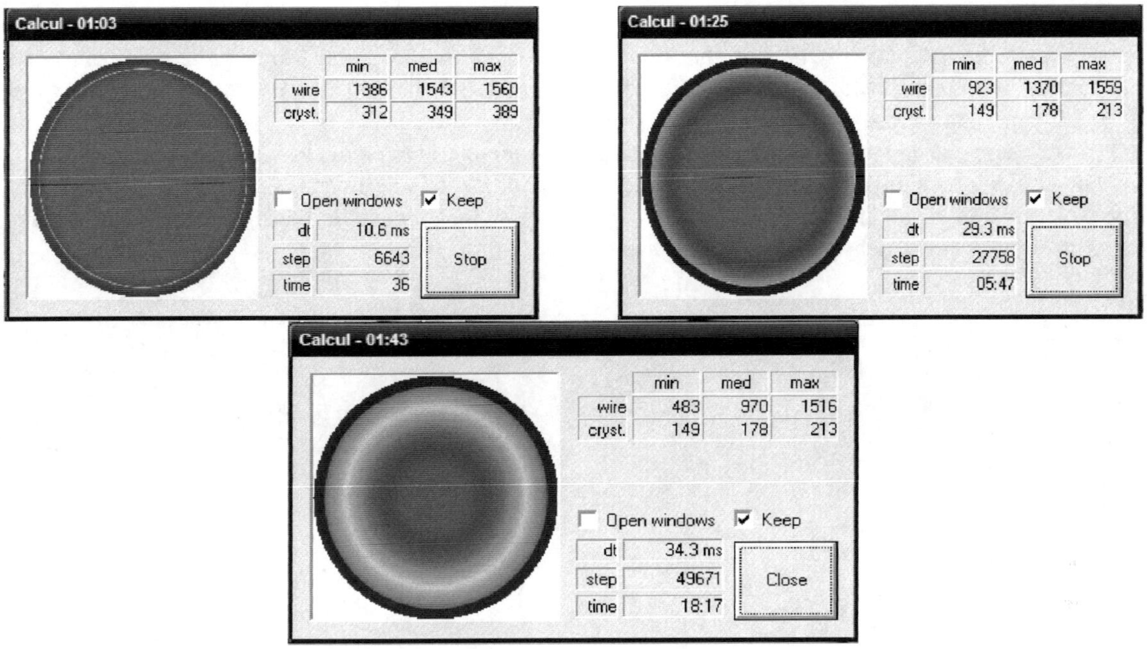

Figure 3. The calculation progress (simulation without micro-coolers addition)

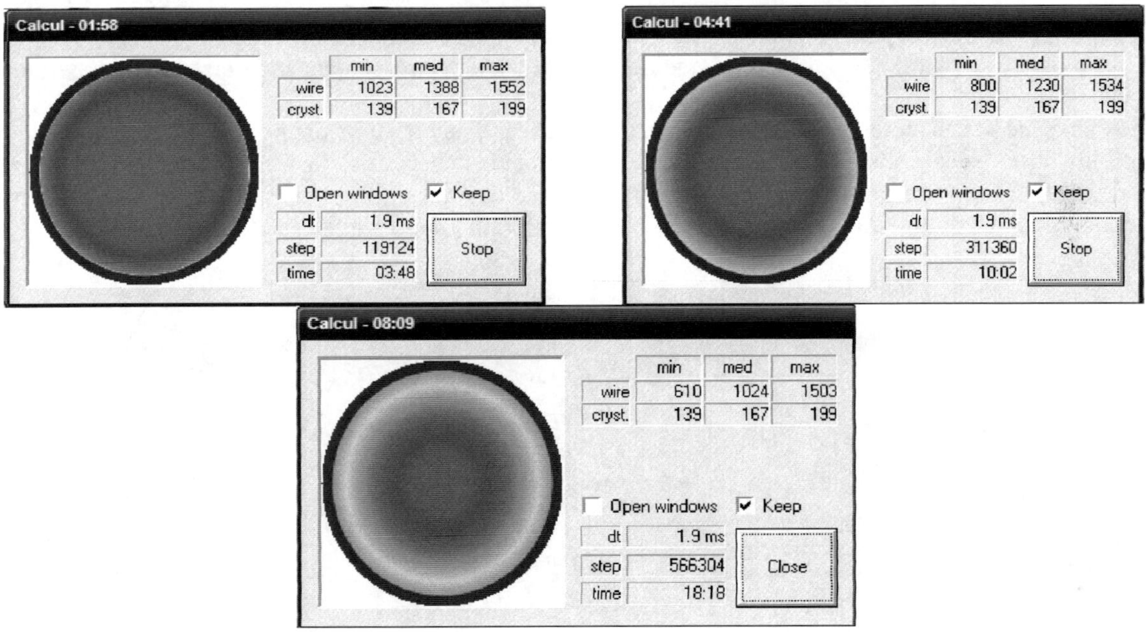

Figure 4. The calculation progress (simulation with 3% micro-coolers addition)

Still referring to the speed of change in time of the solidification front radius, for the few cases studied, we found the following:

- at 3% micro-coolers addition into crystallizer, in the first 100 seconds of casting, result a strong increase in its, an maximum (0.24%/s) is reached after 50 seconds, followed by an intense decrease up to 100 seconds; after about 300 seconds variation is maintained in the constant range (0.075 to 0.035)%/s;

- at 1% micro-coolers added, the rate of change of the solidification front radius is behaving similarly, with the observation that after 120 seconds the variation limits remain constant, the maximum being 0.25%/s, reached after 50 seconds;
- if don't add micro-coolers, the maximum speed variation of the solidification front radius is 0.20%/s, reached after 50 seconds and its limits remain constant after about 100 seconds.

The situation is similar in the version without the addition of micro-coolers, with increased power dissipated in the secondary cooling, with the observation that the value of lower limit respectively superior limit are significantly higher.

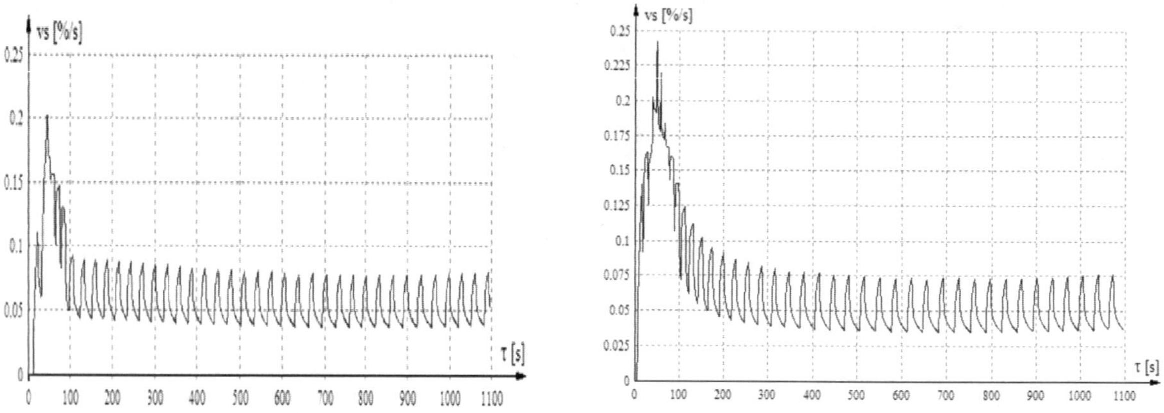

Figure 5. The variation speed of the radius relative of the solidification front (without micro-coolers addition and with 3% micro-coolers addition)

In Figure 6 it is noted that until about 200 seconds is a sharp drop of the relative radius of the solidification front, normal situation given the direct contact between crust and crystallizer interior face, heat transfer occurs mostly through conduction. After detachment the crust, the heat transfer occurs by radiation and there is a linear decrease relative to the radius of the solidification front.

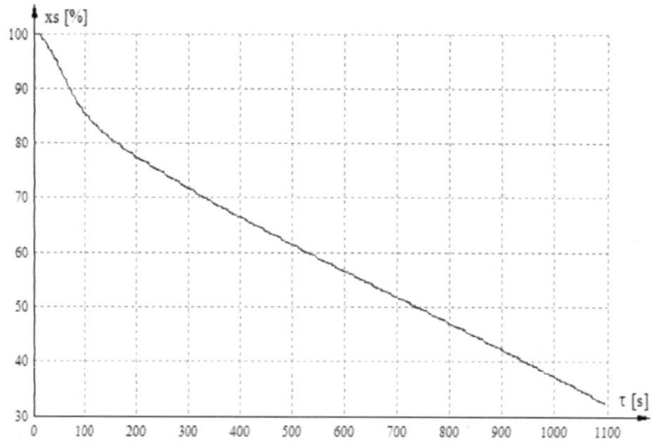

Figure 6. The relative radius of the solidification front (100% represents the wire surface)

From Figure 7 and Figure 8 it is found that in the first 70-80 seconds there is a sharp drop of temperature, during which the crust is formed, so the intense heat disposal is justified. After crustal formation, as a result of the contraction, it detaches from the crystallizer; as a result there is a brief increase of temperature (for about 40-50 seconds) followed by a linear decrease.

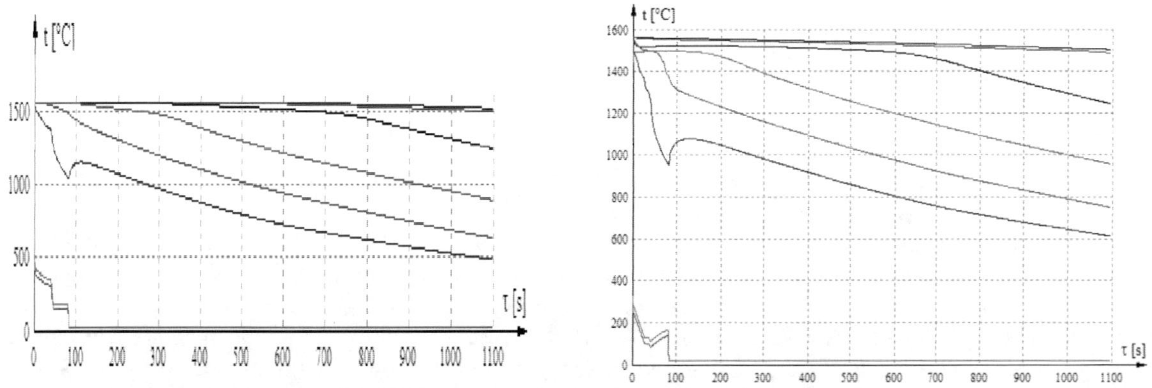

Figure 7. The temperature variation function of time for the polar coordinate φ= 0, at points 0%, 50%, 75%, 90% and 100% of the strand radius, as well as the mid-thickness and its surface crystallizer-coldest
(without micro-coolers addition and with 3% micro-coolers addition)

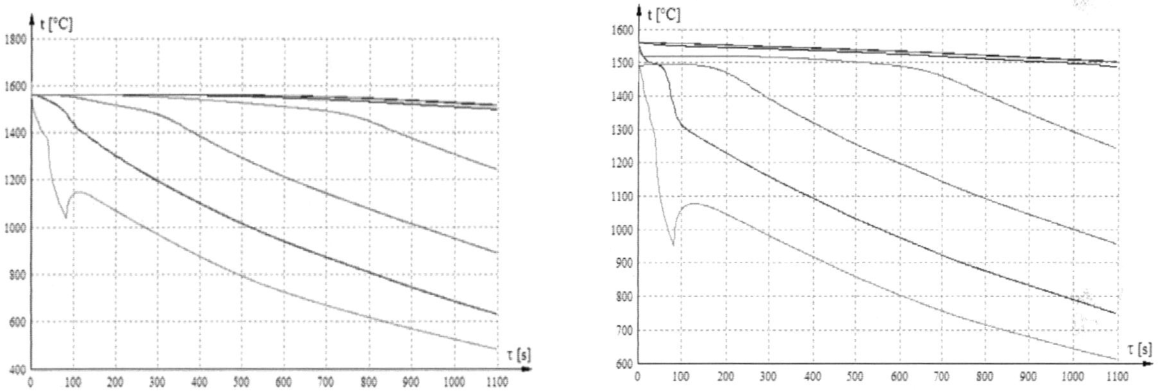

Figure 8. The temperature variation function of time for the polar coordinate j = 0, at points 0%, 10%, 25%, 50% and 100% of the strand radius
(without micro-coolers addition and with 3% micro-coolers addition)

3. Conclusions

By analyzing the obtained simulation results have some conclusions:

- modelling and simulation of temperature variation process on steel in the crystallizer allow highlighting the effect of the addition of micro-coolers on the temperature of solidifying steel;
- from the interpretation of simulation results for crystallizer thermal process intended to manufacturing steel pipes it is found that there is an appropriate heat transfer (which provide a crust thickness adequate to avoid the surface defects) for an extra 3% micro-coolers with a 2-3mm diameter;
- the simulation software can be used for any steel grades of semi-finished steel round with various sections on condition that the parameters is reconfiguration depending by steel grade and billet size;
- the simulation can contribute to the development of suitable casting technologies by which the specific energy consumption and metal are reduced on the one hand, and on the other hand whenever appear certain discrepancies between technological parameters and quality of the molded product, is possible to study through simulation, to clarify these causes.

Acknowledgments
This work was partially supported by the strategic grant POSDRU/159/1.5/S/137070 (2014) of the Ministry of National Education, Romania, co-Financed by the European Social Fund - Investing in People, Sectorial Operational Programme within the Human Resources Development 2007-2013.

References

[1] Ardelean E, Ardelean M, Socalici A and Heput T 2007 Simulation of Continuous Cast Steel Product Solidification, *Revista de Metalurgia* **43**(3) 181-187
[2] Man Y H, Hyun G L and Seung H S 2003 Numerical Simulation of Three-dimensional Flow, Heat Transfer and Solidification of Steel in Continuous Casting Mold with Electromagnetic Brake, *Journal of Materials Processing Technology* **133**(3) 322–339
[3] Santos C A, Spim J A and Garcia A 2003 Mathematical Modeling and Optimization Strategies (genetic algorithm and knowledge base) Applied to the Continuous casting of Steel, *Engineering Applications of Artificial Intelligence* **16**(5–6) 511–527
[4] Vertnik R and Šarler B 2009 Simulation of Continuous Casting of Steel by a Meshless Technique, *International Journal of Cast Metals Research* **22**(1-4) 311-313
[5] Louhenkilpi S, Laitinen E and Nieminen R 1993 Real-time Simulation of Heat Transfer in Continuous Casting, *Metallurgical Transactions B* **24**(4) 685-693
[6] Tieu A K and Kim I S 1997 Simulation of the Continuous Casting Process by a Mathematical Model, *International Journal of Mechanical Sciences* **39**(2) 185–192
[7] Janik M and Dyja H 2004 Modeling of Three-dimensional Temperature Field Inside the Mould During Continuous Casting of Steel, *Journal of Materials Processing Technology* **157–158** 177–182
[8] Fachinotti V D and Cardona A 2003 Constitutive Models of Steel Under Continuous Casting Conditions, *Journal of Materials Processing Technology* **135**(1) 30–43
[9] Meng Y and Thomas BG 2003 Heat-transfer and Solidification Model of Continuous Slab Casting: CON1D *Metallurgical and Materials Transactions B* **34**(5) 685-705
[10] Santos C A, Fortaleza E L, Ferreira C R F, Spim J A and Garcia A 2005 A Solidification Heat Transfer Model and a Neural Network Based Algorithm Applied to the Continuous Casting of Steel Billets and Blooms, *Modelling And Simulation In Materials Science And Engineering* **13** 1071–1087
[11] Saxén H and Sillanpää M 1994 A Model for Decision Support in Continuous Steel Casting, *Modelling and Simulation in Materials Science and Engineering* **2**(1) 79
[12] Popa E, Mihuţ G and Alexa V 2010 Simulation Tendencies in the Continuously Cast Half-products Area, *Acta Techical Corviniensis - Bulletin of Engineering* **1**(III) 95-102
[13] Hardin R A, Liu K, Bechermann C and Kapoor A 2003 A Transient Simulation and Dynamic Spray Cooling Control Model for Continuous Steel Casting, *Metallurgical and Materials Transactions B* **34**(3) 297-306
[14] Šarler B, Vertnik R and Mramor K 2012 A Numerical Benchmark Test for Continuous Casting of Steel, *IOP Conf. Ser.: Mater. Sci. Eng.* **33** 012012
[15] Vakhrushev A, Wu M, Ludwig A, Tang Y, Hackel G and Nitzl G 2012 Modeling of the Flow-solidification Interaction in Thin Slab Casting, *IOP Conf. Ser.: Mater. Sci. Eng.* **33** 012014
[16] Hinata N, Zulaida Y M and Anzai K 2012, Application of Particle Method to the Casting Process Simulation, *IOP Conf. Ser.: Mater. Sci. Eng.* **33** 012114

IOP Conf. Series: Materials Science and Engineering **106** (2016) 012035 doi:10.1088/1757-899X/106/1/012035

Research regarding to behavior on advanced plastic from rolling mills equipment

M Ardelean[1], E Ardelean[1], E Popa[1], A Josan[1] and A Socalici[1]

[1]Politehnica University of Timisoara, Department of Engineering and Management, 5 Revolution Street, 331128 Hunedoara, Romania

E-mail: marius.ardelean@fih.upt.ro

Abstract. New advanced plastic can be used in construction of different equipment's from some industries; due to mechanical properties closer to nonferrous materials. In steel industries uses of this materials are limited because working temperature is too low, related to nonferrous or ferrous material.
In this paper is presented some researches related to replacement of bronze material with advanced plastic in construction of antifriction bearings. For replaces of this material with engineering plastic product, it was calculated analytical and using simulation, forces in node of braking mechanism. Using these loads, it was make simulation regarding behavior of static loads with finite element software. Based on these researches, this bearing can be made from engineering plastic product, in same qualitative and technical condition, and this is a way to reduce maintenance and exploitation cost.

1. Introduction

For the design of cooling bed cannot be specified uniform principles, the construction of this equipment depends primarily on the rolling speed and schedule of rolling, it is also influenced by local conditions such as, for example, the length and width of the hall. How much will become a greater cooling area, the greater will be the cooling capacity. The cooling capacity C, of the cooling bed is related to the productivity of the rolling mill Q and cooling time t_r according to the equation, [1]:

$$C = q \cdot t_r \quad (t) \tag{1}$$

The value of t_r from the equation result of the laminate heat disposal to the environment, mainly through radiation. Temperature of the material to leaving the cooling bed shall be in general under the 150^0C. When, after cooling, the material is directed in the main production flow, the temperature should not exceed 100^0C.

The cooling bed from small profiles rolling mills, is one of the most complicate constructive-functional cooling beds, first of all because it is composed from several mechanisms and all must be correlated in operation. The location of the main mechanisms components of the small section cooling bed rolling mill is shown in Figure 1.

Kinematic analysis of mechanisms aims to determine the positions, speeds and accelerations of mechanism elements (angles of position, angular velocity and accelerations for elements in rotation and displacements, linear velocities and accelerations for items in translation) by the angle position of the driving element for rotation or linear movement for the translational motion of drive element [2].

Kinetostatic analysis is made for structural mechanism groups, considered statically determined, and start from more distant structural element to driving element, that containing effector element. The purpose of this analysis is to determine the kinematic couplers loads, forces and moments [1], [2], [5].

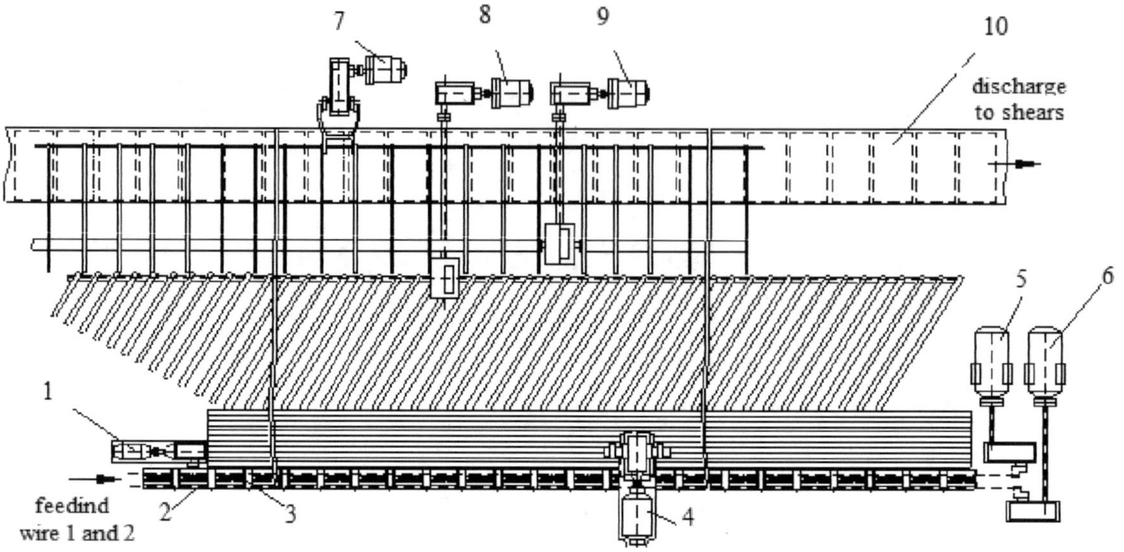

Figure 1. The location of the cooling bed mechanisms
1- separating mechanism; 2,3-supply roller conveyor; 4-the walking drive mechanism with rack; 5,6- brake mechanisms of the wires 1 and 2; 7,9-discharge mechanism; 8- inclined roller drive mechanism; 9- evacuation roller conveyor

2. Presentation of the studied mechanism

Structure diagram of a brake mechanism for the wire 1 is shown in Figure 2. The figure shows that the mechanism contains a rank V, two elements of rank IV, an element of rank III and eight elements of ranks II, which form five closed deformable contours including four class IV and one class V. Joints C and E are double couplers, [1].

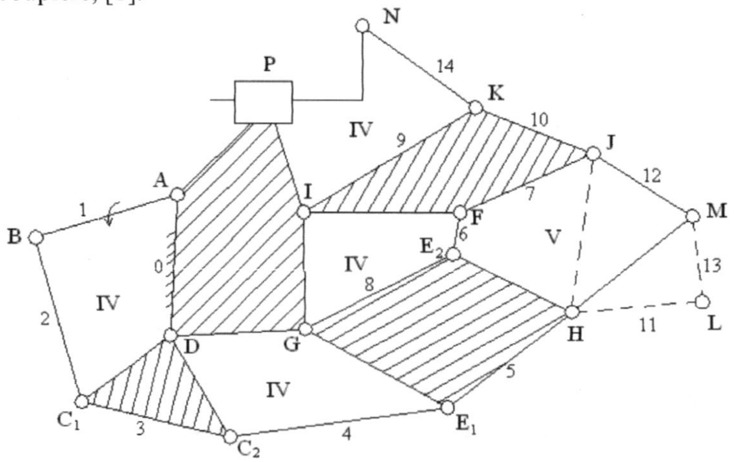

Figure 2. Structure diagram of a brake mechanism

Kinematic scheme of the mechanism is shown in Figure 3. The mechanism is based on a plan reference system, with origin in fixed joint A with its OX axis oriented in a direction parallel to direction of fixed couplers D, G, I.

The mechanism is plan, consists of 11 movable cinematic elements and 15 rotation grade cinematic couplers (forming rectangles A, B, C, D; dyad C, E, G; dyad E, F, I, dyad H (L) MJ; dyad K, NN) and a translational coupling in P, [3].

Family plane mechanisms are mechanisms f = 3, degree of mobility for planar mechanisms is calculated with relation:

$$M_3=3\cdot n-2\cdot C_5-C_4=3\cdot 11-2\cdot 16-0=1 \quad (2)$$

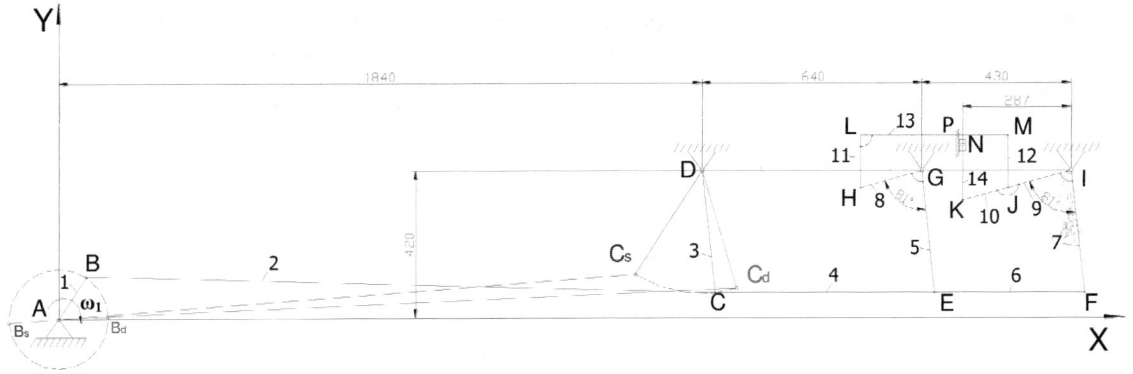

Figure 3. Kinematic scheme of the mechanism

Studied subassembly is part of the braking mechanism of the booth wire, and is shown in Figure 4. Also is presented drawings of three types of bushings used in construction of this subassembly.

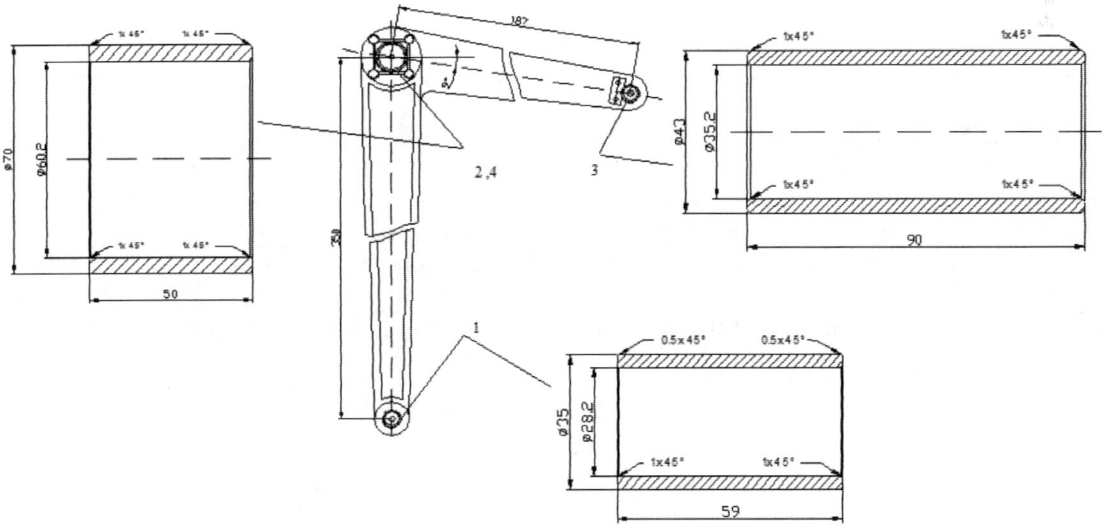

Figure 4. Subassembly of the braking mechanism

For the bushings 1, 2 and 3 maximum loads shall be increased by a factor of safety (c=1.2). Are determined the loads that will make simulation in finite element analysis software Ansys, [4], [5].

$$F_{1max}= F_1\cdot c=162.864\cdot 1.2=162.864\ N$$

$$F_{2max} = F_2 \cdot c = 530.69 \cdot 1.2 = 318.414 \text{ N} \quad (3)$$

$$F_{3max} = F_3 \cdot c = 289.197 \cdot 1.2 = 304.197 \text{ N}$$

3. Simulation and experiments

Based on the results of wear testing (Figure 5), will be presented in this section simulation results to static behavior, focusing on the materials ERTALON 66SA and ERTALYTE, which had the best behavior to wear in the test circumstances, [2], [6], [7], [8], [9].

In figure is presented comparative values of cumulative wear according to the specified pressure for all studied materials, including bronze. In conclusion, the plastic composite ERTALON 66SA performs better than bronze in these test conditions. ERTALYTE perform well at low and medium loads when not wear at all. It is observed in this case a tendency to buckling that leads to an increase of wear. The material ERTACETAL C presents the worst behavior and is observed an accelerated wear, particularly with increasing load, [1].

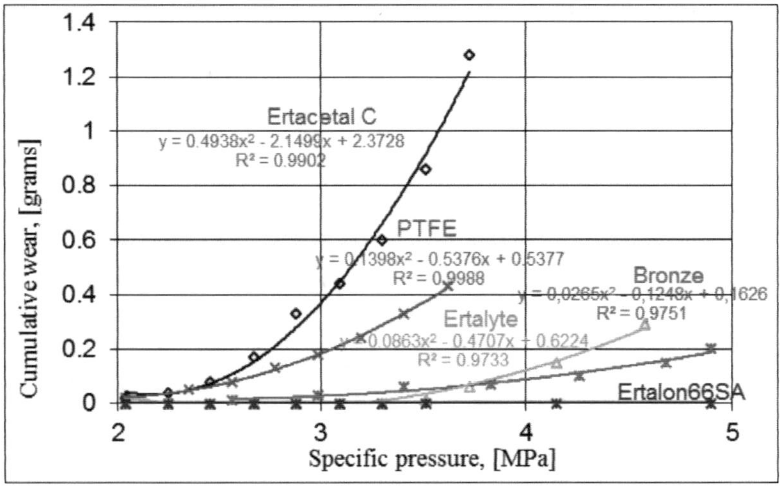

Figure 5. Cumulative wear according to the specific pressure - comparative values

Are established two variants cases of extreme loading, corresponding to a uniform force distribution on the contact surface, or force acting along the contact line – Figure 6 to 9. The loads are applied through the force transfer surfaces or network lines (nodes). From the loading analysis notes that bushing 2 is more loaded and in the following will study only this bushing.

This version represents an extreme load case that would come when it would be extremely worn bushing and replace it as necessary anyway. Based on the presented simulations ERTALON 66 SA is loaded to yield stress (max. 89.714MPa) request to corresponding maximum deformation of 0.019mm.

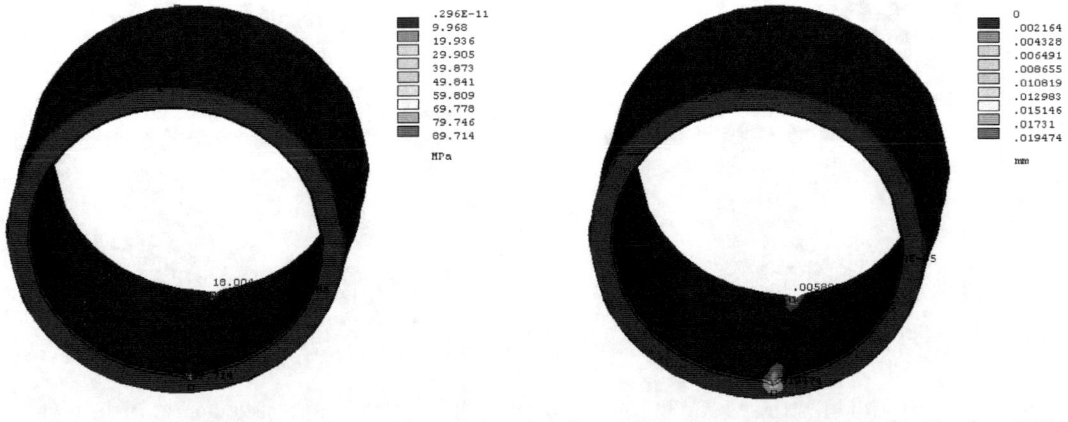

Figure 6. Equivalent stresses and strains at loading of bushing 2, line cases, for Ertalon 66SA

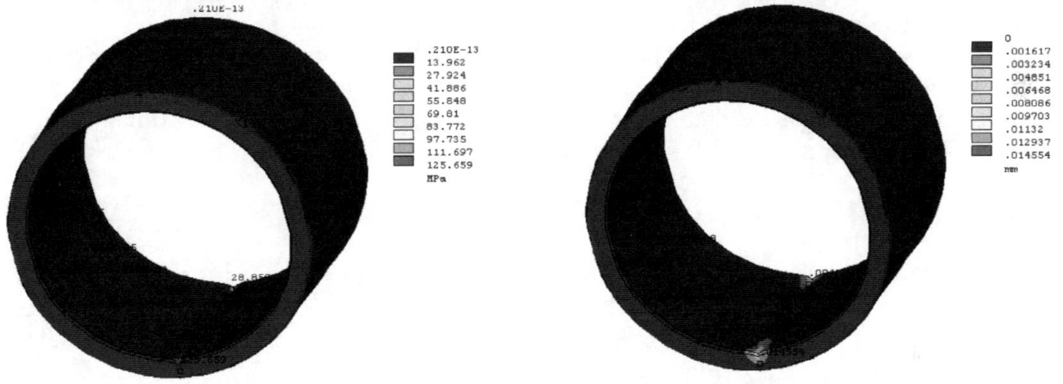

Figure 7. Equivalent stresses and strains at loading of bushing 2, line cases, for Ertalyte

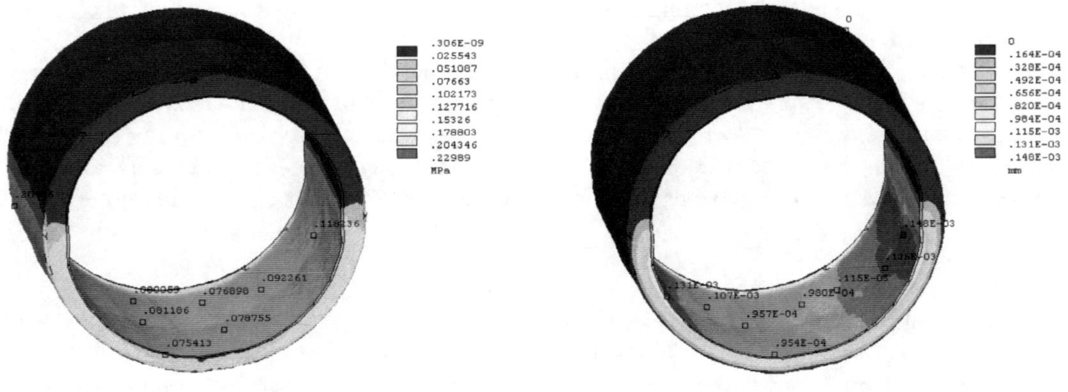

Figure 8. Equivalent stresses and strains at loading of bushing 2, surfaces cases, for Ertalon 66SA

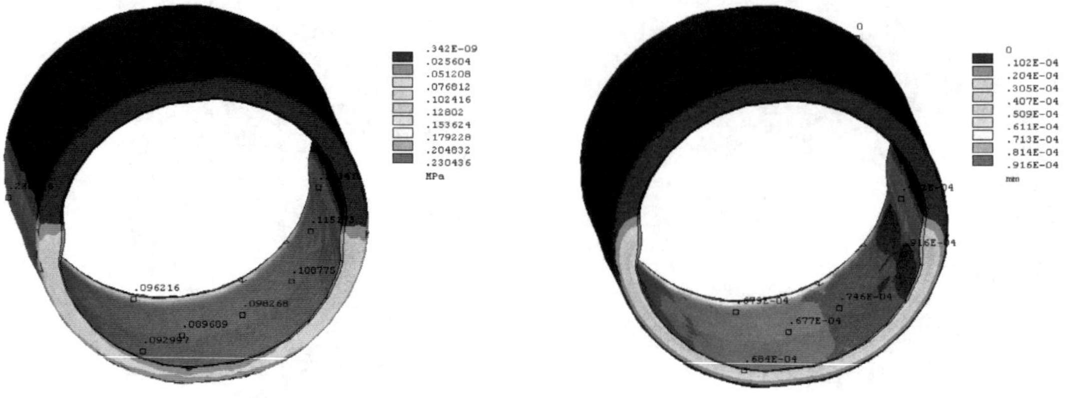

Figure 9. Equivalent stresses and strains at loading of bushing 2, surfaces cases, for Ertalyte

The figures are shown both maximum equivalent stresses and deformations and also stresses and equivalent deformations values at different point's inner or outer bushing.

On this basis we can say that the material ERTALON 66SA is closest to the desired exploitation behavior. ERTALYTE compared ERTALON 66SA behaves weaker showing load more than the yield point of approximately 30%, leading to an accelerated bushing degradation. Maximum equivalent stress were 0.229MPa for material ERTALON 66SA.

4. Conclusion

Using the simulation results in the two versions of load, in conjunction with wear test performed on plastics materials can say that studied bushings may be made of technical plastic material ERTALON 66SA that can replace in optimal condition bronze currently used in the construction of these bearing and at low cost prices. It can be used ERTALYTE (as replacement material) but with worse results than ERTALON 66SA.

Also it can be concluded that in the construction of bushings can be used the same types of materials with the observation that the using of ERTALYTE could be do under conditions of poor lubrication with good results of operating equipment. It should be noted that these bushes can operate at high temperatures (for short time approx. 90-110°C) and even under poor lubrication.

Compared with bronze (which has a purchase price of around 12...14€/kg) these plastic composite materials have a much lower price (about 7...7.5€/ kg for bars and approximately 12... 13€/kg for tubes). Low costs in conjunction with lower density compared to the bronze, 7 times (1.14g/cm^3 for plastics and 8.65g/cm^3 for bronze) leads to significant financial savings. So if processing costs are the same, the economy resulting from the use of plastics is around 0.2 ...4€ / bushing to its size.

References

[1] Ardelean M and Zamfir V 2008 *The Equipment of Cooling Beds Section Mills. Constructive and Functional Analysis*, Mirton, Timisoara, Romania

[2] Starcevic J, Pohrt R and Popov V L 2015 *Plastic Properties of Polytetrafluoroethylene (PTFE) under Conditions of High Pressure and Shear*, Wear, Lausanne, Switzerland

[3] Ardelean M, Ardelean E, Heput T and Socalici A 2009 *Simulation Results Regarding to Cinematic Parameters of Mechanical Systems*, 13th WSEAS International Conference On Systems: Recent Advances In Systems, Book Series: Mathematics and Computers in Science and Engineering, Rodos, Greece, July 22-24, pp 254-259

[4] Kljajin M, Pisačić K, Keček D and Bolarić D 2014 Comparison of Kinematic Analysis Methods at One Demonstration Mechanism, *Technical Gazette* **21**(3) 657-664

[5] Ardelean M, Heputș T, Ardelean E, Socalici A and Vilceanu L 2008 New Methods and Analysis

Functioning Techniques for Cooling Bed Tool of the Profile Laminators, *Metalurgia International* **XIII**(5) 48-53

[6] Demirci MT and Duzcukoglu H 2014 Wear behaviors of Polytetrafluoroethylene and Glass Fiber Reinforced Polyamide 66 Journal Bearings, *Materials & Design* ***57*** *560-567*

[7] Ardelean E, Ardelean M, Hepuţ T and Socalici A 2008 *Aspects Regarding to Dimensioning of Shape Cooling Bed*, International Symposium about Forming and Design in Mechanical Engineering KOD 2008, Novi Sad, Serbia, April 15-16, pp 233-236

[8] Glasscock D, Walter A, Kozielski G and Martens M 2008 *High Performance Polyamides Fulfill Demanding Requirements for Automotive Thermal Management Components*. DuPont Engineering Polymers, available: www.plastics.dupont.com

[9] Kutz M 2015 *Mechanical Engineers' Handbook, Materials and Engineering Mechanics*, Fourth Edition, John Wiley & Sons, Hoboken, NJ USA

IOP Conf. Series: Materials Science and Engineering **106** (2016) 012036 doi:10.1088/1757-899X/106/1/012036

PLCs used in smart home control

C Barz[1], S I Deaconu[2], T Latinovic[3], A Berdie[2], A Pop-Vadean[4] and M Horgos[1]

[1]Technical University of Cluj-Napoca, North University Centre Baia Mare, Electrical Electronic and Computer Department, Victor Babes str., no. 62/A, 300223 Baia Mare, Romania

[2]Politehnica University of Timisoara, Department of Electrical Engineering and Industrial Informatics, 5 Revolution street, 331128 Hunedoara, Romania

[3]University of Banja Luka, Faculty of Mechanical Engineering, 71 Stepanovica Street, 78000 Banja Luka, Republic of Srpska, Bosnia and Herzegovina

[4]Technical University of Cluj-Napoca, Department of Mechatronics and Machine Dynamics, 103-105 Muncii street, 400641 Cluj-Napoca, Romania

E-mail: cristian.barz@cunbm.utcluj.ro

Abstract. This paper presents the realization of a smart home automation using Siemens PLCs. The smart home interface is realized using the HMI Weintek eMT3070a touchscreen, which shows the window for controlling and monitoring the lighting, room temperature, irrigation systems, swimming pool, etc. By using PLCs, the smart home can be controlled via Ethernet and it can be programmed to the needs of tenants.

1. Smart home

The concept of intelligent house (SMART HOME) appeared due to the need for comfort, safety and tranquility in a house. Through it, we can create and manage an entire world with our touch, this being done with a flexible and discrete control system, suitable as well for the inside as for the outside of the house. The Smart Home has the advantage of modular functionalities, which supports the addition of new options or required equipment at any time.

It offers the possibility to create pre-defined scenarios for different equipment or installations in the Smart Home. It also uses information technology to monitor the environment and, according to certain conditions, to control electronic devices and to communicate with the outside world wherein the Smart Home lies. All these are meant to offer more comfort, improved energy efficiency and high security level.

The home automation can include the control of:

- Lighting: lights enabled on different areas of the house depending on usage and necessities, depending on the season or time of the day; illuminating the access ways over night or lighting for surveillance; we can talk about the design of the lights to induce comfort and a pleasant ambience, depending on the scenario previously created through the control console; we can also talk about illumination of the front of the house, the garden or the pool, according to time or in various configurations - party in the garden, on the terrace, watching movies, night mode, etc., the only limit is the imagination of the owner;

IOP Conf. Series: Materials Science and Engineering **106** (2016) 012036 doi:10.1088/1757-899X/106/1/012036

- Heating: personalized temperature control in each area of the smart home or in time intervals, with the possibility of changing the scenario anytime through the local control panel or remotely, through internet or phone;
- Blinds and sunshades (ventilation and air conditioning): obtain the desired level of lighting or shading through their control; they can be automatically lowered or raised depending on the desired scenario or on the command of the owner;
- Irrigation: this means watering your lawn or plants from the intelligent home through automatic irrigation according to schedule and soil moisture level or through manual or remote control;
- Swimming pool: preparing the water temperature in the pool in case you desire to use it after work or during vacations. Remote control of the pool can be done by sending an SMS or over the internet. The automatic dosing system will send status messages for: absence of chemical substances for water treatment, water quality issues, etc;
- Security System: the house can "be taught" to call the owner or close people, if there is an alert situation. These alarms can be sent in case of plumbing issues, gas leaks, electricity outage or cases that would cause material damage [1].

When we suggest an integrated smart home system, we aim primarily to simplify and improve comfort, thus, a simple press of the button it is enough to lower the blinds, to turn off all lights in the house, to set the temperature, to arm the anti-theft system; these facilities will save us time, will increase the comfort level and will also reduce energy consumption. All these facilities are available through remote, as well as through the internet.

2. PLC SIMATIC S7-1200 System control

PLC performs two main tasks of process automation [2-4]:

- inspecting the process by monitoring the status of PLC inputs, using sensors, buttons, limit switches, the state variables in the process, etc;
- processing information from inputs and generating the necessary commands to actuators of the automated process, according to a specific program.

PLC is meant to replace wired circuits with contacts and relays, which are performed by simulating software in the PLC.

The PLC monitors and controls a machine or a process with the help of the STEP7 program, which provides interrogation I/O modules through input addresses (% I) and commands the process through output addresses (% Q). [5], [6]

SIMATIC S7-1200 PLC programming TIA Portal was done in the LAD language program, taking into account various scenarios for the smart home.

STEP 7 Basic Software V12 is programming software for the SIMATIC S7-1200 automation system, and it allows the usage of the following functions to automate the facility [7]:

- The configuration and parameterization of hardware
- Defining communication
- Programming
- Testing
- Develop documentation
- Generate display screens for the SIMATIC basic operating panels.

The TIA Portal programming environment provides configuration, parameterization and programming components such as controller, visualization devices and network elements.

IOP Conf. Series: Materials Science and Engineering **106** (2016) 012036 doi:10.1088/1757-899X/106/1/012036

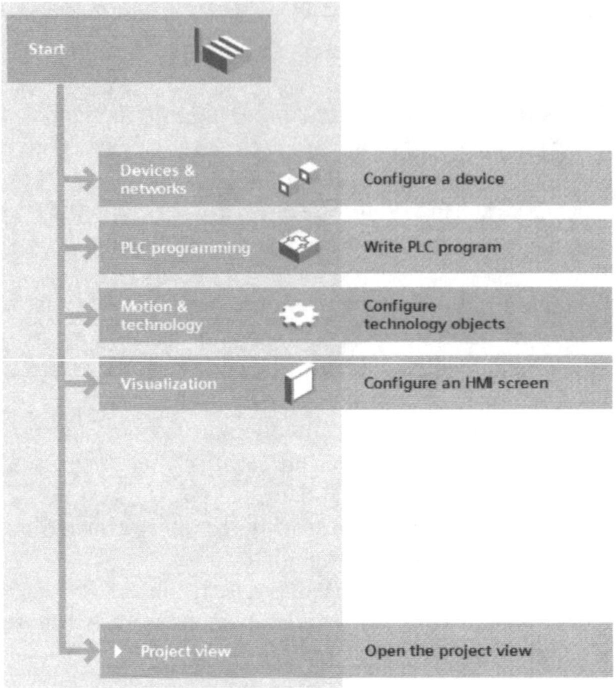

Figure 1. Portal presentation

TIA Portal has two modes of presentation: Portal presentation (provides an overview of tools activity for project processing, Figure 1) and project type (presents an overview of all structural parts of the project).

The PLC used in the paper is the SIMATIC S7-1200 series from Siemens, processor CPU 1215C DC / DC / DC, model 6ES7215-1AG31-0BX0 (Figure 2).

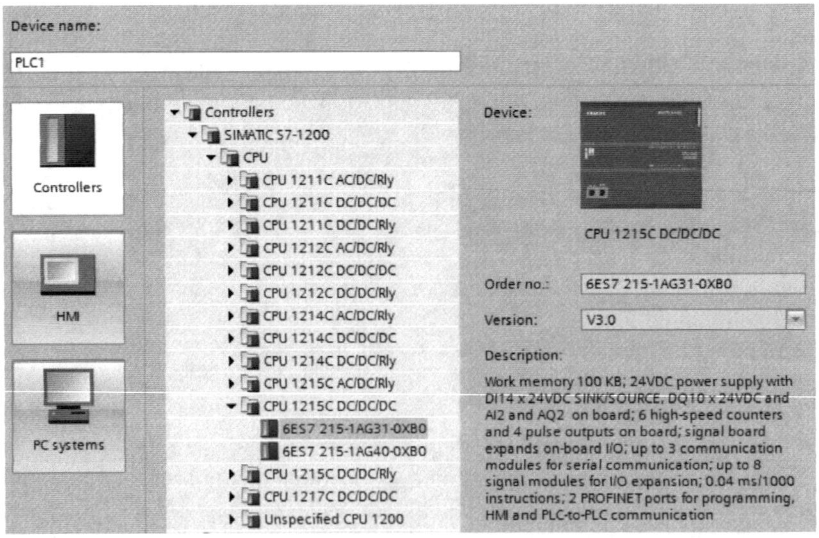

Figure 2. Selecting a CPU model

After the completion of the project configuration and PLC model selection, the TIA Portal programming environment will automatically switch to project presentation from the portal (Figure 3) and the window opens with the hardware configuration.

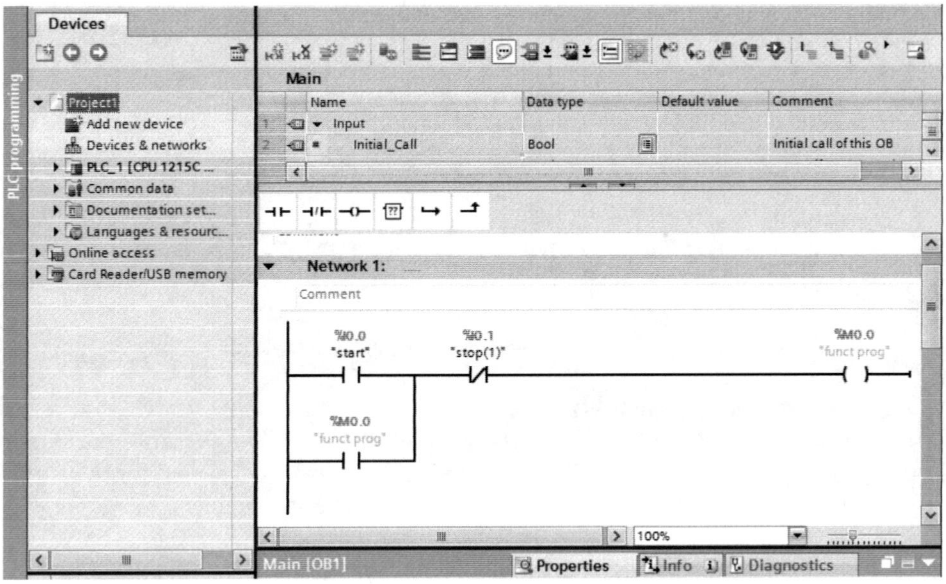

Figure 3. Project type presentation

Here, the actual realization of the smart home program begins, depending on issues that should be monitored and controlled [8].

3. Achieving control programs

3.1. Lighting

Light control can be performed manually or by command switches from digital PLC inputs (% I0.0), and through the motion sensor (% I0.2) conditioned by a crepuscular sensor (% I0.3), which blocks the turning of lights on during the day.

There is a general switch (% I0.1), which will simultaneously stop and start the lighting all over the house, or it can be controlled from the Weintek panel interface.

3.2. Irrigation

It is automatically switched on according to soil humidity measured by a humidity sensor [9]. Its value is sensed by the PLC through analog input % IW78.

If the soil humidity is below the set reference value, it opens an electro valve irrigated circuit (output% Q3.2 of PLC) and grass irrigation shall be performed (Figure 4).

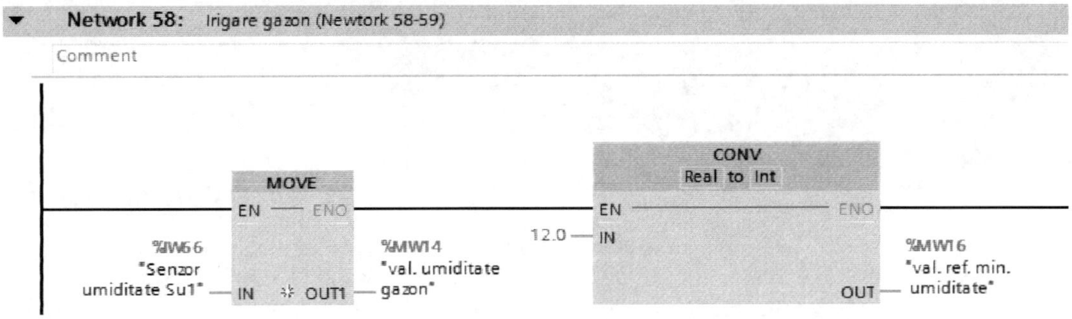

Figure 4. Start up the automatic irrigation

If soil humidity read by the humidity sensor will be higher than the maximum humidity of reference, it will command the closing marker % M3.1, which will break the electro valve circuit El6% Q3.2, electro valve will be closed and the irrigation stopped (Figure 5) [10].

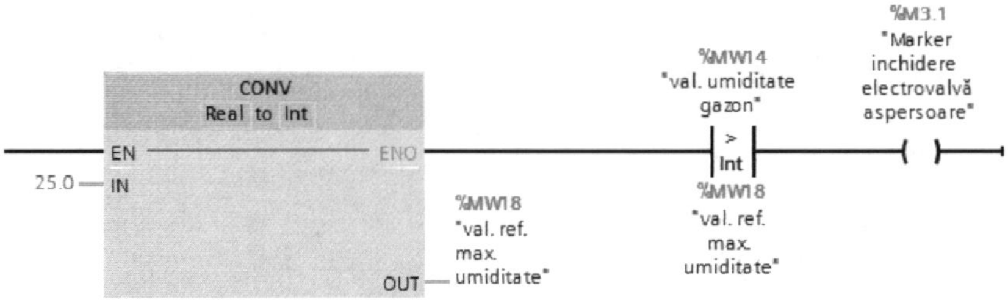

Figure 5. Shutting down the automatic irrigation

4. Creating the control interface for HMI

EMT3070 HMI touchscreen panel is produced by Weintek that facilitates creating a graphical user interface for a large number of PLCs on the market. EMT-3000 Series is designed for industrial use.

Achieving the HMI interfaces from command of the house in Easy Builder Pro for panel Weintek eMT3070 [11]

After establishing the scenario and implementing it in the programming of the PLC in TIA Portal, the command interface of the program is realized and windows are established for each activity. A window is set for each activity (lighting, irrigation, temperature, blinds, pool, etc.), but there also is a main window command.

The main window allows access to all home automation (Figure 6) and creating a username and password permits hierarchical access to the command of certain processes through user classes [12].

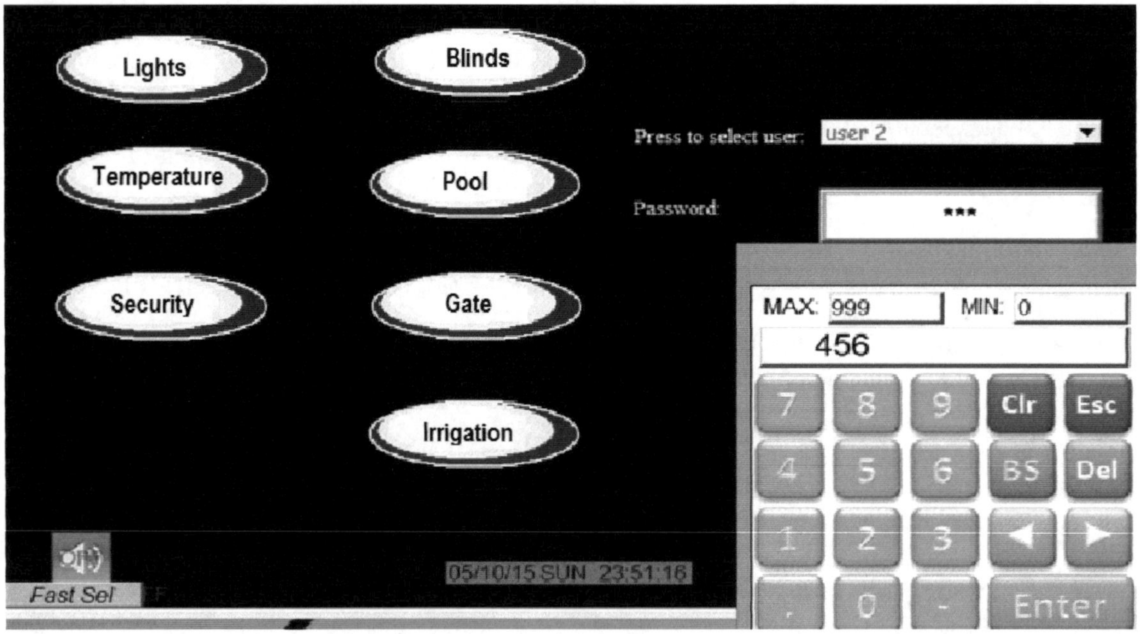

Figure 6. Main window

Similarly, interfaces will be created for all the automatic equipment that is preprogrammed in the TIA Portal for PLC. Examples of the automatic control of the pool (Figure 7) and of the sprinkler irrigation control can be viewed (Figure 8).

Opening or closing the electro valve which starts the sprinklers circuit can be done automatically, through the humidity sensor or manually with an ON / OFF switch.

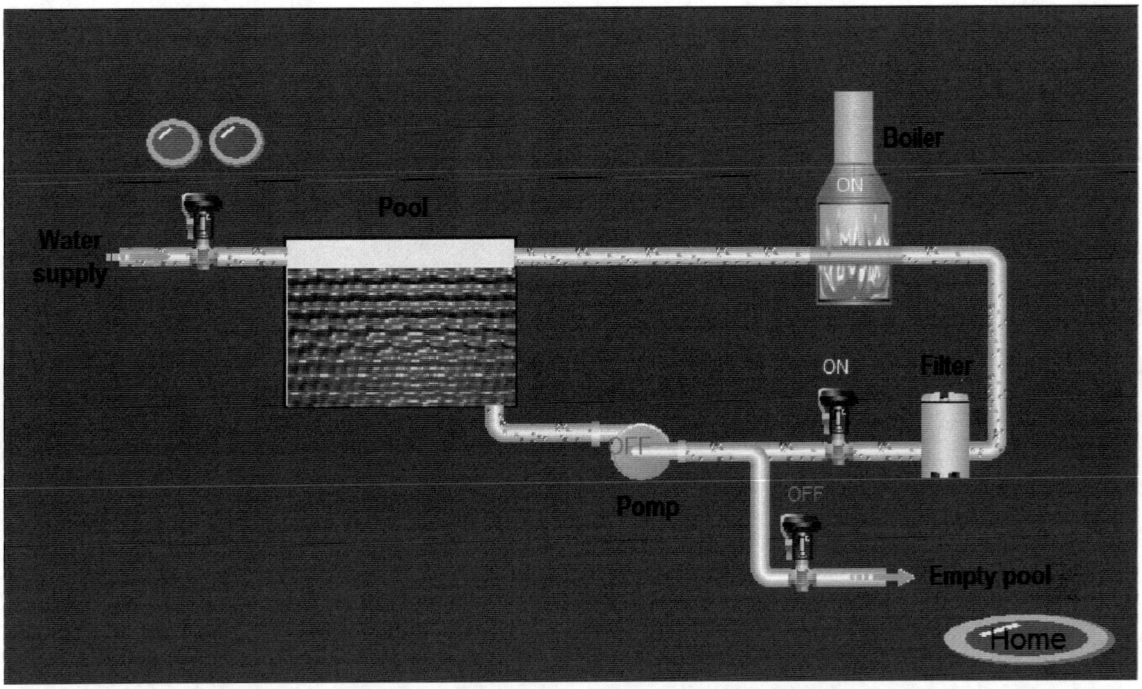

Figure 7. The control interface of pool

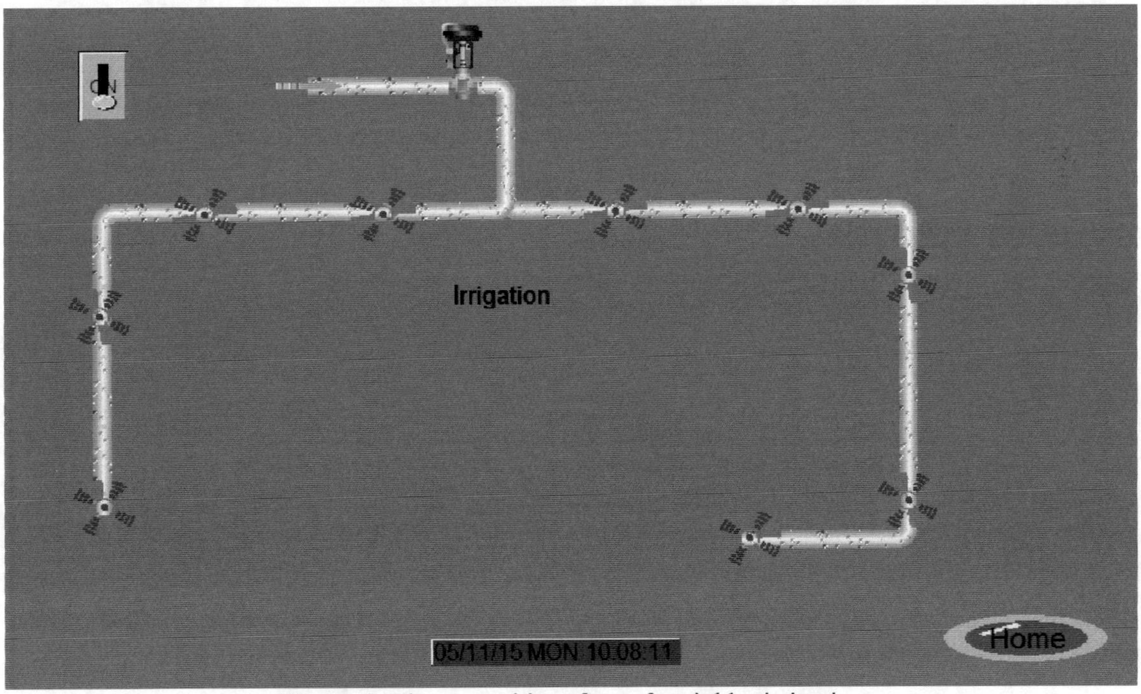

Figure 8. The control interface of sprinkler irrigation

5. Conclusions

In this paper, we intended to highlight how the comfort level can be increased when we choose to implement a smart home system. Using the PLC to implement a smart home system presents a number of advantages, of which the most important are: decrease the number of wiring, lower consumption of energy, fast adaptation to different applications, manage the consumption of energy more efficiently [13].

In this paper we presented a smart home automation using Siemens PLCs and TIA Portal V12 software. PLCs allow controlling the region through Ethernet, but are used for both monitoring and control.

The Smart house control was performed using the Weintek eMT3070a touchscreen and the application was made in EasyBuilder Pro program.

The present study was conducted in the Ladder of TIA Portal language control program.

Future objectives are achieving smart home monitoring and control through web interface and application development on platforms such as iOS, Android, etc.

References

[1] Lung C, Sabou S and Buchman A 2014 *Emergency radio communication network controller implemented in FPGA*, International Symposium for Design and Technology of Electronic Packages SIITME 2014, 20th Edition, Bucharest, Romania, October 23-26, pp 193-196

[2] Mărgineanu I 2005 *Automate Programabile*, Ed. Albastră, Bucharest, Romania

[3] Oprea C and Barz C 2014 *Automate programabile în acționări*, Ed. Risoprint, Cluj-Napoca, Romania

[4] Webb J W and Reis R I 2000 *Programmable logic controllers*, Pentice Hall, New Jersey, USA

[5] Dunning G 2002 *Introduction to Programmable Logical Controllers*, TWI Press Inc., Terre Haute, Indiana, USA

[6] Popescu D 2011 *Automate programabile. Construcție, funcționare, programare și aplicații*, Matrix, Bucharest, Romania

[7] *** *Siemens S7-1200 Programmable controller system manual*

[8] Barz C, Oprea C, Erdei Z, Pop-Vadean A and Petrovan F 2014 *The control of an industrial process with PLC*, International Conference on Applied and Theoretical Electricity (ICATE) Craiova, Romania, October 23-25

[9] Mitar S 2014 *Design and Development of Air Temperature and Relative Humidity Monitoring System With AVR Processor Based Web Server*, 8th International Conference And Exposition On Electrical And Power Engineering (EPE), Iasi, Romania, October 16-18, pp 38-41

[10] Lei S Y 2013 *Servo-control System Design of Automatic Production Line Based on PLC and HMI*, 2nd International Conference on Frontiers of Mechanical Engineering and Materials Engineering (MEME 2013), Hong Kong, October 12-13, pp 1381-1385

[11] Barz C, Latinovic T, Balan A, Pop-Vadean A and Pop P P 2014 Using HMI Weintek in command of an industrial robot arm, *IOP Conf. Ser.: Mater. Sci. Eng.* **85** 012003

[12] ***www.weintek.com

[13] Jinsoo H, Chang-Sic C, Wan-Ki P et al. 2014 PLC-Based Photovoltaic System Management for Smart Home Energy Management System, *IEEE Transactions On Consumer Electronics* **60**(2) 184-189

IOP Conf. Series: Materials Science and Engineering **106** (2016) 012037 doi:10.1088/1757-899X/106/1/012037

RF waves - Measuring their energy (dBm) in three points into Baia Mare, Romania

A Pop-Vădean[1], P P Pop[1], C Barz[2] and C Lung[2]

[1]Technical University of ClujNapoca, Department of Mechatronics and Machine Dynamics, Muncii str., no. 103-105, 400641 ClujNapoca, Romania
[2]Technical University of Cluj-Napoca, North University Center of Baia Mare, Romania

E-mail: adinaluciapopvadean@yahoo.com

Abstract. For the general public - and even for people with a science education - knowing to be subject to Radio Frequency (RF) radiation may arouse a feeling of uneasiness or even fear. Those feelings are instigated by the fact that you cannot see, hear, smell or feel RF radiation. Furthermore, the term „radiation" in itself has become very sensitive from the moment that the hazardous health effects of excessive radioactive decay (ionizing radiation) have become apparent. Media announcements of mostly unconfirmed scientific studies add to the feeling of uneasiness. So, naturally, whenever a new Global System of Mobile communications (GSM) base station is installed in an urban environment, the people living in the neighborhood may get worried. The same applies when we propose to use RF radiation for powering wireless sensors in office and home spaces using harvesting energy sistem (HE).In this paper we determine if the electromagnetic field intensity values, measured at certain points, respect the limits set by law, the protection against risks to human health.

1. Introduction

I chose this topic because of the growing interest for such measurements as RF sources are more and more. In cities and very populated areas there is a large number of potential RF sources: broadcast radio and tv, mobile telephony, wireless networks, etc.

Today over 80% of the population uses mobile phone and radio and television stations have a wider coverage for a growing number of channels. This is possible by increasing the number of base stations that provide a growing number of GSM cellule and by increasing the number of radio and television stations.

This paper discusses methods for measuring the field strength fixed in 30MHz-1 GHz frequency band. Frequency band that uses these methods, is designed band signals broadcasting, television and mobile telephony.

2. Field strength measurement. Mathematical relations

To evaluate the feasibility of the amount of energy that can be purchased (energy harvesting) from the environment (where RF is radio frequency) in the vicinity of a GSM base station BTS and compliance permissible limits the transmission power, we performed measurements of the electromagnetic field intensity fixed located in the University (North University of Baia Mare Centre of Technical University of Cluj Napoca). The mobile phone system measurements are made to determine the electromagnetic field strength in dBm for each radio channel in the frequency spectrum of a base station BTS but it is useful to determine the power density of the electromagnetic field, which is expressed in W/m^2. For this reason the values measured in dBm are converting in W/m^2. For measured values seems that is posible to be enough energy from the environment to supply a sensor.

General Conditions for measuring the electromagnetic field intensity are stipulated by International Regulation. These rules have been adopted as National Standards. The methods used for measuring the electromagnetic field intensity in fixed are those that respect the procedures developed by the National Authority for Management and Regulation in Communications.

The growth trend stations radiofrequency and communications services operators the power increase over the permitted limits, to have greater coverage with fewer sites, can lead to disturbances in other stations or even overcoming limits absorbance power permissible power of the human body.

To avoid such situations, both at commissioning and annual measurements are taken to verify it in the permitted limits. Measurements are performed using methods treated in this paper.

Electronic measurements should be concentrated for the investigation of RF electromagnetic field spectrum broadcasting field intensity and variation in coverage of local channels or cable transmission of radio and television frequencies in the 40MHz and 1GHz range. Specific measurements performed on the RF stations is that for each type of communication signal occupies a certain frequency band. In Table 1 below shows the minimum bandwidth for all types of RF signals transmitted by stations [1], [2].

Table 1. Minimum bandwidth for all types of RF signals transmitted by stations

Signal Type	Minimum bandwidth (kHz)
Double sideband AM	9
Single sideband AM	2.4
FM sound broadcasting	120
Narrow-band FM channel spacing: 12.5 kHz 20 kHz 25 kHz	 7.5 12 12
analog TV	200
GSM	200

E - electric field strength is measured, V_o induce a voltage across the antenna. The relationship between electric field strength and voltage induced in the antenna is a function of frequency. Antenna factor is denoted by K is equal to:

$K=E/V_o$

where:

-K Is the electrical component of the electromagnetic field intensity (μV / m);

-V_o Is the output voltage across its nominal impedance antenna (μV)

An isotropic antenna gain Gi note or a dipole antenna in λ / 2 note to Gd is frequently used in practice than K. Relation between antenna factor of antenna gain and antenna factor is:

- nominal resistance load antenna, RN = 50Ω;

$$K = \frac{f/MHz}{30.81 * \sqrt{Gd}} \qquad K = \frac{f/MHz}{39.47 * \sqrt{Gd}}$$

• nominal resistance load antenna, RN = 75Ω;

$$K = \frac{f/MHz}{37.75 * \sqrt{Gi}} \qquad K = \frac{f/MHz}{48.34 + \sqrt{Gd}}$$

Formulas used to calculate antenna factor can be expressed in logarithmic form:
• nominal resistance load antenna, RN = 50Ω
k=-29.77dB-gi/dB+20log(f/MHz)
k=-31.93dB-gd/dB+20log(f/MHz)
• nominal resistance load antenna, RN = 75Ω
k=-31.54dB-gi/dB+20log(f/MHz)
k=-33.69dB-gd/dB+20log(f/MHz)
where:
- gi is relative to isotropic antenna (dB);
- gd is related to the dipole g λ / 2 (dB);
- g -10logG (dB);
Taking into account a_f ,a_c and field strength can be written in logarithmic form:
e / dB (μV / m) = v_o / dB (μV) + k / dB (m^{-1}) + a_c /dB +a_f/ dB ,
where:
- e is the electric component of electromagnetic field strength (dBμV / m);
- a_c is cable attenuation (dB)
- a_f is the attenuation produced by attenuator, filter and amplifier (dB)

3. Measuring electromagnetic field intensity emitted by a television station

The measurement system has a symmetrical dipole antenna TV suitable parameters frequency range in which measurements are made to capture the electromagnetic field emitted by a television station. The relevant characteristics are measured electromagnetic field spectral analyzer. Measurement results can be processed with computer [3].

The regulations relevant international bodies on the medium and minimum intensity radiofrequency field generated by television transmitters in reception points are shown in Table 2. Radiofrequency field level is expressed in dBμV / m (the intensity) or dBmW (for power), often using those notations dBm and dBμV [4], [5].

Table 2. Average and minimum intensity radiofrequency field generated by television transmitters in points reception

Conditions of reception		Band	I	III	IV	V
Intensity average interference protection	Urban environment	dB[μV/m]	+48	+55	+65	+71
	Rural area without disturbance	dB[μV/m]	+46	+49	+58	+64
Minimum intensity satisfactory TV picture Q = 3	Urban environment	dB[μV/m]	+47	+53	+62	+67
	Rural area without disturbance	dB[μV/m]	+40	+43	+52	+58

For television broadcasting power to determine the amplitude and the image carrier, the carrier sound, luminance signal component and chromite. The longer determine the bandwidth and frequency difference between the sound signal and image signal. Complex video signal bandwidth of 8MHz, Norm CCIR-G, D / K [6].

IOP Conf. Series: Materials Science and Engineering **106** (2016) 012037 doi:10.1088/1757-899X/106/1/012037

4. Field strength measurement for GSM BTS station

GSM mobile phone system is designed as a cellule radio network, each cell is covered with radio signal from a base station BTS (Base Transceiver Station). Each BTS consists of a set of transceiver equipment for each assigned radio channel. Frequency band allocated to mobile radio communications is 890-915 MHz for mobile phone communication from the BTS and BTS 935-960MHz for communication from the mobile phone. Each transmission direction is allocated a frequency band of 25 MHz. Thus obtained 124 radio channels, which can achieve 124 simultaneous mobile communications. Channels of communication are based on a structure in time division multiplex TDMA / CDMA radio implemented on a carrier frequency. For this reason in GSM telephone system taking measurements to determine the strength of the electromagnetic field in dBm for each radio channel in the frequency spectrum of a base station BTS. Because at high frequencies, close to 1 GHz, there are minimum requirements to protect against risks to human health, established by legislation, it is useful to determine the power density of the electromagnetic field, which is expressed in W / m^2. For this reason in the following tables converting measurement of electromagnetic field strength for each radio channel in the frequency spectrum of a base station measured in dBm, the values of the electromagnetic field power density, expressed in W / m^2.

5. Results

Given the specifics of the study we conducted measurements of electromagnetic field intensity for GSM BTS station in fixed locations located in the University of Baia Mare. Measurements were performed with Agilent E4402B spectrum analyzer and measure Log-periodic antenna, model 3148 EMCO, channels network operators Orange, Vodafone show in Tables 3, 3.1, 4 and 4.1 [7], [8].

Table 3. Results

F	G	H	I	J	K	L
	channel width 0.2 MHz		Ka (antenna factor) 24		958.4	
					956.2	
					957	
			Ac (cable attenuation) 3		951	
					frecv [MHz]	949.6
						953.6
					951.4	
measuring points	Placement of the measured				951.8	
					954.4	
1	Str. Victor Babeş, no.41				952.6	
2	Platouliceu nr.1					
3	Platoucămin					
4						
5	SC Orange România SA					
Values measured in dBm						
F MHz	Pct. 1	Pct. 2	Pct. 3			
958.4	-50	-64	-70			
956.2	-53	-65	-58			
957	-49	-61	-67			
951	-62	-69	-77			
949.6	-52	-63	-52			
953.6	-60	-61	-51			
951.4	-69	-69	-60			
951.8	-68	-52	-59			
954.4	-59	-57	-63			
952.6	-57	-67	-59			

Table 3.1. Orange measurements

	A	B	C	D
1	measuring point no. 1 –Str. Victor Babeş no. 41			
2	Freq[MHz]	dBm/0.2 MHz	PFD dB(W/m2)	PFD w/m2
3	958.4	-61	-72.77	0.000000052845
4	956.2	-53	-64.77	0.000000333426
5	957	-49	-60.77	0.000000837529
6	951	-62	-73.77	0.000000041976
7	949.6	-52	-63.77	0.000000419759
8	953.6	-60	-71.77	0.000000066527
9	951.4	-69	-80.77	0.000000008375
10	951.8	-68	-79.77	0.000000010544
11	954.4	-59	-70.77	0.000000083753
12	952.6	-57	-68.77	0.000000132739
13	measuring point no. 2 – Platouliceu nr. 1			
14	Freq[MHz]	dBm/0.2 MHz	PFD dB(W/m2)	PFD w/m2
15	958.4	-64	-75.77	0.000000026485
16	956.2	-65	-76.77	0.000000021038
17	957	-61	-72.77	0.000000052045
18	951	-69	-80.77	0.000000008375
19	949.6	-63	-74.77	0.000000033343
20	953.6	-61	-72.77	0.000000052845
21	951.4	-69	-80.77	0.000000008375
22	951.8	-52	-63.77	0.000000419759
23	954.4	-57	-68.77	0.000000132739
24	952.6	-67	-78.77	0.000000013274
25	measuring point no. 3 –Platoucămin			
26	Freq[MHz]	dBm/0.2 MHz	PFD dB(W/m2)	PFD w/m2
27	958.4	-70	-81.77	0.00000006653
28	956.2	-58	-69.77	0.00000105439
29	957	-67	-78.77	0.00000013274
30	951	-77	-88.77	0.00000001327
31	949.6	-52	-63.77	0.00000419759
32	953.6	-51	-62.77	0.00000528445
33	951.4	-60	-71.77	0.00000066527
34	951.8	-59	-70.77	0.00000083753
35	954.4	-63	-74.77	0.00000033343
36	952.6	-59	-70.77	0.00000083753

Table 4. Results

F	G	H	I	J	K	L
	channel width 0.2 MHz		Ka (antenna factor) 24		935,2 935,4 935,6	
			Ac (cable attenuation) 3		frecv [MHz]	935,8 936 936,2
					936,4	

Measuring points	Placement of the measured		
1	Str. Victor Babeş, no.41		
2	Platouliceu nr.1		
3	Platoucămin		
4			
5	SC Vodafone SA		
Values measured in dBm			
F MHz	Pct. 1	Pct. 2	Pct. 3
935.2	-42	-52	-48
935.4	-63	-60	-52
935.6	-50	-57	-50
935.8	-49	-59	-62
936	-61	-49	-70
936.2	-49	-53	-56
936.4	-57	-56	-52

Table 4.1. Vodafone measurements

1.	A	B	C	D
2.	measuring point no. 1 –Str. Victor Babeş no. 41			
3.	Freq[MHz]	dBm/0.2 MHz	PFD dB(W/m2)	PFD w/m2
4.	935.2	-45	-56.77	0.000002103778
5.	935.4	-63	-74.77	0.000000033343
6.	935.6	-50	-61.77	0.000000665873
7.	935.8	-49	-60.77	0.000000837529
8.	936	-61	-72.77	0.000000052845
9.	936.2	-49	-60.77	0.000000837529
10.	936.4	-57	-68.77	0.000000132739
11.	measuring point no. 2 – Platouliceul nr. 1			
12.	Freq[MHz]	dBm/0.2 MHz	PFD dB(W/m2)	PFD w/m2
13.	935.2	-52	-63.77	0.000000419759
14.	935.4	-60	-71.77	0.000000066527
15.	935.6	-57	-68.77	0.000000132739
16.	935.8	-59	-70.77	0.000000083753
17.	936	-49	-60.77	0.000000837529
18.	936.2	-53	-64.77	0.000000333426
19.	936.4	-56	-67.77	0.000000167109
20.	measuring point no. 3 – Platoucămin			
21.	Freq[MHz]	dBm/0.2 MHz	PFD dB(W/m2)	PFD w/m2
22.	935.2	-40	-59.77	0.000001054387
23.	935.4	-52	-63.77	0.000000419759
24.	935.6	-50	-61.77	0.000000665273
25.	935.8	-62	-73.77	0.000000041976
26.	936	-70	-81.77	0.000000006653
27.	936.2	-56	-67.77	0.000000167109
28.	936.4	-52	-63.77	0.000000419759

IOP Conf. Series: Materials Science and Engineering **106** (2016) 012037 doi:10.1088/1757-899X/106/1/012037

6. Other possible applications of RF

6.1. What is Rectenna

In cities and very populated areas there is a large number of potential RF sources: broadcast radio and tv, mobile telephony, wireless networks, etc. The problem is collecting all these disparate sources and converting them in useful energy. The conversion is based on a rectifying antenna also called rectenna.

A rectenna is a special type of antenna that is used to convert microwave energy into direct current electricity. They are used in wireless power transmission systems that transmit power by radio waves. A simple rectenna element consists of a dipole antenna with an RF diode connected across the dipole elements. The diode rectifies the AC current induced in the antenna by the microwaves, to produce DC power, which powers a load connected across the diode. Schottky diodes are usually used because they have the lowest voltage drop and highest speed and therefore have the lowest power losses due to conduction and switching. Large rectennas consist of an array of many such dipole elements.

Nowadays there is an active research area investigating a number of alternatives to extract energy energy from the environment and convert it to electrical energy to power an electronic device. Taken one by one, each alternative still provides a low level of energy compared to the power requirements of today's electronic devices, except for specific applications. However, it is expected that energy harvesting will have an important role in future microelectronic devices for a number of reasons.

6.2. RF Harvesting

Some RF harvesting converts electromagnetic radiation into electricity. This can be done in two ways.

• Make use of existing electromagnetic radiation (GSM, FM, WiFi)

• Broadcast an electromagnetic signal at a specific wavelength in order to power a wireless node. The first solution has the advantage that radiation can be used that is already present. The disadvantage is the low energy density (typically µW/cm2). Furthermore, it is not always desirable or even legal to block radiation (e.g. for emergency calls). Therefore, a solution is to use dedicated broadcasting device, which can power sensors in the neighborhood.

A point of concern is the maximum power that is allowed to be transmitted into the environment, which is typically around 100mW. A company called Powercast uses this principle to charge mobile devices. Using 3W of transmitted power at an operating frequency of 900MHz and a distance between device and transmitter around 30 cm, around 100mW of power is received. Larger distances mean larger radiation power, but 3W is already much higher than the allowed energy in Europe (around 100mW). However, the received power decreases very rapidly with distance. Furthermore, 3W as transmission power is not allowed in Europe. There is room for improvement, though, in transmission, receiving (improved antenna design) and the conversion efficiency. For example, at a transmission power of 100mW, values of 1.5mW at 20 cm [9] have been reported. Other techniques still in the research phase are developed by MIT ("non-radiative-resonant-energy transfer"), for distances less than a few meters. The technology relies on copper coils tuned to resonate in identical magnetic fields [10].

In addition, researchers at the University of Tokyo have developed a four-layer plastic sheet with printed coils, organic transistors, and MEMS switches that use inductive coupling to power devices fitted with receiver coils [11]. Both of these technologies are still undergoing research, and commercialization is at least a few years away.

In Table 6 the estimated figures of energy output values per harvesting principle are listed [12].

IOP Conf. Series: Materials Science and Engineering **106** (2016) 012037 doi:10.1088/1757-899X/106/1/012037

Table 6. Estimated power output values per RF harvesting principle [12]

SOURCE	SOURCE CARACTERISTICS	PHYSICAL EFFICIENCY	HARVESTED POWER
RF GSM	900MHz 0.3-0.03μW/cm² 1800MHz 0.1-0.01μW/cm²	50%	0.1μW/cm²

These numbers are just first order indications, and they have to be considered to have a large error margin. However, they do give some indication of the typical expected power levels.

7. Conclusion

Given the great distance from sources of electromagnetic radiation (GSM antennas group), values measured do not pose a risk in the University area, the Group Aurel Vlaicu school and student residence.

References

[1] ***Recomandation ITU-R SM.443-3, Bandwidth measurement at monitoring stations

[2] ***Recomandation ITU-R BT.417-5, Minimum field strength for wich protection may be sought in planning terrestrial television service

[3] ***Recomandation ITU-R BT.665-7, Radio Frequency protection radio for AM terrestrial television

[4] ***Recomandation ITU-R BS.412-9, Planning standards for terrestrial FM sound broadcasting at VHF

[5] ***Recomandation ITU-R SM.1268, Method of Measuring the Frequency of FM Broadcast Emission at Monitoring Stations

[6] ***Recomandation ETSI ETR 132

[7] ***Agilent AN 1315 Optimizing RF and Microwave Spectrum Analyzer Dynamic range

[8] ***Agilent spectrum Analysis Basics. Application Note 150 AN 1315

[9] Vullers R J M, Visser H J, Veld B and Pop V 2008 *RF Harvesting Using Antenna Structures On Foil*, Proc. of Power MEMS 2008, Sendai, Japan, November 10-11, pp 209-212

[10] Fildes J *Wireless energy promise powers up*, BBC News, http:// news.bbc.co.uk/2/hi/technology/6725955.stm

[11] Takamiya M, Sekitani T, Miyamoto Y, Noguchi Y, Kawaguchi H, Someya T and Sakurai T 2007 *Design Solutions for a Multi-Object Wireless Power Transmission Sheet Based on Plastic Switches,* IEEE Int. Solid-State Circuits Conf. (ISSCC 2007) Dig. Tech. Pap., San Francisco, USA, February 11-15, pp 362-609

[12] van Hoof C 2008 *Micro-Power Generation using Thermal and Vibrational Energy Scavengers. 12 Years of Energy Harvesting – An overview*, presented at IEEE Int. Solid-State Circ. Conf. (ISSCC 2008), San Francisco, USA, February 3-5

IOP Conf. Series: Materials Science and Engineering **106** (2016) 012038 doi:10.1088/1757-899X/106/1/012038

Particular bi-fuel application of spark ignition engines

S Rațiu[1], V Alexa[1] and I Kiss[1]

[1]University Politehnica Timisoara, Faculty of Engineering Hunedoara, Department of Engineering and Management, 5 Revolution street, 331128 Hunedoara, Romania

E–mail: sorin.ratiu@fih.upt.ro

Abstract. This paper presents a comparative test concerning the operation of a spark-ignition engine, make: Dacia 1300, model: 810.99, fuelled alternatively with gasoline and LPG (Liquefied Petroleum Gas). The tests carried out show, on the one hand, the maintenance of power and torque performances in both engine fuelling cases, for all the engine operation regimes, and, on the other hand, a considerable decrease in CO and HC emissions when using poor mixtures related to LPG fuelling.

1. Introduction

The term "bi-fuel" refers to the alternative fuelling of an internal combustion engine with two fuels (in this case, one is liquid – the gasoline, and one gaseous – the LPG), and should not be confused with the term "dual-fuel" which designates an operating principle based on the simultaneous burning of two fuels (liquid and gaseous). An example relative to the second term would be a diesel engine operating with LPG, based on the pilot injection principle [1].

The legislative norms on environmental pollution, increasingly stringent, require, among others, alternatives to the fossil-fuel combustion, by partial or full use of certain types of unconventional fuels, which are less polluting. An example in this regard is the LPG. In general, the "liquefied petroleum gases" (LPG) are the commercial propane and butane (as well as their mixture). The propane (C3H8) and butane (C4H10) are chemical compounds consisting of carbon (C) and hydrogen (H), which justifies their designation of "hydrocarbons". For a car engine, in the current conditions when the LPG distribution network is quite limited, the possibility to ensure the use two fuels, i.e. gasoline and LPG (the "bi-fuel" variant), appears as a rational solution.

The use of LPG to the internal combustion engines does not involve special design changes, especially in case of spark-ignition engines equipped with carburettor. The simplest fuelling solution consists of passing the liquefied gas stored in the tank (cylinder) through a vaporizer, and then entering the vapours into the inlet system by means of a mixer, which doses the air-fuel mixture in an open-loop control system. This is the variant studied in the experiment proposed herein. The solution involving a closed-loop control system of the LPG injection (in liquid phase) into the air column, located on the inlet route, is currently applied. This latter solution is the most advanced, enabling higher power and efficiency performances, because the noxious emissions are reduced compared to the solution involving a mixer [2-5], [7].

This study aims to highlight the peculiarities of using the simplest solution for adapting the engine designed to operate with gasoline, by mounting the mixer upstream to the standard carburettor. So, the flowing features of the fluid through the fuel inlet system are changed, fact that influences the filling ratio of the cylinders and, therefore, the performances, the engine power yield being mostly influenced.

2. Presentation of the experimental equipment

The study was conducted on an experimental stand found in the Laboratory of Internal Combustion Engines of the Faculty of Engineering Hunedoara, University Politehnica Timisoara, Department of Engineering and Management, Specialty: Road motor vehicles. The stand is equipped with a 4-stroke spark-ignition engine, make: Dacia 1300, model 810.99, enabled to be alternatively fuelled with gasoline and LPG [6]. The engine loading was made using hydraulic disc brakes, with the possibility of measuring the engine speed and torque, the fuel and air consumption, the advance when triggering the electric spark, and the composition of exhaust gases.

Figure 1. Overview of the test stand

As mentioned in the introduction, we opted for the solution that involves the mounting of the LPG mixer upstream to the standard carburettor diffuser, as shown in Figure 2. This is a mechanical device that, using the Venturi principle, ensures the correct air-fuel ratio at all operation regimes of the engine. The depression created by the mixer allows the controlled gas absorption according to the throttle position.

Figure 2. The LPG mixer

The carburettor type is CARFIL 32 IRMA, up-draught, with one mixing chamber and a balanced float chamber balanced with the air filter downstream area.

The liquefied gas stored in the tank in the form of cylinder is driven towards the reducer-vaporizer (Figure 3.a), which is an electro-pneumatic device equipped with dual pressure control that enables the evaporation of the liquefied gas in such a way as to ensure a stable supply of the engine with a gas amount adequate to the various operation regimes. From the evaporator, the gas is sucked through the mixer (shown in Figure 2) into the carburettor diffuser, allowing the gas to mix, at the same time, with the air flow. The following items are also included in the plant: a LPG solenoid valve (Figure 3.b), which is a device mounted between the tank and the reducer-vaporizer that stops the gas flow towards the reducer when the engine is fuelled with gasoline. It has also the role of gas impurity filter; a gasoline solenoid valve (Figure 3.c) mounted between the gasoline pump and carburettor stops the gasoline flow when the engine is fuelled with gas. The switching from gasoline to LPG is made at the push of a button.

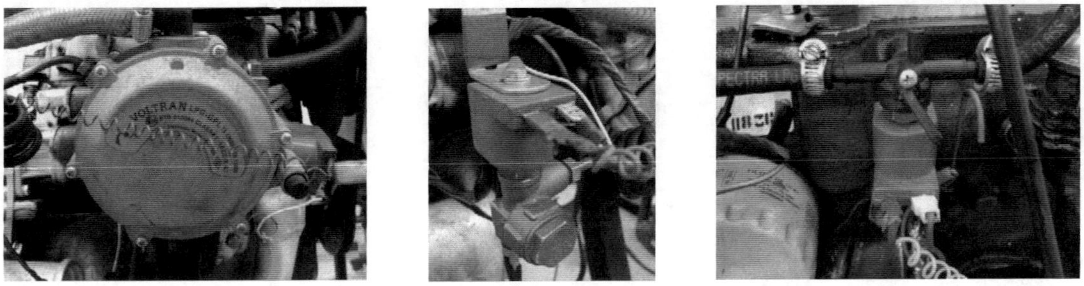

a. b. c.

Figure 3. Composition of the LPG plant [6]:
a – reducer-vaporizer, b – LPG solenoid valve, c – gasoline solenoid valve

3. Testing session and the experimental methodology

Each steady operation regime has been determined by the engine speed, the pressure in the intake manifold, and the shutter position. In both cases of fuel supply, the engine operated with the classic electric ignition equipment, consisting of accumulator battery, induction coil, breaker diffuser and standard spark plugs.

The experimental measurements were made at various operation regimes which involved variable loads and a wide range of speeds:

- Full load (fully opened shutter) and speed variable between 2000 and 5000 rev/min, in order to check if the standard power and torque performances are affected in case of LPG fuelling;
- Partial loads for two speed values (1600 and 3500 rev/min), representing specific regimes for a car driving in urban and, respectively, in extra-urban traffic conditions, with constant speed.

For each steady operation regime, we determined the flue gas composition using the analytical and diagnostic station “CAPELEC CAP 3200 GO”.

4. Results and conclusions

The full load regimes: The data obtained in these cases have shown that the dynamic performances of the engine are reduced when using LPG, compared to those obtained when using gasoline (Figures 4 and 5). This is primary due to the lower calorific value of LPG than the gasoline, and secondly to the solution chosen to adapt the engine to the operation with LPG: the mixer mounted upstream the carburettor diffuser (Figure 2) involves the introduction of an additional gas-dynamic resistance which adversely affects the process of filling the engine cylinders. This results in a decrease in the filling factor and thus a reduction in the engine power. This phenomenon is more obvious at high speeds, when the gas-dynamic losses, due to this additional resistance, are higher. In addition, this additional gas-dynamic resistance, introduced by mounting the mixer, entails a change in the flow conditions, with impact on the dosage feature and the filling factor when using gasoline.

The trend of engine power reduction when using LPG, due to the above phenomena, is, to some extent, diminished by the increase of the specified yield when using gasoline. This can be attributed to several factors: improving the combustion and increasing the amount of heat released by the reaction thanks to the presence of the required air and a better mixture with the gaseous fuel; reduction of the specific heat of the flue gases mixture evolving in the expansion stroke; possible reduction of heat transferred to cylinder walls.

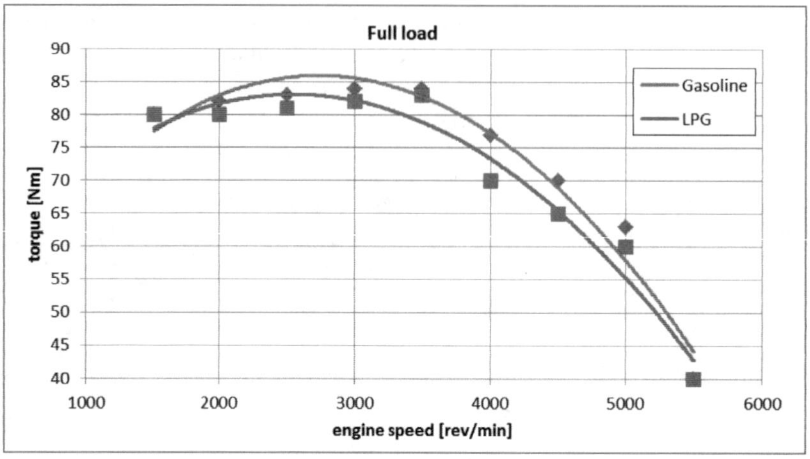

Figure 4. Torque versus speed, at full load

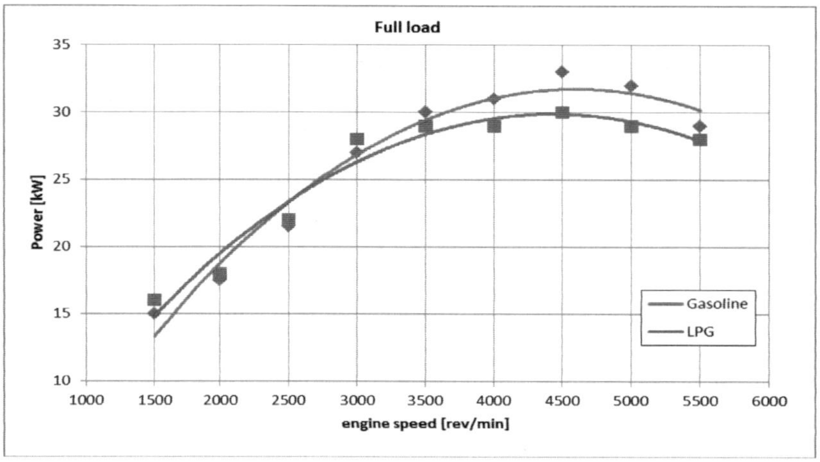

Figure 5. Power versus speed, at full load

The operation with poorer mixtures (LPG variant) led to changes in the concentration of exhaust emissions. The amounts of carbon monoxide (CO) and unburnt hydrocarbons (HC) decreased compared to the standard fuelling variant (gasoline). These changes reflect, on the one hand, the effect of combustion improvement and, on the other hand, the effect of increasing the oxygen concentrations involved in the formation of nitrogen oxides.

The partial-load regimes: The measurements carried out for these regimes revealed the following:

- A significant increase of the indicated yield compared to the gasoline fuelling variant. This allowed that, in case of an equal volume of mixture allowed inside the cylinder (the same position of the shutter) and having a lower calorific value (poorer mixture) to obtain the quasi constant maintenance of the power.
- We noticed the same general trends to reduce the emissions of CO and HC and increase the emissions of NO (Figures 6 and 7).

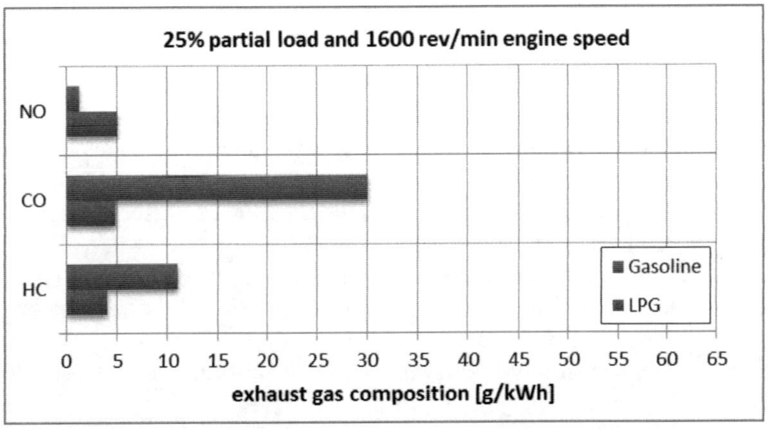

Figure 6. The exhaust gases composition at the load of 25% and speed of 1600 rev/min

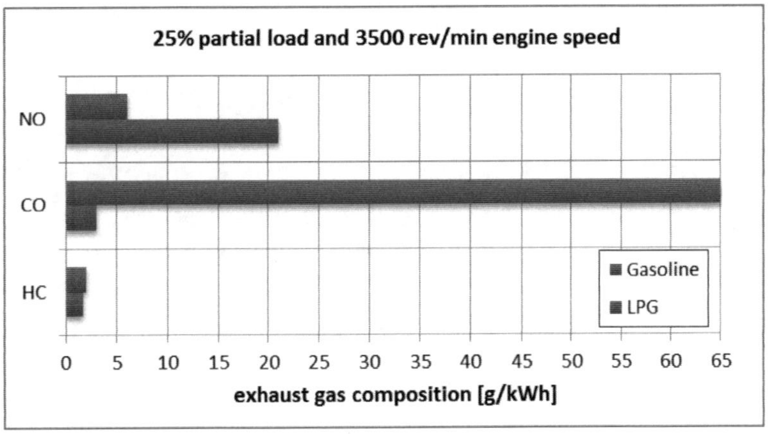

Figure 7. The exhaust gases composition at the load of 25% and speed of 3500 rev/min

As final conclusions, we can say:

- This experimental study showed that by a simple transformation of a classic "Dacia 1300" engine, consisting of a minimum adaptation of the carburettor and installation of an appropriate fuelling equipment, we can provide the engine operation with LPG in advantageous conditions, even if the standard power and torque performances have suffered, especially at high speeds and loads;
- There has been an improvement in the effective yield by increasing the specified yield, for the entire range of engine operation regimes;
- The CO and HC emissions have been reduced, but this entailed the disadvantage of significant increase of NO emissions;
- The experimental study focused on highlighting the possibilities offered by the use of LPG to an engine that was designed to be fuelled with gasoline and has to maintain its operating capacity;
- Finally, the „bi-fuel" operation of an engine is also justified by the lower price of LPG compared to gasoline.

References

[1] Robert L E 1986 *Automotive Engine Alternatives*, Proceedings of the International Symposium on Alternative and Advanced Automotive Engines, Vancouver, Canada, August 11-12

[2] ***1998 World LPG-as Association *Automotive LPgas – Today's fuel for a cleaner tomorow*

[3] Hollemans B, Conti L and de Kok P 1995 *Propane the "clean" fuel as the next century for light and heavy duty vehicles*, TNO Paper VM9504

[4] Lui B R 1998 *The effects of fuel composition, system design and operating conditions on in – system vaporization and hot start of a liquid – phase LPG injection system*, SAE Paper 981388

[5] Van der Steen M 1996 *Gaseous fuels: past experience and future expectations*, TNO Paper VM9608

[6] Raţiu S 2009 *Internal combustion engines for road vehicles – Processes and Features – Laboratory experiments (in Romanian)*, Publisher Mirton, Timişoara, Romania

[7] ***2011 *Bosch Automotive Hand–book*, 8th Edition

Institute of Physics
Dirac House, Temple Back
Bristol BS1 6BE UK

ISSN: 1757-8981
ISBN 978-1-5108-1957-3

CURRAN ASSOCIATES INC.
proceedings.com